Concrete Gravity and Arch Dams on Rock Foundation

Concrete Gravity and Arch Dams on Rock Foundation

Bronstein Vadim Izrailovich,
Vainberg Alexander Isaakovich,
Gaziev Erast Grigorievich,
Landau Yuri Alexandrovich, and
Mgalobelov Yuri Borisovich

Edited by
Landau Yuri Alexandrovich and
Mgalobelov Yuri Borisovich

Translated by Mgalobelov Ju.B.

CRC Press
Taylor & Francis Group
Boca Raton London New York

CRC Press is an imprint of the
Taylor & Francis Group, an **informa** business

Cover image: The Inguri Dam, Georgia (Shutterstock)

First published in Russian as:
"Бетонные гравитационные и арочные плотины на скальном основании", published in Russian: Ассоциация «Гидроэнергетика России» М.: 2019. – 628 p"

CRC Press/Balkema is an imprint of the Taylor & Francis Group, an informa business

© 2021 Taylor & Francis Group, London, UK

Typeset by codeMantra

Library of Congress Cataloging-in-Publication Data
Names: Bronstein, Vadim Izrailovich, 1936-2011, editor.
Title: Concrete, gravity and arch dams on rock foundation / edited by Vadim Izrailovich Bronstein, Alexander Isaakovich Vainberg, Erast Grigorievich Gaziev, Yuri Alexandrovich Landau, Yuri Borisovich Mgalobelov.
Other titles: 880-01 Betonnye gravitatsionnye i arochnye plotiny na skal'nom osnovanii. English
Description: Boca Raton : CRC Press, 2021. | Includes bibliographical references and index.
Subjects: LCSH: Concrete dams.
Classification: LCC TC547 .B44813 2021 (print) | LCC TC547 (ebook) | DDC 627/.82—dc23
LC record available at https://lccn.loc.gov/2020045159
LC ebook record available at https://lccn.loc.gov/2020045160

Published by: CRC Press/Balkema
                Schipholweg 107C, 2316 XC Leiden, The Netherlands
                e-mail: Pub.NL@taylorandfrancis.com
                www.crcpress.com – www.taylorandfrancis.com

ISBN: 978-0-367-60831-6 (hbk)
ISBN: 978-0-367-74237-9 (pbk)
ISBN: 978-1-003-15671-0 (ebk)

DOI: 10.1201/b22629
DOI: https://doi.org/10.1201/b22629

This book was published following the results of the contest "The Best Edition on Hydropower", conducted by the "Hydropower of Russia" Association in 2019.

Contents

About the authors

Bronstein Vadim Izrailovich (19.01.1936–07.12.2011) graduated from the Moscow Institute of Water Management in 1959 as a hydraulic engineer. He was Doctor of Technical Sciences, Academician of the Russian Academy of Water Economics. He was one of the leading experts in the field of arch dam engineering, earthquake resistance of power structures, and control of their reliability and safety. He is known as the chief engineer of the Inguri arch dam design and the head of design work to stabilize the slope of the Zagorsk SPP. From 1962, he worked at JSC Hydroproject Institute and from 1996 to 2011 as the Deputy Director for Safety and Monitoring of Structures of the CSGOES. He took an active part in the design of Hudoni arch dam in Georgia, the Toktogul HEP in Kyrgyzstan, the Rogun HEP in Tajikistan, the Kaseb HEP in Tunisia, Kapanda in Angola, and other HEP, SPP, and NPP in Russia and abroad. As a leader and direct executor, he participated in the development of programs for calculating the earthquake resistance of concrete and soil dams and underground structures and conducting full-scale studies of the earthquake resistance of HEP in Russia and abroad. He made a contribution to the comprehensive justification of the strength of arch dams and developed new design solutions and technological methods for regulating the SSS of structures. He was Member of ICOLD and ISRM, scientific and technical councils of the JSC Hydroproject Institute and RusHydro PJSC and the author of more than 140 publications in domestic and foreign editions, with five copyright certificates for inventions and computer programs. He was Honored Power Engineer of the CIS.

Vainberg Alexander Isaakovich was born on 15.06.1941 and graduated from the Ukrainian Institute of Water Engineers in 1963 as a hydraulic engineer. He is Doctor of Technical Science, Professor. After graduation, he served in the Soviet Army from 1966 to 1986 and was a graduate student, teacher, and assistant professor at the Ukrainian Institute of Water Management. Since 1986, he worked at the Ukrhydroproject Institute (Kharkov) as the chief specialist in calculating the strength and stability of HS; since 2005, he has been working as the Deputy General Director of Ukrhydroproject PJSC. His design and scientific activities are associated with the justification of complex HP in Ukraine

(Dnepr HEP, Dnester SPP), Russia, China, Vietnam, and other countries. He made a great contribution to the development of the method for calculating the SSS of dams, taking into account the phasing of their construction. He developed new methods for solving probabilistic problems of assessing the reliability of HS within the framework of the parametric and systemic theory of reliability. He is the author of a biexponential law, which in the best way to determine the extraordinary values of discharges in river Dnepr. He published the monograph "Formation of stresses in gravity dams", "Reliability and safety of HS. Selected problems", and others. At the same time, he is the head of the Department of HS of Kharkov State Technical University of Construction and Architecture.

Gaziev Erast Grigorievich was born on 13.12.1931 and graduated in 1953 from the Hydrotechnical Department of the Moscow Power Engineering Institute (MPEI) and then from graduate school at the Department of Hydrotechnical Structures. In 1958, he defended his doctoral dissertation at MPEI and in 1980 at the Moscow Civil Engineering Institute defended his doctoral dissertation on the mechanics of rock masses. In 1958–1959, he worked as an engineer and lead engineer at the Institute of Geological Sciences of the Academy of Sciences of Armenia. In January 1960, he joined the Hydroproject Institute, where he was engaged in research on HS and their rocky foundations. He was the head of the research department of arch dams of the NIS Hydroproject. He participated in the research of many dams, including the Inguri arch and Sayano-Shushensk arch-gravity dam. In 1963, he completed an internship in Paris at the Coin and Bellier bureau. From 1966 to 1969, he worked as a UN expert in Mexico, where he organized research laboratories at the Ministry of Water Resources and the Ministry of Public Works. Under his leadership, research was carried out on a number of dams designed at that time, including the La Amistad dam (Mexico-USA) on the river Rio Grande. In 1994–2002, he was researcher at the Institute of Engineering and professor at the Mexican Autonomous University in Mexico City, where he published the monograph "Rock Mechanics in Construction" and "Stability of Rock Formations and Methods of Fixing Them". He is the author of more than 200 publications, including 17 books in Russian, Spanish, and Serbian-Croatian. He was Chief Specialist of GSGOES RusHydro PJSC since 2002, Member of ISRM since 1972, and the Chairman of the Russian Association of Geomechanics in the period of 2004–2018.

Landau Yuri Alexandrovich was born on 21.03.1939. He graduated from the Ukrainian Institute of Water Engineering in 1961 as a hydraulic engineer and graduated from postgraduate studies at VNIIG named after B.E. Vedeneev. He was a candidate of technical sciences since 1969, doctor of technical sciences since 1999, full member of the Academy of Construction of Ukraine, and member of the International Energy Association (IEA). Since 1961, he has been working in the Ukrainian branch of the Hydroproject Institute JSC, where he has gone from an engineer to the Deputy Director of Ukrhydroproject PJSC (Kharkov).

Its activities are related to the scientific justification of HP, including environmental protection. He was the head of work on the scientific justification of new solutions for HS, including the study of stress state and optimization of the technology of erection of HS. He led the development of the state program for ensuring the environmental safety of HP. Under his leadership, innovative solutions were developed and implemented in the design and construction of the Kiev and Kanev HEP, the pumping stations of the feces Dnepr-Donbass, Tashlyksk SPP and the Aleksandrovsk HEP, the HEP Guaninge, and Chontenhe in China, Thakmo, Yali, and Nam Chien in Vietnam. He is one of the leading scientists in the field of hydropower, a member of the Ukrainian State Academy of Railway Transport, a scientific editor, and one of the authors of the encyclopedic publication "Energy: History, Present and Future". He is the author of more than 70 printed works in domestic and foreign publications, including monographs on concrete dams, hydropower, and environmental protection, with 125 copyright certificates and patents for inventions and three poetry collections. He is a member of ICOLD.

Mgalobelov Yuri Borisovich was born on 04.01.1942 and graduated from the Leningrad Polytechnic Institute as a hydraulic engineer. He is a candidate of technical sciences since 1973, a doctor of technical sciences since 1990, and a full member of the Russian Academy of Water Economics. In the years 1964–1980, he worked at the Hydroproject Institute as an engineer, team leader, chief specialist of the concrete dam department, and Deputy Chief Engineer of the project of the Middle Yenisei HEP. He took an active part in the design of the high-altitude arch dam of the Inguri, the arch-gravity dam of the Sayano-Shushensk HEP, the Toktogul concrete dam, and arch dams: Kasseb (Tunisia), Mansur Eddahbi (Morocco), and Kirdzhali (Bulgaria). In 1980–1989, as the head of the department of concrete structures, he led the scientific substantiation of the projects of arch dams of the Katun HEP and the Hudoni HEP in Georgia and concrete dams of the HP Kapanda (Angola), Kurpsay and Tashkumyr (Kyrgyzstan), Hoabin (Vietnam), Tashlyksk and Konstantinov (Ukraine), and Sangtudin (Tajikistan). He participated in the development of BR of the Gosstroy of the USSR for the design of concrete dams and their foundations in terms of new calculation methods for studying arch dams and their foundations in spatial conditions taking into account nonlinear effects, strength, and stability criteria, as well as in the development of engineering measures for strengthening rock foundations and slopes. Since 1989, he worked at the Hydroproject Institute as a chief specialist in the technical department, as a chief engineer, and as the head of the Center for Project Scientific Justification; he currently works as a scientific adviser to the Director General of Institute Hydroproject JSC. During this time, he led the justification of the projects of the HP Al-Baghdadi (Iraq), Teri (India), Merove (Sudan), Kudankulam NPP (India), Shon La, Lai Chau, and Nam Chien (Vietnam), and Plyavinsk (Latvia). Since 1992, he has concurrently been the Deputy General Director of the International Institute of Geomechanics and Hydrotechnics (IIGH) and Member of ICOLD and ISRM. He is the author of more than 160 publications, including 7 books and 4 certificates for inventions.

Annotation

This book, on the basis of a generalization and critical analysis of materials on constructed concrete dams, accumulated experience of their operation, and current trends, discusses a set of problems related to design and construction of concrete dams. There are expound modern principles of designing gravity and arch dams; the fundamental principles of validation of their safety in Russia as compared to the USA standards were formulated. Much attention was paid to dams made of rolled concrete with consideration of their specific features; ways of increasing the efficiency of dams due to the improvement of layout and structural solutions, calculation methods, and more complete consideration of the features of natural conditions are considered. The book presents and analyzes the designs of erected concrete dams, which allow understanding better the approaches and principles of decision-making when designing dams, taking into account the specifics of natural, construction, and other conditions, and analyzes a number of new solutions that reflect various ways of engineering search in the direction of further improvement of concrete dams.

This book may be useful to hydraulic engineers involved in the design, construction, and operation, as well as computer studies of concrete dams; the book can be used as a textbook for university students studying in the specialty of hydraulic engineering.

The authors hope that the use of materials and recommendations of this work in the design, construction, and operation of concrete dams will increase their reliability and efficiency.

Moscow 2020

Introduction

"A structure is a kind of body, which, like other bodies, consists of outlines and matter, the first being created by the mind, and the second – taken from nature. The first requires mind and thought, the second – preparation and choice."

P. Alberti (1404–1472), Italian architect

The construction of dams has the same ancient history as human civilization.

The need for a fuller use of renewable highly efficient hydropower resources, a strong increase in water demand for water supply and irrigation, protection against catastrophic floods, and the development of shipping caused the intensive construction of hydraulic projects (HPs) with integrated reservoirs in the whole world in the XX century. The construction of HPs played a huge role in the socioeconomic development of the world community.

According to ICOLD data, more than 50 thousand dams with a height of more than 15 m were built in the world, which made it possible to accumulate water for water supply, hydropower, and irrigation. In the year of 2000, 2,650 billion kWh electricity was generated by hydroelectric plants (HEPs) with a capacity of 670 million kW (about 19% of world production). In 2015, the capacity of HEP increased up to 1,120 million kW, and generation increased by 47% compared to 2000 and reached 3,900 billion kWh.

It is reservoirs that make it possible to redistribute and use rationally water resources, which becomes an urgent task today in almost all countries of the world; words of the famous writer *Antoine de Saint-Exupery* about water sound prophetic today: "One cannot say that you are necessary for life, you are Life itself… You are the greatest wealth in the world."

Forecasts of the socioeconomic development of society in the XXI century show a steady increase in energy and water consumption, maintaining the dominant role of the electric power in ensuring the development of the economy and improving the living conditions of people. At the same time, the global problem of our civilization is the problem of maintaining a safe state of the environment for the life of society, overcoming the growing environmental crisis. Among the measures taken by humankind to overcome the environmental crisis, the transition to low-carbon electricity with the wide use of renewable energy sources, including hydropower resources, plays an important role.

China has become the world leader in the use of hydropower resources: in 2015, the capacity of HEPs reached 320 million kW and the generation of 1,126 billion kWh. By 2020, China plans to increase the capacity of its HEPs to 350 million kW.

In accordance with the "General Scheme for the Development of the Electric Power Industry", Russia will generate about 15 million kW power at HEPs and pumped storage plants (SPPs) by 2030. According to the "New Energy Strategy for the Period until 2035" in Ukraine, HEPs and SPPs may generate about 5 million kW. Kyrgyzstan and Tajikistan are planning to develop their large hydro-energy resources; further construction of HEPs is also envisaged in Kazakhstan, Uzbekistan, and Georgia.

Data on the hydropower potential and its use as of 2015 in countries with the largest hydropower resources are given in the table below [265].

Table. Countries with the largest hydropower resources

Country	Hydropower potential on production		Development of hydropower potential		
	Technical (billion, kWh)	Cost effective (billion, kWh)	On capacity (million, kW)	On production	
				Billion (kWh)	% of cost-effective
China	2720	1753	3200	1126	66
Russia	1670	852	47.9	160	19
Brazil	1250	763.5	96	415	54
Canada	981	536	76	388	72
Congo	774	400	2.6	8	2
India	660	442	46	136	30
USA	528	376	79.5	276	73
Tajikistan	439	263	5.2	17	7
Peru	395	260	4.2	23	9
Nepal	367	220	0.8	3	1.3
Japan	285	136	28	102	90
Ethiopia	260	162	4.3	11	7
Norway	300	213	31	143	70
Venezuela	260	100	14.5	90	90
Turkey	216	165	25.8	67	41
Colombia	200	140	9	50	36
Chile	162	-	6.2	24	-
Argentina	130	-	11	40	-
Cameroon	115	105	0.7	4	4

Characteristics of global trend are construction of hydraulic projects at the foothills and in the mountains and increase of dam height.

Concrete dams on rock foundations are one of the most common and effective types of dams, as they were built in all countries of the world under various natural conditions.

The advantages of concrete dams include the compact layout of HP, minimum volume of work, time, and costs of construction, favorable conditions for the erection and operation of facilities and the admission of construction discharges, and reducing time and cost of construction, as well as minimizing the negative impact on the nature.

The experience in the construction and long-term operation of concrete dams on rock foundations built and operated in a variety of natural conditions has shown their effective operation and high reliability, also in areas with high seismic activity and with severe climatic conditions.

It is worth to note that of all the dams with a height between 150 and 200 m constructed in the whole world in 2000, more than 60% were concrete dams (35% of these were arch dams); of the highest dams (with a height above 200 m), 78% were concrete dams (60% of these were arch dams).

The following concrete dams could be considered as an outstanding achievement of world dam engineering of the XX century:

- arch-gravity Hoover (1936) in the USA, 222 m high (see Figure 1.11),
- gravity Grande Dixence (1961) in Switzerland, 285 m high (see Figure 1.19),
- multi-arch Daniel Johnson (1968) in Canada, 214 m high (see Figure 1.39),
- gravity Toktogul (1978) in Kyrgyzstan (project of the USSR), 215 m high (see Figure 1.23),
- arch Inguri (1980) in Georgia (project of the USSR), 271.5 m high (see Figure 1.30),
- arch-gravity Sayano-Shushensk (1981) in Russia, 242 m high (see Figures 1.28 and 2.6),
- the massive buttress Itaipu (1982) in Brazil-Paraguay, 196 m high (see Figure 1.43).

By the beginning of the XXI century, the height of concrete dams reached 300 m or more. In China, the highest arch dams Jinping-1, 305 m high (see Figure 8.16) and Xiowan, 294.5 m high (see Figure 8.15) were built.

Taking into account that accidents of such unique engineering structures as high concrete dams can lead to catastrophic consequences for the population, economy, and the environment, ensuring their reliable operation and safety is the most important task at all stages of design, construction, and operation of dams.

Serious changes have taken place in the construction of concrete dams over the past decades. The technology of the dam construction becomes the most important factor, often determining their effectiveness and competitiveness. The practice of construction required urgently a qualitative leap in the technology of erecting concrete dams. Introduction of the new rolled concrete compacted (RCC) has dramatically increased the efficiency of construction of concrete dams.

This progressive method has been widely developed in countries where construction of large HC has a big scope, such as China, the USA, Japan, Brazil, Colombia, South Africa, Spain, and others. By 2015, about 600 dams in 56 countries of the world were built with this technology. Their height reached 200 m and above. So, in Colombia, the Miel-1 gravity dam was built 188 m high (see Figure 2.20), in China – the Longtan gravity dam, 216.5 m high (see Figure 2.21) and Guangzhao, 201 m high; in Ethiopia – the Gibe-III gravity dam, 249 m high. RCC technology was also used in the construction of arch dams.

Design solutions of modern concrete dams and the technology of their construction are linked inextricably, and their further improvement is largely determined by the construction technology.

In recent decades, certain successes have been achieved in improving the design solutions of concrete dams, in strengthening their foundations, and in creating high-strength hydraulic concrete. Experience in design, construction, and operation of concrete dams together with computer simulations of dams with rock foundations has created the prerequisites for a more complete use of the bearing capacity of concrete and rock foundations, which in turn led to a further increase of the dam height, their construction in sites with difficult natural conditions, increased reliability, and efficiency of concrete dams. In China, the most powerful HEP in the world, Three Gorges, with an installed capacity of 22.5 million kW was built (see Figure 1.45).

Following the statement of the famous French philosopher *Michel Montaigne (1533–1592)*: "it is not enough to accumulate experience, it is necessary to weigh and discuss it, to digest it and think it over to extract all possible arguments and conclusions from it." – world experience in the construction of concrete dams, current trends, and scientific and technological achievements should be used and creatively developed, taking into account the peculiarities of the natural conditions of sites when making optimal decisions in the design and construction of concrete dams. At the same time, security remains a top priority.

A number of monographs and textbooks are devoted to the design and construction of concrete dams. Among others, the guide by *Grishin M.M.* [65] "Concrete dams (on rock foundations)" (1975); monograph by *Sudakov V.B.* and *Tolkachev L.A.* [160] "Modern methods of concreting high dams" (1988); textbook "Features of the design and construction of hydraulic structures in hot climates" (1993), edited by *Rozanov N.P.* [61]; critical review of concrete dam structures on rock foundations (constructed and invented) by *Mgalobelov Yu.B.*, *Landau Yu.A.* "Non-Traditional concrete dam construction on rock foundation" [233], A.A.Balkema Publisher (1997); separate sections of the textbook "Hydraulic structures" (1996 and 2009), parts 1 and 2, edited by *Rasskazov L.N.* [59, 60].

They mainly reflect the experience of building concrete dams in the 60–80 years of the XX century. Aspects of technology and constructive solutions of gravity dams, based on the modern experience of their construction, were covered in the monograph by *Lyapichev Yu.P.* [93] "Design and construction of modern high dams" (2009).

However, a monograph covering a wide range of problems in the design and construction of massive gravity and arch dams based on a synthesis of world, USSR, and Russia experience and modern trends in the construction of concrete dams doesn't exist. With this book, the authors are hoping to fill this gap.

The book can be useful to hydraulic engineers involved in the design, construction, analysis research, and operation of concrete dams, and can be used as a textbook for university students studying in the specialty of hydraulic engineering.

The book is based on a generalization and critical analysis of materials of constructed concrete dams and accumulated experience of their operation and current trends and describes the complex of problems related to design and construction of concrete dams. Modern principles of designing gravity and arch dams and the fundamental principles of validation of their safety compared to the USA standards are formulated. Special attention was paid to dams made of rolled concrete with consideration of their specific features; the ways of increasing the efficiency of dams due to the improvement of layout and structural solutions, methods of analysis, and more complete consideration of the features of natural conditions were considered. The book presents and analyzes the designs of erected concrete dams, which allow one to understand better the approaches and decision-making principles when designing dams, taking into account the specifics of natural, construction, and other conditions, and analyzes a number of new solutions that reflect various ways of engineering search in the direction of further improvement of concrete dams.

The authors hope that the use of materials and recommendations of this work in the design, construction, and operation of concrete dams will allow increasing their reliability and efficiency.

The book contains eight chapters, list of references, and two appendices.

The first chapter provides information on the history of the construction of concrete dams (gravity, buttresses, and arch), starting from ancient times until the beginning of the XXI century.

The second chapter describes the layout of HPs with concrete dams.

The third chapter provides basic information on rock foundations, including the nature of rocks, their classification, physical and mechanical characteristics, and the effect of reservoirs on state of the rock foundations and concrete dams.

The fourth chapter describes the features of connection of concrete dams with rock foundations, preparation of rock foundations, strengthening, and waterproof measures.

The fifth chapter gives basic information about hydraulic concrete and its physical and mechanical properties.

The sixth chapter describes the loads and effects on the dam, including temperature and seismic effects, information on methods for dam analysis.

The seventh and eighth chapters devoted to gravity and arch dams provide constructive solutions to dams and their elements and the features of dam analysis together with the foundation and analyze the conditions of their operation and features of the stress-strain state (SSS) of the dam system "dam foundation".

The list of references contains published materials used in writing the book.

Appendix 1 contains links to sources from which photographs of dams and HEPs are borrowed. The appendix 2 contains accepted abbreviations.

This book was written by a team of authors consisting of doctors of technical sciences Bronstein V. I., Vainberg A. I., Gaziev E. G., Landau Yu. A., and Mgalobelov Yu. B.

The authors express their gratitude to the world community of hydraulic engineers: designers, researchers, builders, and operators, who create perfect structures of concrete dams in difficult natural conditions, many of which are outstanding achievements of the civilization.

Chapter 1

History of the development of concrete dams

"You can't really master by any scientific discipline not knowing the history of its development."

Auguste Comte (1798–1857), French philosopher

1.1 Masonry dams from ancient civilizations to the XX century

At all times, water has been the most important factor determining the life of people and the development of productive forces. The foci of ancient civilizations are associated with the rivers: Tigris and Euphrates in Mesopotamia and the Nile in Egypt. Already at the dawn of civilization for 4000–3000 BC, the first dams and reservoirs were built. The oldest in the world are considered the Kosheish dam, 15 m high, built in Egypt under the pharaoh Menes (about 3000 BC) and the Sadd-El-Kefara dam, built in Egypt between 2950 and 2750 BC in the valley of the Wadi Garavi rock desert Jebel Galala between the Nile and the Red Sea.

The height of the Sadd-El-Kefara dam was 14 m, the length along the crest was 110 m, and the angle of inclination of the downstream slope was 35°–45° and the upstream −30°. The structure included two dams in the distance 32 m from rubble masonry with 29 m width at the foundation (upper dam) and 37 m (downer dam) separated from each other, and the core between them was 32 m wide at the foundation. The core is made of a mixture of sand and gravel and the lining slopes of the dam – from well-cut limestone slabs. The purpose of the dam remains unclear to date.

The dam was built during the pharaoh Soti-1 (1319–1304 BC) and still performs its functions [28,56]. The construction of masonry dams was the forerunner of modern concrete dams and required high engineering skills. Their purpose was the creation of reservoirs for water supply, irrigation, and flood control.

Dam Sadd al-Arim built between 1000 and 700 BC near the city of Marib (now Yemen), 10 m high and about 600 m long, was considered one of the wonders of the ancient world. In Persia, during the era of King *Darius* (VI century BC), a series of dams were built on the rivers Dzharahi, Kar, etc. BC Shuao dam built on the Yellow River in China is currently operating.

In the code of laws of the king of Mesopotamia *Hammurabi* (XIX century BC), much attention was paid to the rules for the safe and lossless operation of reservoirs. One of the articles of the law read: "Article 56. If someone dumps water and his neighbor is flooded with water, then he must measure him 10 guru (150 dm^3) of bread for each bur (6.3 ha) of flooded land".

The Assyrian king *Sinaherib* (705–680 BC) was the first who carried out the transfer of runoff from the basin of several rivers to another river, building a dam on the Khosr River and a canal 16 km long, through which water was supplied to the capital Nineveh.

The Chavdarnishar Dam near Kutahya (VI–IV centuries BC, the region of Kariya, the territory of modern Turkey) was built to protect Aizana from floods; the height of the dam is 10 m, and the length along the crest is 80 m. Near the city of Corum, an Oryukai dam was built with a crest length of 40 m and a height of 16 m. Both were earth dams with a crest width of 4.5–5.5 m and slopes made of stone blocks.

In the VII century BC, in the territory of present-day Yemen, not far from the city of Sana, the Maribo HP was built with an earth dam 18 m high and 600 m long along the crest. The hydraulic plant included two outlets and an automatic spillway. Water discharge regulation was carried out using removable stone beams.

Already in those days, analysis was carried out to determine the size of dams, and there were rules governing water use. Thus, a cuneiform tablet, discovered during excavations of the city of Ur, the capital of the Sumerian state in Mesopotamia (about 1800 BC), contains an example of calculating the dam [65].

The dam constructed in antiquity was gravity, perceiving water pressure with its weight. The idea of using the arched effect by giving the dam a curved shape was expressed by the Byzantine emperor *Justinian* I (527–565), who intended to build a dam in Dara near the Syrian-Turkish border. "He will build this dam not in the form of a straight line, but in the form of a curve curved against the flow, for a better perception of head" wrote *Prokopiy* from Caesarea in 560.

In Spain, during the time of Ancient Rome (in the I–IV centuries AD), more than 70 dams were built, of which the highest dam Almonacid de la Cuba (98–117), 34 m high, after periodic reconstructions, is preserved to the present. This dam was originally made in the form of three arches, resting on the foundation and massive buttresses, and later with further reconstructions was turned into a gravity dam. Another large Proserpina dam, 21.6 m high, was made by a combined structure from masonry on the upstream side, to the downstream face of which an earth dam adjoins (Figure 1.1). In the XVII century, the dam was repaired. The dam is still in operation [187].

Figure 1.1 Proserpina combined dam, 21.6 m high (Spain, I–IV c.). General view.

Today we may be amazed at the engineering skills of the ancient builders, given that some of these dams are still operational. These hydraulic structures (HS) are outstanding achievements of ancient civilizations.

The arch dam Shah Abbas was erected in the Khorasan Province of Iran 700 years ago during the reign of the Persian *Safavid* dynasty. The height of the dam is 60 m, and the thickness on the crest is 1 m. Due to the 60-m height for 550 years, the dam was called the highest, thinnest, and oldest arch dam in the world (Figure 1.2).

In the XV century and later in Western Europe, dams of various types were built with a height of mainly up to 40 m on relatively large rivers. Some of these dams are still in operation today, and the profiles of gravity dams are close to the profiles of modern dams. At this time, the first design and construction criteria were developed, which did not change much until the end of the XIX century.

In 1384, one of the first arch dams from masonry Almansa Dam, 23 m high, was built in Spain (Figure 1.3) and in 1537, Ponte Alto Dam in Italy [141].

In 1594, Tibi Dam was built in Spain from masonry 42 m high and 34 m wide at the foundation (Figure 1.4), which for a long time was the highest dam in Europe. Despite being over 400 years old, it is still working.

The La Sierra de Elche dam built in 1632 in Spain on the Vinadopo River is considered the first arch dam built in Europe after the fall of the Roman Empire (Figure 1.5). Its historical significance is in the use of the geometric configuration in the form of an arc for the construction of an arch dam used in the construction of bridges and buildings. The original purpose of this dam was to curb the rivers during floods and torrential rains and the use of accumulated water for irrigation. But over the years, the dam became unusable from clogging, its ability to retain water decreased, and repeated cleanings were required to restore its work. In 1995, after a severe flood, a landslide occurred, which devastated the reservoir. In 2008, work began on the restoration of the reservoir, and at the beginning of 2009, it was again filled with water and still serves people today.

Figure 1.2 Shah Abbas arch dam, 60 m high (Iran, 1320).

Figure 1.3 Almansa arch dam, 25 m high (Spain, 1384).

Figure 1.4 Tibi Dam, masonry, 46 m high (Spain, 1594).

In 1799, in Spain, the El Gasco dam was built 90 m high of a combined construction of masonry from the upstream and an earth dam adjacent to its face in downstream. In the XIX century, masonry arch dams were built: Zola Dam (1843) 38 m high in France (see Figure 8.1); Carpa 16 m high in 1875 in Peru; Beer Valley 15 m high in 1884 in the USA (see Figure 8.3); and Parramatta Lake 15 m high in 1857 in Australia.

Figure 1.5 La Sierra de Elche arch dam, 22 m high (Spain, 1632).

The gravity dams of Eschboh, (1891) 24 m high, Einsiedel, (1896) 28 m high, and Fuelbecke, (1896) 27 m high, were built in Germany. Beetaloo, (1890) 37 m high, and Victoria, (1890) 25 m high, in Australia; Lago Lavezze, (1883) 40 m high, and Lago Lungo, (1891) 47 m high, in Italy; Maigrauge, (1872) 24 m high, in Switzerland; and Burrator, (1898) 35 m high, in England are still in operation [141].

In the XVIII–XIX century, the quantity of masonry gravity dams was strengthened by installing buttresses from masonry on a mortar, for example, the Pabellon dam in Mexico, which has been preserved to date [141].

Outstanding achievements of the XIX century, the invention of cement, concrete, and reinforced concrete, ushered in a new era of dam construction.

The distinguished English physicist *J. Thomson* wrote: "Three factors are involved in technological progress: knowledge, energy and material."

Gravity, arch, and buttress dams began to be built exclusively from concrete, which ensured their wide distribution and rapid progress.

1.2 Concrete dams in the XX century

1.2.1 Dam construction in the first half of the XX century

Since the beginning of the XX century, the number of dams has increased sharply, which was caused, first of all, by the development of the electric power industry, which ensured the possibility of efficient use of hydropower resources. In the first half of the XX century, concrete dams were built as part of integrated HPs in the USA and Canada, in the countries of Western Europe, Central and South America, Asia, the USSR, and Australia. HEP capacity reached hundreds and then thousands of MW; reservoir capacities sharply increased, and the height of the dam on rock foundations reached 100 m and more. In various natural conditions, different types of concrete dams were built on rock foundations: gravity, arch, and buttresses.

The Schrah gravity dam, 112 m high, built in 1924 in Switzerland became the highest dam in the early 1990s of the year XX century.

During this period, due to a sharp increase of dam height and reservoir volumes, the problem of ensuring the reliable operation of dams became urgent, the solution of which was hampered by the lack of the necessary experience in the design and construction of concrete dams.

The first concrete dams were largely imperfect: the concrete was compacted (rammed) manually, there were no temperature-shrinkable joints and drainage, and the foundation was not reinforced, which led to cracking of the concrete, to violation of the strength and stability of the dam, and, as a result, to numerous accidents with catastrophic environmental consequences, including loss of life and huge material losses. As result of the destruction in 1923 of the multiarch Gleno dam, 75 m high, in Italy, about 500 people died. The destruction of the arch-gravity dam of St. Francis, 59 m high, in the USA killed about 400 people (see Figure 8.8).

The practical needs for improving reliability and safety led to the improvement of structures and technology for the construction of concrete dams and the development of methods for studying the static work of dams, rock mechanics, and hydraulics.

In order to simplify the technology of concrete laying in the USA, a method was developed for concreting with highly plastic, the so-called, cast concrete, but it had low strength, significant shrinkage, and insufficient frost resistance. In the 1930s of the XX century, to convey concrete to the dam, belt conveyors were used. The concrete was compacted by vibration using sufficiently powerful vibrators, which made it possible to put low plastic concrete mixtures in dams and to obtain hydraulic concrete that is waterproof with the necessary strength and frost resistance.

Great success was achieved during the construction of the largest Hoover HEP in the world in 1936 in the USA with a capacity of 1.34 million kW with the highest arch-gravity dam, 222 m high, with a concrete volume of 2.5 million m^3. The experience accumulated in the USA made it possible to develop methods for combating temperature cracking during hardening of concrete and to create rational concrete compositions and an efficient concreting technology using cement with relatively low heat generation, with plasticizing and air-entraining additives and artificial pipe cooling of laid concrete. Due to the cementation of construction joints between blocks after cooling of the laid concrete and completion of its shrinkage, the reliability and durability of the structure were ensured, the conditions were improved, and the construction time was shortened. Hoover dam concreting technology has long been widely used in the construction of concrete dams with minor modifications.

The problem of ensuring the monolithic nature of concrete dams was solved by reducing the volume of concrete laid at the same time, which allowed accelerating the dissipation of exothermic heat. The following cutting schemes for concrete blocks were developed (Figure 1.6):

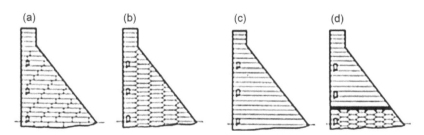

Figure 1.6 The scheme for cutting dams into concrete blocks: (a) in dressing, (b) columnar, (c) sectional, and (d) mixed.

- sectional cutting, when the block is laid within the dam section between temperature transverse joints;
- in the dressing, i.e., with overlapping vertical joints between blocks of a given row by blocks of an overlying row;
- columnar, when the blocks are laid in the "pillars" with the formation of longitudinal vertical joints, subsequently cemented;
- sectional cutting, when the block is laid within the dam section between temperature transverse joints;
- mixed when the first blocks are laid with "pillars", and the subsequent ones over the entire length of the section.

During the construction of gravity dams, mainly dressing (Dnepr HEP, 1932) and columnar (Hoover, Grand Coolie dams in the USA, Figure 1.7, see Figure 1.25) schemes were used.

Of the total number of 410 concrete dams with a height of more than 30 m that were built in the world between 1900 and 1940, gravity dams amounted to 69%, arch to 22%, buttresses to 3%, and multiarch to 6% [56].

In the first half of the XX century, significant experience was gained in the design and construction of gravity, arch, and buttress dams; their structures and construction technology were improved, and their reliability increased, which made it possible to build them extensively in the first half of the XX century.

Gravity dams. This period is characterized by the construction of gravity dams in many countries of the world, including the USA, Canada, Japan, countries of Western Europe, the USSR, and South Africa. Large HPs with high gravity dams were built: in Mexico – La Boquilla dam (1916), 74 m high; in Switzerland – Schrah (1924), 111 m high; in France – Le Chambon (1934), 136 m high; in Australia – Mundaring (1902), 71 m high, and Burrinjuck (1928), 91 m high.

In the USSR, in 1932, the Dnepr HEP dam (Figure 1.8) was built with a height of 62 m, a crest length of 760 m, and a concrete volume of 0.73 million m^3 (Figure 1.8), which according to its technical solutions was at the level of the best achievements of world dam engineering of its time.

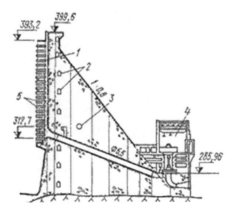

Figure 1.7 The profile of the gravity dam, 167.6 m high, of the Grand Coulee HEP (USA): 1-groove flat gate, 2-galleries, 3-cable gallery, 4-building of HEP, 5-trash lattice.

Figure 1.8 **General view of the Dnepr HEP with a dam 62 m high (former USSR, now Ukraine).**

The dam was designed with a curved outline in the plan, which allowed increasing the length of the spillway front and providing a design flood pass of 40 thousand m³/s.

In Algeria, in 1934, the Sherfa dam was built with a height of about 30 m (Figure 1.9) – one of the first dams with the use of prestressed anchors for compressing concrete.

Arch dams. In the first half of the XX century, arch dams were built in many countries in sites with favorable topographic and geological conditions including in the USA, Japan, France, Portugal, Italy, Switzerland, Spain, Austria, Norway, Greece, USSR, Turkey, Bulgaria, Romania, South Africa, Australia, and New Zealand.

The first high arch dams Buffalo Bill (1910), 107 m high, and Arrow-rock (1915), 107 m high, were built in the USA.

Initially, simple cylindrical forms of arch dams with a vertical upstream face were used, which simplified the production of works, due to the absence of overhanging consoles.

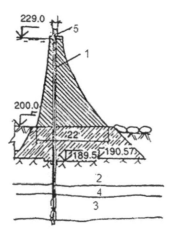

Figure 1.9 **Profile of the Sherfa dam about 30 m high (Algeria) with prestressed anchors: 1 – anchor cord; 2 – sandstone; 3 – limestone; 4 – clay marl; and 5 – extension for anchor lock.**

In 1937, the Gergebil cylindrical arch dam (see Figure 1.26) was built in Dagestan (USSR, $h = 69.5$ m, $L/h = 1.5$; $\beta = 0.28$; where h is the height of the dam, L is the length along the crest, and the β-ratio of the thickness of its bottom to the height is the so-called "shapeliness coefficient").

However, later on, dams of a more complex shape with a "constant central angle" but more economical especially in the conditions of V-shaped sections became widespread. A high dam of this type, Pacoima Dam ($h = 116$ m, $L/h = 1.5$; $\beta = 0.258$), was built in the USA in 1929 (Figure 1.10).

To ensure a favorable stress-strain state (SSS) and to reduce the volume of concrete, dome dams with vertical curvature (the so-called "double curvature" dams) were built, which later became widespread. The first dams with a noticeable curvature of the cantilevers include the Ceppo Morelli dam ($h = 46$ m) built in Italy in 1930 and the Mareges Dam ($h = 90$ m) in France in 1935.

The first domed dam was the Osigletta dam ($h = 76.8$ m) built in Italy in 1939. At the suggestion of the famous Italian engineer *Guido Oberti* in 1935, a perimeter joint was used for the first time in this dam, the purpose of which was to improve the stress state of the dam by reducing tensile stresses on the upstream face, aligning the support contour of the dam and imparting symmetry to its dome part. Such dams are widespread, especially in Italy.

The desire to increase the use of high-strength properties of compression concrete due to the spatial effect prompted USA designers to develop Hoover arch-gravity dam, 222 m high, which was built in 1936 and was the highest dam in the world at that time (Figure 1.11).

Buttress dams. In the 1920–1940s of the XX century, buttress dams were built mainly of two types: with flat pressure (Figure 1.12a) ceilings (a type developed by engineer *Ambursen*) and multiarch (Figure 1.12b and c). These thin-walled heavily reinforced (reinforcement consumption up to 40 kg and more per 1 m³ of concrete) dams were small in height, mainly up to 35 m. Similar dams were built in the USA, Japan, Mexico, Italy, France, Norway, and other countries.

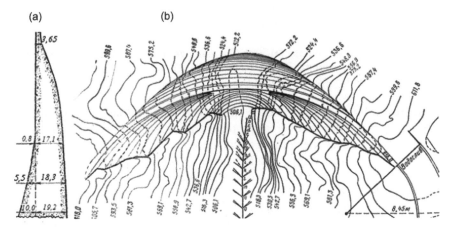

Figure 1.10 Pacoima arch dam, 116 m high (USA): (a) section along the key console and (b) plan.

Figure 1.11 Hoover arch-gravity dam, 222 m high (USA). View from the downstream.

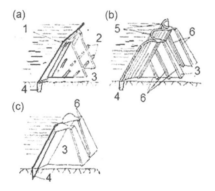

Figure 1.12 Buttress dams: (a) with flat ceilings; (b) multiarch, (c) massive buttress: 1– flat ceiling, 2 – beam stiffness, 3 – buttress, 4 – tooth, 5 – arch ceiling, and 6 – edge stiffeners.

In buttress dams, slide stability is ensured by using a load of water acting on the inclined upstream face and decreasing the filtration pressure (uplift) on the bottom of the dam, which reduces the volume of concrete.

The constructed high dams with flat pressure ceilings include the Rodriguez Dam (span $\ell = 6.7$ m, $h = 76$ m) in Mexico (1935) and the Possum Kingdom ($h = 57.8$ m, $\ell = 12.2$ m) in the USA (1941) – Figure 1.13 [65].

In order to ensure stability on longitudinal bending of relatively thin buttresses, stiffening beams were erected between them. Concrete savings in such dams compared to gravity ones reached 25%–45%. Of the high multiarch dams, the Bartlett dam ($h = 87.5$ m, $\ell = 21$ m) in the USA (1939) and the Beni Badel ($h = 61$ m, $\ell = 20$ m) in Algeria (1941) can be noted.

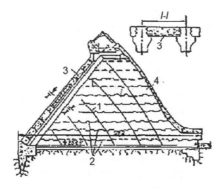

Figure 1.13 Possum Kingdom buttress dam (USA), 57.8 m high: 1 – temperature-shrink joints, 2 – construction joints, 3 – flat ceiling, and 4 – buttress (thickness 2.53–2.74 m).

A feature of the Beni Badel dam (Figure 1.14) built on a complex foundation of sandstones and weak marl shales is that buttresses have persistent abutments located on relatively strong sandstones. Between the buttresses and persistent abutments, an "active joint" with flat jacks was arranged. For the perception of tensile stresses on the upstream face in the upper zone of the buttresses, prestressed reinforcement was used.

A significant impact on the design of such dams is exerted by the seismic conditions of the site.

Concrete savings in multiarch dams compared to gravity dams can be about 30%–60% and more. However, such dams usually require more reinforcement, their design is more complicated, and therefore they have not received significant application.

In 1929, the first massive buttress dam was constructed in the USA with single buttresses (designed by engineer *F. Netzley*) with a height of 39 m (Figure 1.12c), after which such dams became widespread in many countries of the world, including Italy,

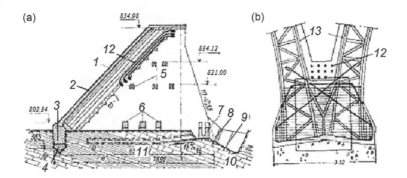

Figure 1.14 Beni Badel buttress dam (Algeria), 61 m high: (a) transverse section; (b) connection of arches with buttresses; 1 – reinforced concrete arch (span $\ell = 20$ m); 2 – waterproofing from bitumen, 3 cm thick; 3 – thrust array of arches; 4 – tooth; 5 – beam stiffness; 6 – spacers; 7 – bottom spacer frame; 8 – abutment; 9 – flat jacks; 10 – sandstone; 11 – marl shale; 12 – prestressed buttress reinforcement (design force 10 MN); and 13 – reinforcing bars.

Sweden, Romania, Japan, Iran, and Uruguay. This is due to the simplification of the design, reducing the consumption of reinforcement, improving construction conditions, favorable working conditions in harsh climatic conditions, and earthquake resistance.

1.2.2 Dam construction in the second half of the XX century

Acceleration of socioeconomic development of society in the second half of the XX century required an increase in both the production of electricity and the use of water resources. The importance of HPs has sharply increased as the most efficient of renewable energy sources, which is also the basis for the integrated use of water resources.

The need to increase electricity production, provide water to a rapidly developing industry and an increasing population, provide irrigation, and prevent catastrophic floods has led to a significant increase in the construction of HPs around the world (Table 1.1) [62,90].

Water reservoirs with a volume of more than 100 million m³, most of which have HEP, account for more than 95% of the total volume of all reservoirs.

During this period, the generation of electricity at HEP increased eight times, reaching 2,650 billion kWh in 2000 (about 19% of world electricity production) with a HEPs capacity of 670 million kW, and water consumption increased by almost six times (Table 1.2).

In the years 1950–1970 of the XX century, much attention was paid to lightweight types of dams: buttresses, gravity dams with extended joints, anchored to the foundation, etc., the prospects of which were justified by a significant decrease in concrete volume and more efficient use of its strength properties.

Nevertheless, despite the fact that the largest number of constructed concrete dams were massive gravity dams, an opinion was formed that the strength properties of concrete are not used in them, and they are not effective due to the large volume of concrete, even taking into account the simplification of the design.

Table 1.1 Dynamics of growth in the number and volume of water reservoir with a volume of more than 100 million m³ in countries of the world

Period of water reservoirs creation	Number of reservoirs/Total volume of water reservoirs (km³)						
	Europe	Asia	Africa	North America	Central and South America	Australia and Oceania	Total
Before 1900	9/3	5/2	1/0	25/9	1/0	-/-	41/14
1901–1950	104/122	46/18	15/15	342/344	22/18	10/11	539/528
1951–1985	404/491	526/1,068	89/870	516/623	179/623	63/66	1,777/4,982
Total	517/616	577/1,628	105/885	883/1,678	202/641	73/77	2,357/5,552

Table 1.2 The dynamics of water consumption in the world

Name	Water consumption full/irrevocable (km³/year)			
Year	1900	1950	1975	2000
Water consumption (with rounding)	400/270	1,110/650	3,000/1,800	6,000/3,000

In many countries, e.g., the USA, Japan, Western European countries, USSR, China, and others, arch dams were built in sites with favorable natural conditions, the number of which increased significantly especially among high dams.

With the improvement of the methods of static and dynamic analysis of arch dams, the methods of modeling them including research on strength, and geomechanical models with modeling of the foundation of the geological structure, it was possible to effectively use high-strength properties for compression of concrete and rock foundations. The field observations and the monitoring of the dams and foundation state, including the SSS, of the filtration regime during construction and operation with the help of installed control and measuring equipment (CME) were greatly developed. All of this made it possible to obtain a more complete picture of the works of the dams together with the foundations, which led to the improvement of the design solutions of concrete dams and an increase in their height with an appropriate justification for their reliability.

Thanks to the improvement of the design of concrete dams and the technology of their construction, the refinement of analysis methods to substantiate their strength and stability, and the increase in the general level of knowledge about the static works of dams together with the foundation, the relative number of dam failures has significantly decreased. However, catastrophic accidents occurred in the second half of the XX century. In 1959, as a result of the accident at the arch dam Malpasset Dam (France), 66 m high, the city of Frejus was destroyed, and more than 500 people died (a detailed description of the disaster causes is given in Section 3.3.4).

A significant increase in dam height was made possible due to the transfer of HC to foothill and mountainous regions, where the most effective conditions were for creating large regulatory water reservoirs, increasing the capacity of HEP and reducing the negative impact of construction on the environment.

High dams began to be built in areas characterized by severe climatic conditions and in sites with difficult geological conditions, high seismicity, which in many cases were previously considered unacceptable for construction. The height of the concrete dams was close to 300 m. The height of the world's largest gravity dam Grande Dixence (see Figure 1.19) built in Switzerland in 1961 was 285 m, and the highest arch dam was Inguri Dam (see Figure 1.30), 271.5 m high (1980), built in the USSR (now Georgia).

The construction of dams in difficult geological conditions required significant work to strengthen the rock foundations, for example, the arch dams Mauvoisin, 236 m high, in Switzerland (1958, see Figure 8.48), Montenard, 155 m high, in France (1961), and Nagowado, 156 m high, in Japan (1969) in an area with a seismicity of 10 points (see Figure 1.36b) and arch dam Chirkei, 233 m high, (1978, see Figure 1.32) in Russia and Inguri.

The accumulated experience and success in the design and construction of high concrete dams, as well as significant progress in the selection of concrete compositions, allowed us to move on to increasing concrete classes and, accordingly, to increase the allowable compressive stresses in dams from 3.5–5 to 10 MPa and more.

However, the use of technology for the construction of concrete dams developed mainly in the first half of the XX century. The associated large labor costs and long construction periods began to slow down the further development of concrete dams, adversely affecting their competitiveness.

In 70–80 years of the XX century, it became clear that gravity dams do not withstand competition with soil dams, despite the best conditions for the layout of structures, the passage of construction and operating discharges with the device spillway in the dams, the location of HEP structures (water inlets, water pipes, and HEP buildings), and other advantages. Even in the sites most suitable for the construction of concrete dams, in many cases, soil dams were built, which was caused by increase of their competitiveness due to the success achieved in the design and technology of erecting soil dams. It is worth noting that the construction of stone-filled dams with cores, reinforced concrete screens, and asphalt-concrete diaphragms first of all used high-performance equipment and in-line high-speed construction methods, little dependent on local environmental conditions, which ensured a decrease in their cost and a reduction in construction time. It was the technological factor that became decisive in the dominance of soil dams, reducing the competitiveness of concrete dams.

"To pose a question well means to solve it by half," wrote the famous Russian scientist *D. I. Mendeleev.*

To overcome the crisis, it was necessary to search for new effective solutions with a significant change in the technology of erecting concrete dams, using modern methods of high-speed construction, which reduce labor worth, cost, and construction time.

The development and implementation of a new technology for the construction of dams from rolled concrete (RCC) ensured the high efficiency and competitiveness of gravity dams and gave a powerful impetus to its wide construction in many countries of the world.

RCC dams are constructed from a rigid concrete mixture with a low cement content and a high content of pozzolana (fly ash), layer-by-layer compacted by vibration rollers. The basis for the development of this new technology was the achievements in global dam engineering, the vast accumulated experience, and the creative approach of engineers to solve the problem.

Already in 1961–1964 and in Italy at the construction of the high Alpe Gera Dam with a metal screen, $h = 174\,\mathrm{m}$, $L = 520\,\mathrm{m}$, and $V = 1.7$ million m^3 (Figure 1.15), a new technology was successfully applied. Hard concrete was laid without vertical joints with layers of thickness 0.8 m, with sealing mounted packages of vibrators mounted on the tractor and cutting temperature-sedimentary joints by steel vibrating knives. This allowed to significantly reduce construction time and reduce the cost by 10%–12% compared with traditional technology.

A new direction with the use of RCC was formulated in the reports of *J. Rafael & R.W. Cannon* at the Asilomar conferences in 1970 and 1972 in the USA devoted to ways to accelerate construction and increased efficiency of concrete dams [243–246].

RCC was first used at the Tarbela HP in Pakistan, where in 1975, 0.34 million m^3 of such concrete was laid in 44 days in a washout pit up to 80 m deep in the rock foundation of a water pool in a spillway tunnel.

The first gravity dam made of RCC, the Shimajigawa Dam, 89 m high, was built in Japan in 1981. Since then the construction of RCC dams has developed rapidly, their technology and designs have improved, and the height has increased. Extensive construction of such dams was carried out in many countries of the world, including Japan, the USA, China, Brazil, Mexico, Spain, Greece, France, South Africa, Morocco, Australia, Vietnam, and Thailand in various natural conditions.

Using RCC along with massive gravity dams, arch-gravity dams were built, for example, Wolverdans, 70 m high, and Knell Port, 50 m high, in South Africa

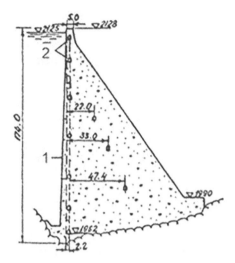

Figure 1.15 Alpe Gera gravity dam, 174 m high (Italy): 1– metal screen and 2 – galleries.

(1989–1990), as well as arch dams Puding (see Figure 8.71), 75 m high (1994), and the highest Shapai (see Figure 1.34), 132 m high (2001), in China.

The introduction of RCC technology into construction practice helped to effectively solve the problem of heat dissipation of exothermy and ensured the use of flow methods and the comprehensive mechanization of concrete work, which caused a sharp increase in the efficiency and competitiveness of concrete dams. So, at many HPs, changing the project from a stone-filled dam with a reinforced concrete screen to RCC dam allowed reducing the construction cost by 25%, for example, Salto Caxias in Brazil (1999) [93,162].

Experience in the construction and operation of RCC dams has shown their high reliability. By 1995, the number of operated and constructed RCC dams in the world with a height of more than 15 m reached 160, and by 2000, it reached 260 (including 7 arch and arch-gravity dams), of which 43 were in China, 42 in Japan, 35 in USA, 29 in Brazil, 5 in Mexico, 21 in Spain, 6 in France, 14 in South Africa, 9 in Morocco, and 9 in Australia [263,264].

High gravity dams from RCC were built: Miel-1 – $h = 188$ m and concrete volume $Vr = 1.67$ ($V = 1.73$) million m^3 in Colombia (2002). Ralko – $h = 155$ m and $Vr = 1.6$ ($V = 1.65$) million m^3 in Chile (2004); Miyegase – $h = 155$ m and $Vr = 1.54$ ($V = 2$) million m^3 (2000); and Urayama – $h = 156$ m and $Vr = 1.29$ ($V = 1.86$) million m^3 in Japan (1999); where Vr and V are the volume of RCC and the total volume of dam concrete, respectively.

On large dams, the average packing intensity of RCC reached 10 thousand m^3 per day, and the growth rate of the dam was 10 m in height and more per month, which ensured a sharp reduction in construction time.

In the last quarter of the XX century, lightweight gravity and buttress dams were built extremely rarely. Due to the complex construction technology, large worth, and long construction periods, they could not withstand competition with massive gravity dams.

In general, the second half of the XX century was characterized by great achievements in the construction of concrete dams. The scope of their application has expanded significantly. Extensive construction of concrete dams was carried out in a variety of natural conditions, including harsh climatic and complex geological conditions and high seismicity. Prerequisites were created for a further increase in dam height to 300 m and above.

Of the 140 dams built and constructed in the world in 2000 in a height of 150 m and above, more than 60% were concrete dams, including arch 35%, and among the highest dams 200 m and above, concrete dams accounted for about 78%, including arch – 60%.

The difficulties caused by the lag in the technology of erecting massive gravity dams were overcome, a new highly efficient technology for erecting RCC dams was widely developed. Concrete dam construction has reached a new level, ensuring high competitiveness and efficiency, Which significantly increased the reliability of concrete dams.

Favorable conditions have been created for further development and widespread construction of gravity and arched dams in the XXI century.

Gravity dams. Along with the massive gravity dams of the classical type, the lightweight types of gravity dams received some development, including (Figure 1.16) the following:

- dams with expanded joints and cavities at the foundation, allowing to reduce filtration pressure on the bottom of the dam;
- with cavities in the dam body, filled with ballast or used to place technological equipment (for example, HEP);
- anchored with preliminary compression of concrete from the upstream face with tie anchors, embedded into the foundation.

The use of lightweight types of gravity dams by reducing the uplift at the foundation and creating artificial compression of the dam concrete and the contact zone by prestressed anchors made it possible to reduce the concrete volume by 10%–15%; however, it complicated the dam design and the conditions of its construction.

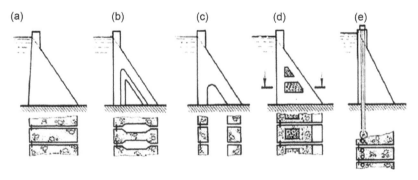

Figure 1.16 Types of gravity dams: (a) massive classic type; (b) with extended joints; (c) with cavities at the foundation; (d) with cavities in the dam filled with ballast or to accommodate technological equipment; and (e) with anchoring in the foundation.

In massive gravitational dams, transverse flat permanent joints between the dam sections are often arranged, and in narrow sites, joints have been used to ensure work of the dam sections together [65]:

- extruded or articulated;
- monolithic (due to which the dam works in spatial conditions with the transfer of part of the hydrostatic load to the banks of the gorge).

One of the first dams with extended joints was the Räterichsboden dam, 94 m high, in Switzerland (1954), where concrete savings were 9% compared to massive. An example of a dam with a cavity at the foundation, which was rarely used, is the Müldorfer dam (Figure 1.17), 40 m high, in Austria with a concrete saving of about 10% compared to a massive one [56].

In England, a lightweight dam, Olt Leyridge gravity dam, 25 m high, was built, and anchoring the dam into the foundation made it possible to compress the dam profile (Figure 1.18).

Despite some competition in 1950–1960 of the last century of light buttress dams, the gravity massive dams of the classical profile continued to occupy a dominant position, which was associated with the simplicity of construction and simple conditions of construction.

During this period, the highest gravity dams were built. The massive Grande Dixence dam (Figure 1.19), 285 m high, 695 m in length along the crest, and 6 million m^3

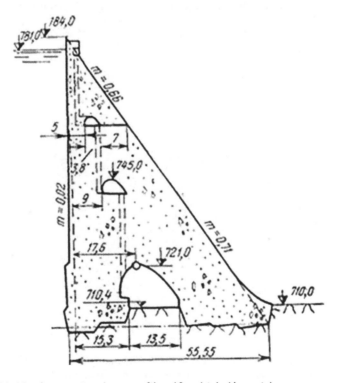

Figure 1.17 Muldorfer gravity dam profile, 40 m high (Austria).

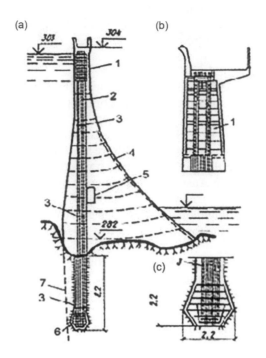

Figure 1.18 Anchored Olt Leyridge dam, 25 m high (England): (a) profile; (b) crest; and (c) anchor anchoring into the foundation; 1 – head dam, 2 – anchor, 3 – screw joints of rods, 4 – surface reinforcement, 5 – viewing gallery, 6 – end broadening of the anchor, 7 – cementing curtain.

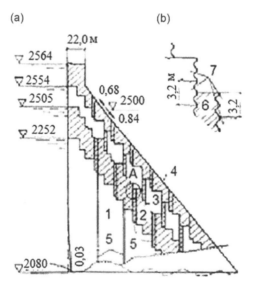

Figure 1.19 Grande Dixence gravity dam, 285 m high, (Switzerland): (a) profile and (b) knot A; 1, 2, 3, and 4 – 1st, 2nd, 3rd, and 4th stages of the dam, respectively, 5 and 6 – intercolumn joints, and 7 – filling concrete joints.

concrete volume was erected by Switzerland (1961), in four stages with cutting with columnar blocks and subsequent cementation of inter-column joints, which also allowed filling the water reservoir with stages to intermediate levels with an incompletely constructed dam.

The Bhakra massive gravity dam in India (1963), $h = 226$ m, $L = 518$ m, and $V = 4.1$ million m^3, was built in extremely unfavorable geological conditions at the foundation, weakened by tectonic disturbances and three packs of clay rocks, cool, inclined towards the downstream and located between sandstones and conglomerates. In this regard, the dam profile in the lower part was broadened (Figure 1.20), and engineering measures were taken to strengthen the dam foundation, including clearing clay layers, tectonic disturbances filled with concrete, etc.

In the USA, in 1974, Dworshak massive dam was built with a height of $h = 219$ m and a concrete volume of $V = 5$ million m^3 and in Canada, Revelstoke – $h = 175$ m and $V = 1.6$ million m^3.

High dams were built in Russia in severe climatic conditions: massive Krasnoyarsk (Figure 1.21) – $h = 124$ m, $L = 1,072$ m, and $V = 4.9$ million m^3 (1970); with extended joints in Bratsk (Figure 1.22), $h = 123$ m, $L = 1,430$ m, and $V = 4.4$ million m^3 (1965), and Ust-Ilim, $h = 105$ m, $L = 1,562$ m, and $V = 4.3$ million m^3 (1976).

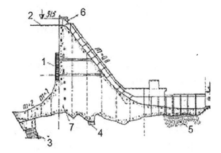

Figure 1.20 Bhakra gravity dam, 226 m high (India): 1 – spillway lattice; 2 – reservoir drawdown level; 3, 4, and 5 – packs of clay rocks; 6 – segment gate; and 7 – gallery at the foundation.

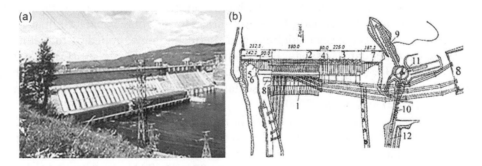

Figure 1.21 Krasnoyarsk HEP (USSR, now Russia): (a) view from the downstream and (b) plan: 1 – building HEP, 2 – station dam, 3 – spillway dam, 4, 5, 6, and 7 – deaf, bedding, right-bank, and left-bank dams, 8 – mounting place, 9 and 10 – upstream and downstream shipping routes, 11 – rotary device, and 12 – ship elevator.

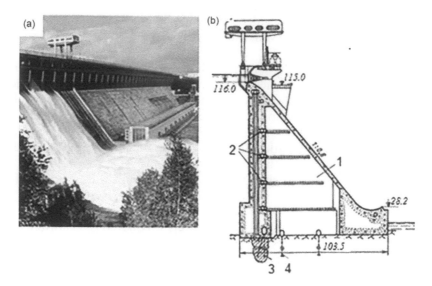

Figure 1.22 Bratsk gravity dam, 126 m high (USSR, now Russia): (a) view from the downstream and (b) dam profile: 1 – extended intersection joint, 2 – viewing galleries, 3 – cement curtain, and 4 – drainage wells.

In Kyrgyzstan, Toktogul massive dam was erected in a narrow gorge (Figure 1.23) – $h = 215$ m, $L = 300$ m, and $V = 3.2$ million m^3 (1977) – in complex geological conditions and extremely high seismicity (about 10 points).

To transfer the main load from hydrostatic pressure to the bottom of the gorge and partial to the slopes, the dam is cut by transverse joints, diverging fan in plan. At the Toktogul dam, a fundamentally new technology of single-layer (0.5–1 m thick) concrete was developed and introduced by large blocks with an area of up to 2.5 thousand m^2 and using hard low-cement (cement consumption 115–150 kg/m^3) concrete, compacted by hinged vibrators.

In the RCC dam Shimajigawa, (volume $Vr = 65$ thousand m^3 with a total volume of $V = 317$ thousand m^3) erected in the highly seismic region of Japan, the areas on the upstream and downstream faces and in the lower contact part are made of vibrated concrete. The design solutions and the technology applied at this dam were also used at other dams in Japan. For example, Sakaigawa Dam – $h = 115$ m, $L = 298$ m, and $Vr = 0.37$ ($V = 0.72$) million m^3 (1993), was also erected in the region with high seismicity and at the highest, the Miyagase dam (Figure 1.24) – $h = 155$ m, $L = 400$ m, and $Vr = 1.54$ ($V = 2$) million m^3 (2000).

On the first RCC Yellow Creek dam in the USA built entirely, $h = 52$ m, $L = 543$ m, and $Vr = V = 331$ million m^3 (1982), due to the formation of cracks and unacceptable filtration, cementation of the dam body was required. In 1987, the highest RCC Upper Stillwater Dam in the USA was built – $h = 97$ m, $L = 815$ m, and $Vr = 1.12$ ($V = 1.28$) million m^3. In the dams under construction, negative factors identified during the construction of the first RCC dams were taken into account.

The first RCC Kengkou Dam in China was built in 1986 with a height of $h = 53$ m, $L = 123$ m, and $V = 62$ thousand m^3, after which the

Figure 1.23 Toktogul gravity dam, 215 m high (USSR, now Kyrgyzstan). View from the downstream.

Figure 1.24 Miyegase RCC gravity dam, 155 m high (Japan). View from the downstream.

construction of the RCC dams was widespread. In 1995, large dams were built: Shuikou – $h=101$ m, $L=791$ m, and $Vr=0.6$ ($V=1.71$) million m^3 and Guaninge[1] – $h=82$ m, $L=1,040$ m, and $Vr=1.24$ ($V=1.81$) million m^3 and in 1999, the Jiangya Dam – $h=131$ m, $L=370$ m, and $Vr=1.1$ ($V=1.38$) million m^3.

In Brazil (1998), Salto Caxias dam was built in the composition of a hydraulic plant with a capacity of 1.24 million kW – $h=67$ m, $L=1.1$ km, and $Vr=0.9$ ($V=1.44$) million m^3. A spillway dam with a length of 290 m ensured a flood discharge of 48.3 thousand m^3/s. In the technical and economic comparison, the RCC dam turned out to be 25% more economical than a stone-fill dam with a reinforced concrete screen with the same construction time of 30 months [259]. In Chile (1996), the Pangue dam was built – $h=113$ m, $L=410$ m, and $Vr=0.67$ ($V=0.74$) million m^3, and in Colombia (2000), the Porce 2 dam – $h=118$ m, $L=425$ m, and $Vr=1.3$ ($V=1.45$) million m^3.

In Greece (1999), Platanovryssi dam was built – $h=95$ m, $L=305$ m, and $Vr=0.42$ ($V=0.44$) million m^3, and in Spain (2000), Riabl dam – $h=99$ m, $L=630$ m, and $Vr=0.98$ ($V=1.02$) million m^3 with a minimum volume of vibrated concrete.

In Morocco (1991), Aoulouz dam was built – $h=75$ m, $L=480$ m, and $Vr=0.61$ ($V=0.83$) million m^3. In Algeria (2000), in the region with a seismicity of eight points, Beni Haroun dam was built – $h=118$ m, $L=714$ m, and $Vr=1.6$ ($V=1.9$) million m^3 with a vertical upstream face, from the side of which a layer of vibrated concrete 1 m thick was laid. An upstream cofferdam was included in the dam profile [93].

Successful construction of massive gravity dams in the last quarter of the XX century using the new highly efficient RCC technology provided favorable conditions for their further widespread construction in the XXI century.

In 1985, the construction of the Grand Coulee HEP with a capacity of 6,809 MW with a gravity dam 165.8 m high and 1,552 m long along the crest was completed in the USA (Figure 1.25).

Arch dams. In the second half of the XX century, the construction of arch dams was most intensively carried out in Western Europe, especially in France, Italy, Switzerland, Portugal, and Spain, as well as in the USA, Japan, South Africa, the USSR, China, Turkey, and Iran.

Figure 1.25 Grand Coulee HEP (USA).

As part of the oldest Gergebil HEP station in Russia, an arch-gravity dam, 69.5 m high, was built (the height of the arched part is 60 m, the height of the plug is 9.5 m (Figure 1.26), and the length of the crest is 76 m.

The arch dam of the Gunib HEP was built on the Karakoisu river above the Gergebil HEP. The full height of the dam is 73 m, of which 33 m is the arch part and 40 m is the plug (Figure 1.27); the length of the dam along the crest is 58.7 m. The thickness of the arch in the crest is 4 m, at the contact with the plug – 6 m, and the thickness of the plug is 20 m. A deep outlet with a cross section of 4.3 × 4.5 m has a throughput capacity 448 m³/s. The foundation of the dam was solid limestones with the inclusion of silt-stones, mudstones, and sandstones. Rocks in the channel are overlain by alluvial boulder-pebble deposits with a depth of up to 20–25 m. The seismicity of the HEP location region is nine points on the MSK-64 scale.

In the 1950s, several mostly thick cylindrical dams were built in relatively narrow sites. Tignes dam was built in France (1952) with a height of $h = 180$ m, a crest length of $L = 296$ m, a shapeliness ratio of $\beta = 0.245$, and concrete volume $V = 0.65$ million m³.

Monticello dam with $h = 90$ m, $L = 310$ m, and $\beta = 0.289$ was built in the USA (1953) under difficult geological conditions (sandstones with marls and shales). Montenard dam with $h = 155$ m, $L = 217$ m, and $\beta = 0.267$ was built in France (1961) on the foundation of shale limestones with large tectonic cracks up to 0.6 m wide, whose planes are almost parallel to the gorge; Kukuan with $h = 86$ m and $L = 135$ m was built on island Taiwan (1962) on the foundation of strongly fractured quartzites and shales.

In some cases, arch dams with a cylindrical upstream face were built in wide sites: Moulin Ribou – $h = 16.2$ m, $L = 162$ m, and $\beta = 0.11$ in France (see Figure 8.47) and Sayano-Shushensk arch-gravity dam (Figure 1.28) – $h = 243$ m, $L = 1,070$ m, $\beta = 0.436$, and $V = 8$ million m in Russia.

(a)

(b)

Figure 1.26 Gergebil arch-gravity dam (USSR, now Russia): (a) view from the downstream and (b) top view.

(a)

(b)

Figure 1.27 **Gunib arch dam (USSR, now Russia): (a) view from the downstream and (b) top view.**

Arch-gravity dams built in relatively narrow sections are known, which was associated with specific conditions. In the USA (1964), in the body Glen Canyon dam with $h = 216$ m, $L = 458$ m, $\beta = 0.424$, and $V = 3.65$ million m^3, pipelines of HEP were designed. Pipelines in a significant degree predetermined the form of arch-gravity dam. Kirdzhali dam with $h = 103$ m, $L = 340$ m, and $\beta = 0.379$ was built in Bulgaria (1962) in very difficult geological conditions (chloride-biotite shales with talc zones included in the left-bank abutment weakened sliding surfaces with fall towards the downstream).

Among the dome dams built during this period, the highest concrete dams in the world are as follows:

- Inguri (Figures 1.29 and 1.30) with $h = 271.5$ m, $L = 752$ m, $\beta = 0.228$, and $V = 3.90$ million m^3 in Georgia (1980);
- Vaiont with $h = 266$ m, $L = 190.1$ m, $\beta = 0.087$, and $V = 0.35$ million m^3 (Figure 1.31) in Italy (1961) with perimeter joint;
- Mauvoisin with $h = 236$ m, $L = 520$ m, $\beta = 0.224$, and $V = 2.03$ million m^3 (see Figure 8.48) in Switzerland (1957);
- Chirkei with $h = 232.5$ m (the height of the arch part – 184.5 m), $L = 280$ m, $\beta = 0.27$ (arch part 0.175), and $V = 1.39$ million m^3 (Figure 1.32) in Russia (1978);

Figure 1.28 Sayano-Shushensk arch-gravity dam, 242 m high (USSR, now Russia). View from the downstream.

- Contra with $h = 220$ m, $L = 380$ m, $\beta = 0.13$, and $V = 0.67$ million m^3 (see Figure 8.45) in Switzerland (1965);
- New Bullards Bar with $h = 197$ m and $L = 789$ m (see Figure 8.62, 1969) and Mossy-rock with $h = 185$ m, $L = 502$ m, $\beta = 0.206$, and $V = 0.98$ million m^3 (1958) in the USA (see Figure 8.60) without perimeter joint.

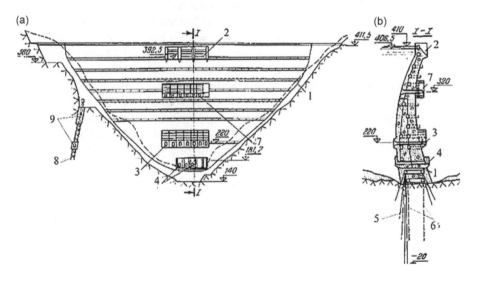

Figure 1.29 Inguri arch dam (USSR, now Georgia): (a) view of the downstream face and (b) section along the center console; 1 – perimeter joint, 2 – surface spillway, 3 – depth outlets, 4 – construction outlets, 5 – cement curtain, 6 – drainage curtain, 7 – spillway (project from which refused), 8 – tectonic fault, and 9 – fault filling.

Figure 1.30 Inguri arch dam (USSR, now Georgia): (a) view from the downstream and (b) top view.

Dome dams were also built in wide sections with a ratio of $L/h = 4.5$, for example, Valley di Ley – $h = 141$ m, $L = 690$ m, $L/h = 4.9$, and $\beta = 0.2$ in Switzerland (1961, see Figure 8.59); Kariba – $h = 128$ m, $L = 579$ m, $L/h = 4.5$, and $\beta = 0.159$ between Zambia and Zimbabwe (Figure 1.33, 1959); and Pangola – $h = 89$ m, $L = 449$ m, $L/h = 5.04$, and $\beta = 0.21$ in South Africa (see Figure 8.31, 1968) [65].

There are known cases of the construction of dome dams in difficult geological conditions with significant discontinuity and heterogeneity of the rock foundation in regions of high seismicity, which required engineering reinforcement measures.

At the foundation of the Inguri arch dam, a tectonic fault and four large fissures were sealed with concrete, the saddle was cut with a joint system (see Figure 8.37) to ensure the strength of the dome part of the dam during possible movements along the fault under seismic effects of eight points. Engineering measures to strengthen the rock foundation were carried out in France on the Mauvoisin dam (at the foundation are siliceous shales alternating with layers of clay-carbonaceous shales) and also in Japan at Kurobe-4 dam (1960) – $h = 186$, $L = 492$ m, and $\beta = 0.213$ (granites in the upper part of the foundation with severe tectonic disturbances, seismicity of 10–11 points) and Nagowado (1969) – $h = 155$ m, $L = 356$ m, and $\beta = 0.226$ (at the foundation are shales with weak interlayers, seismicity of 10–11 points, see Figure 4.4).

Since the late 1980s of the XX century, during the construction of arch and arch-gravity dams, the RCC technology was also used. In this case, the dams were made of a simpler form with a vertical or close-to-it upstream face.

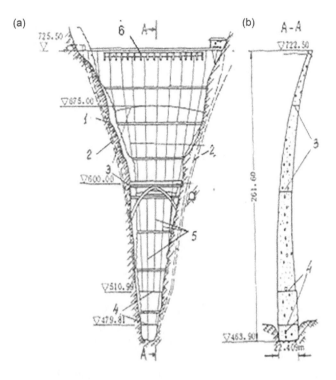

Figure 1.31 Vaiont arch dam (Italy), 266 m high: (a) scan along the downstream face and (b) profile of the dam; 1 – contour of the foundation, 2 – perimeter joint, 3 and 4 – subhorizontal joints, 5 – intersection joints, and 6 – spillway.

Figure 1.32 Chirkei arch dam (former USSR, now Russia), 232.5 m high. View from the downstream.

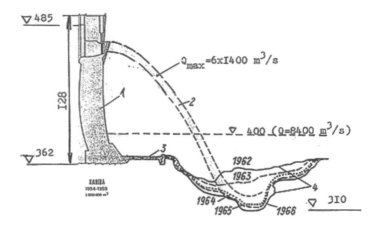

Figure 1.33 Kariba arch dam, 128 m high (between Zambia and Zimbabwe). The erosion funnel in the foundation gneisses after passing of the flood discharge: 1 – dam, 2 – jet trajectory, 3 – apron, and 4 – erosion surface in years.

The first RCC arch-gravity dams built in South Africa in 1989–1990 were Wolverdans – $h = 70$ m, $L = 268$ m, and $Vr = 180$ ($V = 210$) thousand m^3 and Knellport – $h = 50$ m and $L = 200$ m, which were made cylindrical with a laying of the downstream face 0.5.

The first RCC arch dam Puding (see Figure 8.71) built in China in 1994 is cylindrical – $h = 75$ m, $L = 196$ m, and $Vr = 103$ ($V = 145$) thousand m^3 with laying of the downstream face 0.35. The highest RCC arch dams made in China are Shimenzi (see Figure 8.70a) – $h = 109$ m, $L = 176$ m, and $Vr = 188$ ($V = 211$) thousand m^3 built in severe climatic conditions in 8-point seismic region (2001) and Shapai (Figure 1.34) – $h = 132$ m, $L = 250$ m, and $Vr = 365$ ($V = 392$) thousand m^3 (2002) in which water tightness is ensured by laying from the upstream face of vibrated concrete 2–4 m thick and a geomembrane made of PVC tape.

Figure 1.34 Shapai arch dam (China). View from the right bank.

Arch and arch-gravity dams were mainly constructed by spillways, with surface spillways and deep outlets located in the central part of the dam. Surface spillway dams were usually carried out with the deflection of a freely falling jet from the head located on the crest of the dam. In spillway dams, a developed spillway surface was extremely rare, for example, in the arch-gravity dam of the Aldeadavila (see Figure 8.5), $h = 139.5\,$m in Spain (1964), and in the arch dam Ova-Spin (Figure 1.35), $h = 73\,$m, $L = 130\,$m, and $\beta = 0.137$ in Switzerland (1968) with a spillway over the HEP building.

In some cases, surface spillways were established near bank abutments, for example, Hitoshuze dam (see Figure 8.42), $h = 148\,$m, in Japan (1963) or in the piers and Santa Eulalia dam, $h = 74\,$m (see Figure 8.21), in Spain (1966).

Arch dams with deep outlets were usually erected with a rejection of the jet, for example, the Boundary dam (see Figure 2.28), $h = 104\,$m, $L = 226\,$m, and $\beta = 0.086$, in the USA (1967), designed for flow discharge 7.1 thousand m^3/s and Kariba (Figure 1.33) with a very large specific flow rate equal to $176\,m^2$/s with a significant jet departure length. When the flood discharge over a period of 5–7 years was up to 9 thousand m^3/s, an erosion funnel formed up to 50 m deep in the channel behind the dam, composed of highly fractured quartzites and biotite gneisses; subsequently, the depth of the funnel increased to 70 m and was stabilized.

At many HPs behind the arch dam, there was a HEP building with a water intake and pipelines of the HEP in the dam. To reduce their impact on the SSS of the dam and provide more favorable conditions for construction, water inlets were usually arranged from the upstream side outside the dam contour, and head pipelines were installed on the upstream or downstream faces, for example, the thick domed dam of Nagowado, 156 m high (Figure 1.36b), in Japan (1969); the arch-gravity dam of the Sayano-Shushensk HEP (Figure 1.28); and the thick domed Mossyrock, 184 m high (see Figure 8.60), in the USA (1964) with the intersection of the dam body in the lower part.

Figure 1.35 **Ova Spin arch dam (Switzerland), 73 m high.**

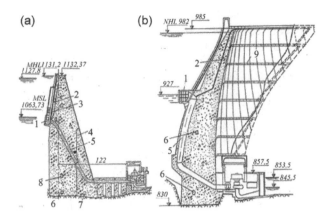

Figure 1.36 (a) Glen Canyon arch-gravity dam (USA) and (b) Nagowado dome dam (Japan): 1 – water intake head, 2 – aeration pipe, 3 – gate, 4 – construction seam, 5 – head pipeline, 6 and 7 – cementation and drainage gallery, 8 – viewing gallery, and 9 – service bridges.

In some cases, in arch-gravity and thick domed dams, pipelines were placed in the body of the dam, for example, in the arch-gravity Glen Canyon dam (Figure 1.36a), 216 m high (1964), and in the New Bullards Bar thick dome dam (see Figure 8.62), 197 m high (1969), in the USA. The arrangement of spillways, water intake, and pipelines of HEP in dams had a significant impact on the type and design of the dam, but in general, such a combination provided favorable operating conditions and a decrease in the total cost of the hydraulic plant.

The design of arc dams was also influenced by the traditions of national engineering schools. So, if Italian engineers in almost all dome dams arranged contour (perimeter) joints, then most French and American engineers consider that the device of all kinds of joints to be ineffective, since they reduce the static indeterminacy of the dam and thereby reduce its bearing capacity. However, in a number of American dams, joints and cuts were used in the lower zone of the dam on the upstream face.

Dome arch dams were built in narrow and relatively wide sites. When designing them, much attention was paid to the selection of rational outlines of the axes of the arches.

Arches with more complex compared with circular outlines: three-centered outlined in a parabola, logarithmic spiral, and other curves, often with a broadening to the heels, were selected based on the desire to provide the most optimal stress distribution in the dam and more favorable working conditions of rock massifs of bank abutment.

Thanks to the progress made in the design of arch dams and the use of concrete of higher classes, arch dams have become more economical. Usually, depending on the width of the site, the volume of concrete of arch dams on average was 50%–70% of the volume of concrete of gravity dams in the same sites.

Arch dams proved to be very reliable structures. A number of dams built in regions with high seismic activity (8–10 points) remained almost intact after strong earthquakes, for example the Carfino Dam, 40 m high, in Italy.

The Vaiont arch dam, 266 m high, withstood enormous overload when a landslide with a volume of about 300 million m³ of rock fell into the reservoir, which led to the formation of a wave with the height of the overflow layer of water through the crest of the dam 110–135 m (see Figure 2.3).

Of the highest dams (200 m and higher) built and under construction in 2000, arch dams accounted for 60%. Achievements in the construction of arch dams in the second half of the XX century and their high competitiveness and effectiveness ensured their further widespread construction in the XXI century.

Buttress dams. In the second half of the XX century, multiarch and buttress dams were further developed, and the construction of dams with flat head ceilings practically stopped.

In multiarch dams, the distance between buttresses was increased to 35–50 m, due to which a decrease in concrete volume was achieved. So in the Granvaal dam with a height of $h = 88$ m and a length along the crest $L = 330$ m in France (1960) and Neber, $h = 65$ m and $L = 450$ m in Tunisia (1955), the distance between the buttresses was 50 m and in the Erragen dam, $h = 80$ m and $L = 500$ m, in Algeria (1958) was 35 m.

Certain difficulties were encountered when arranging spillways in multiarch dams, for example, Granvaal dam in which a chute spillway was constructed in the central buttresses and a HEP building was located between the buttresses (Figure 1.37).

Sometimes, multi-arch dams were erected with "fork" buttresses to ensure the stability of buttresses in the transverse direction, for example, in the Norano dam, 27 m high, in Norway and an 18-m arch span.

In France (1961), a combined dam, the Roseland Dam, of an original design was built with a central arch 150 m high and 200 m span and side buttresses with flat ceilings (Figure 1.38).

The highest multiarch dam Daniel Johnson (Manikuagan-5) with a height of $h = 215$ m, a crest length of $L = 1314$ m, and a concrete volume of $V = 2.23$ million m^3 (1970) was erected in Canada with a central arch with a 161.5 m span and with 13 arches of 76.2 m span (Figure 1.39). On the heels of the central arch, there are two spillway openings with a toe-springboard at the exit to reject the jet [61].

Multiarch dams, despite a significant decrease in concrete volumes, to 60% or more, have significant drawbacks associated primarily with the complexity of the design, large reinforcement, which led to an increase in labor costs, to an increase in the cost

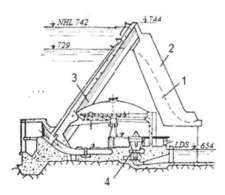

Figure 1.37 Granvaal multiarch dam, 88 m high (France): 1 – chute spillway, 2 – buttress, 3 – arch ceiling, and 4 – HEP building.

Figure 1.38 Roseland combined dam with a central arch, 150 m high, and side buttresses with flat ceilings (France).

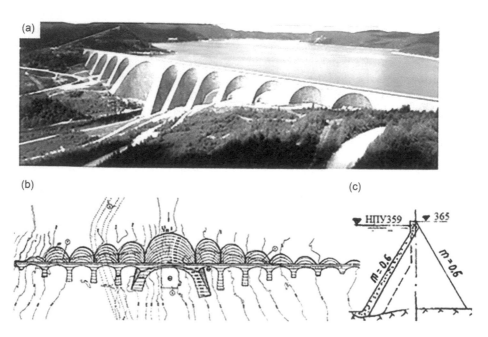

Figure 1.39 Daniel Johnson (Manicoagan-5) multiarch dam, 215 m high (Canada): (a) view from the downstream; (b) plan; and (c) section along the bank arch.

and time of construction, and ultimately to a decrease in efficiency. In this regard, such dams were rarely built in the last quarter of the XX century.

Massive buttress dams. Massive buttress dams, compared to other types of buttress dams, are more widely developed due to simplified construction, limited reinforcement, and improved construction conditions.

Massive buttress dams up to 125–180 m high, mainly spillway dams, were built in severe climatic and difficult geological conditions often in regions with high seismic activity (up to 9 points and higher). Along with dams with single buttresses, dams with double buttresses are known (proposal by engineer *C. Marcello*), which were used in Italy; the toe of the buttresses was flat, with a circular or polygonal outline (Figure 1.40).

In Scotland, low dams were constructed with single buttresses with a slightly inclined flat upstream face and concreting of temporary joints between the toe of the buttresses, for example, dam Loch Sloy, 55 m high (Figures 1.40g and 1.41).

In Japan, in areas with high seismicity (8–9 points), high massive buttress dams were built such as Khatanagi, $h = 125$ m and $L = 275$ m, with single and double buttresses and the distance between them $\ell = 16$ m and $\ell = 22$ m (1962) and Ikawa, $h = 103.6$ m, $L = 240$ m, and $\ell = 14$ m, with single buttresses (1959). In Italy, a dam with double buttresses was erected (1952), $h = 111.5$ m, $L = 253$ m, and $\ell = 22$ m.

Dams with single buttresses were built in the former USSR: Andijan, $h = 115$ m, $L = 1,180$ m, and $V = 2.9$ million m^3 (now Uzbekistan, 1980), in an area with a seismicity of 9 points and Zey (Figure 1.42), $h = 115.5$ m, $L = 714$ m, and $V = 2.1$ million m^3, in severe climatic conditions in a region with high seismicity (now Russia, 1978).

The highest massive buttress dam Itaipu (Figures 1.43 and 1.44), 196 m high, was built in Brazil-Paraguay (1982) as part of the hydraulic plant with the largest HEP in the world at that time with a capacity of 12.6 million kW, which further increased to 14 million kW.

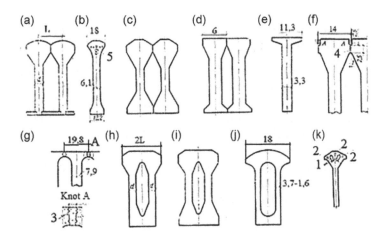

Figure 1.40 Schemes of horizontal sections of buttresses of massive buttress dams: (a) and (l) with a rounded outline of the head face [(l) Olef dam, 54 m high]; (b–f) with a polygonal outline of the head face [(b) Giaveretto dam, 83 m high]; (l) with a cut head: 1 – hollows; and 2 – holes for drainage of filtration water; (g) with a flat head face of the Scottish type dam Loch Sloy, 55.4 m high; 3 – joint; (d) and (e) Swedish type with massive and thin buttresses [(d) Hidalgo-Elv dam and (e) Lengbnori dam, 32 m high]; (f) with a cut in the joint at the pressure face – Menjill dam, 105 m high.

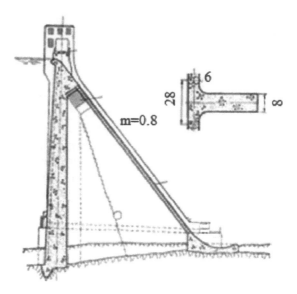

Figure 1.41 Massive buttress dam Loch Sloy (Scotland), 55 m high.

The second half of the XX century is the time of prosperity and "extinction" of buttress dams. As the German philosopher *G.W.F. Hegel* wrote: "Truth is born as heresy and dies as prejudice."

1.3 Prospects for the construction of concrete dams in the XXI century

At the level of 2000 with the capacity of all HEPs, 670 million kW and annual electricity production of 2,650 billion kWh, the development of cost-effective potential in the world amounted to 31%, in Europe 71%, Central and North America 70%, South America 33%, Asia only 22%, and Africa only 6%. Therefore, the development of hydropower resources in the countries of Asia, South America, and Africa is of the greatest interest.

By 2010, the capacity of all HEPs increased by 38% and amounted to 926 million kW, the annual production of 3,550 billion kWh.

According to forecasts of socioeconomic development of the world in the XXI century, electricity consumption will increase by two times by 2030 and by four times by 2050 compared with 2000, and the demand for water for public water supply, industry, agriculture, and irrigation will also increase sharply.

In the conditions of a crisis state of the environment, it is planned, first of all, to further develop renewable hydropower resources with the construction of HPs with integrated reservoirs.

Currently are being built HEPs with a capacity of about 160 million kW in the world, of which in Asia – 126 million kW.

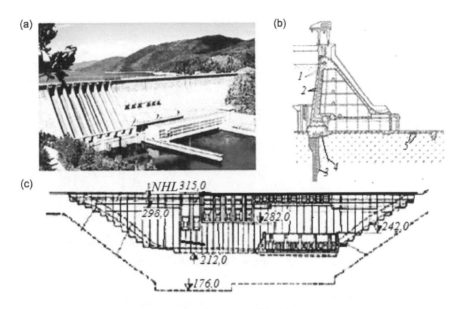

Figure 1.42 **Zeya** massive buttress dam (Russia): (a) view from the downstream; (b) section through spillway dam; and (c) view from the upstream.

Figure 1.43 **Panorama hydraulic plant Itaipu (Brazil-Paraguay).**

More than 400 million kW is planned for construction.

The construction of HPs moves mainly to the mountainous and premountainous regions, where there are favorable conditions for the construction of concrete dams on rock foundations.

Of the dams built and under construction in the world in 2010 with a height of 200 m and above, 75% are concrete dams, including 55% arch dams and 20% gravity dams (Table 1.3). By 2015, the construction of the highest Jinping-1 arch dam with a height of 305 m (see Figure 8.16) was completed; Xiowan arch dam with a height of 294.5 m

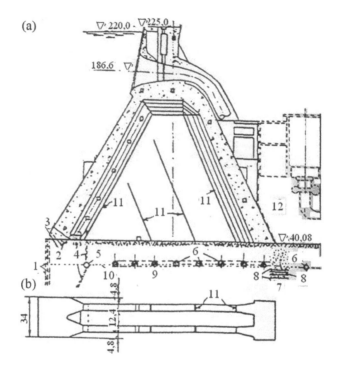

Figure 1.44 Itaipu massive buttress dam, 196 m high (Brazil-Paraguay): (a) section along the spillway dam and (b) horizontal section on the buttress; 1 – cement curtain, 2 and 3 – strengthening cementation, 4 – drainage halls, 5 – drainage tunnel, 6 – antislide concrete dowels, 7 – dense basalt, 8 – breccia, 9 – porous basalt, 10 – heterogeneous rocks, 11 – temperature-shrink joints, and 12 – HEP building.

(see Figure 8.26), Xiluodu with a height of 278 m (see Figure 8.14), and Laxiva 250 m high were built in China; and Kishau gravity dam, 236 m high, was built in India. The largest Three Gorges HEP in the world was built in China with a capacity of 22.5 million kW, with a gravity dam 181 m high, a 2,335 m crest length, and 16.1 million m^3 concrete volume (Figure 1.45).

In 2002, the construction of a high RCC gravity dam Miel-1, 188 m height, 340 m crest length (see Figure 7.6), and 1.75 million m^3 concrete volume, was completed in Colombia region with a seismicity of 8 points; in 2009, China completed the construction of the highest RCC gravity dam Longtan, 216.5 m high (see Figure 2.21)

In total, in 2010, the number of dams built from RCC reached about 600 in the world including 18% of dams with a height of more than 100 m. Of the 55 dams built in 2015, more than 100 m high, 80% are gravity (from RCC 70%) and 20% arch (from RCC 2%) [194,243,262].

Considering the evolution of concrete dam engineering, it is noted that a variety of creative approaches, struggle of ideas, and various ways of engineering search aimed

Table I.3 Dams with a height of 200 m and above being in exploitation and construction

No.	Dam name	Country	Year of construction	Height (m)	Type of dam
1	Jinping-1	China	Under construction	305	A
2	Nurek	Tajikistan	1980	300	RE
3	Lianghekou	China	Under construction	295	RE
4	Xiowan	China	2012	2,945	A
5	Grand Dixence	Switzerland	1962	285	G
6	Xiluodu	China	2010	278	A
7	Inguri	Georgia	1980	2,715	A
8	Vaiont	Italy	1961	266	A
9	Nuozbadu	China	2012	261	RE
10	Teri	India	2002	2,605	RE
11	Laxiva	China	2010	250	A
12	Dirinir	Turkey	Under construction	247	A
13	Mayco	Canada	1973	243	RE
14	Guavio	Colombia	1990	243	RE
15	Gibe-III	Ethiopia	Under construction	245	RCC
16	Sayano-Shushensk	Russia	1981	242	AG
17	Changheba	China	Under construction	241	RE
18	Ertan	China	1999	240	A
19	Kishau	India	2005	236	G
20	Mauvoisin	Switzerland	1957	236	A
21	Orovil	USA	1967	235	RE
22	Irminik	Turkey	2002	235	A
23	Goupitan	China	Under construction	233	A
24	Shuibuya	China	2009	233	RE
25	Chirkei	Russia	1978	2,325	A
26	Karun 4	Iran	Under construction	230	A
27	Ta Shang	Burma	Under construction	2,275	RCC
28	Bhakra	India	1963	226	G
29	Hoover	USA	1936	222	AG
30	Jingpinghe	China	2010	221	RE
31	Contra	Switzerland	1965	220	A
32	Dworshak	USA	1973	219	G
33	Longtan	China	2009	2,165	G (RCC)
34	Glen Canyon	USA США	1964	216	AG
35	Toktogul	Kyrgyzstan	1978	215	G
36	Daniel Johnson	Canada	1968	215	MA
37	Portugués	Puerto Rico	Under construction	210	A (RCC)

A, arch; RE, rock-earth; G, gravity; RCC, rolled concrete compacted; AG, arch-gravity; MA, multi arch.

at improving structures and construction technologies and increasing reliability and competition of various types of dams. Concrete dam engineering as well as other areas of technology are characterized by the emergence and accumulation of contradictions, the resolution of which due to the emergence of new solutions and inventions gives a quantum leap and opens up new effective directions.

The needs of practice require new more effective solutions that can be found in a constant engineering search taking into account the features and the variety of natural conditions of HP.

Figure 1.45 The panorama most powerful in the world of the Three Gorges HEP with a gravity dam 181 m high (China).

Note

1 Analysis of the thermal stress state of the dam during the construction period was performed by specialists of UkrHydroproject (Kharkov) and IIGH (International Institute of Geomechanics and Hydrostructures).

Chapter 2

Layouts of hydraulic projects with concrete dams

"There is supreme courage: courage ... creation, where the plan is vast embraced by creative thought ..."

A.S. Pushkin (1799–1837), Russian poet

2.1 General

The HPs includes a complex of hydraulic structures (HS), the purpose of which is to use a water flow to generate electricity, water supply, irrigation, and water transport, as well as flood protection.

HS must meet safety requirements, ensure normal operating conditions, and have the ability to control their condition during construction and operation.

The needs of practice require new, more effective solutions that can be found in a constant engineering search, taking into account the peculiarities and the variety of natural conditions of HS.

The main HPs include structures whose destruction or damage leads to the termination or violation of the normal operation of the HP, HEP (SPP), water intakes, shipping structures, and flooding of the protected area.

The main HSs by functional attribute are divided into the following:

- water-retaining and spillway structures (dams and spillways) intended for the formation of a front head of a HP, creating a head at HEP and ensuring necessary water discharges are passed in downstream, including flood, ice, and sediment flushing;
- HEP (SPP) structures designed to generate electric energy including water inlets and water conduits supplying water to turbines in the building HEP and diverting water to the downstream;
- shipping structures – locks and ship elevators, designed for passage of vessels through the HP;
- intakes for water supply and irrigation.

Secondary HSs include dividing walls, piers that are not part of the head front, and fish protection structures.

The main structures of large HPs are objects of increased responsibility, the malfunction of which leads to significant damage, and their accidents can have disastrous consequences. Therefore, ensuring their reliable operation and safety is the most important factor in the design, construction, and operation.

The main HSs depending on their height and type of foundation soil, the socioeconomic responsibility of HP, and the possible consequences of their destruction are divided into four classes [9].

Class I includes the following:

- HSs of nuclear power plant (NPP) regardless of capacity;
- dams with a reservoir volume of more than $1\,km^3$ and capacity of HEP and thermal power stations (TPSs) more than 1 million kW;
- concrete dams on a rock foundation at a height of more than 100 m.

Class II includes concrete dams on a rock base at a height of 60–100 m, class III – from 25 to 60 m, and class IV – less than 25 m.

The class of the main structures that make up the head front is established by the structure assigned to a higher class.

HPs with concrete dams on rock foundations are built on plain and mountain rivers in a variety of natural conditions.

The advantages of concrete dams include ensuring the compact layout HS, favorable conditions for the construction, and operation of spillway structures, as well as passing of construction discharges.

When choosing the layout of HP structures, it is necessary to ensure the reliability of structures at all stages of their construction and operation and maximum economic efficiency.

When designing several competing HP, sites are considered that differ in topographic, engineering-geological, construction, and other conditions. For each site, a rational layout is developed, technical and economic indicators are determined, and based on their comparison, the site is selected. In the selected site, taking into account the data of more detailed geological surveys and studies, several layout options are developed on the basis of a more detailed design with clarification of the types and designs of structures. So, options can be developed with concrete gravity and arch dams (under appropriate conditions) that differ in the types and location of HEP structures, spillways, and passing construction discharges.

The choice of layout is a complex integrated engineering task related to the multifactorial nature and variety of initial conditions, many of which vary over time and are of an uncertain or probabilistic nature. These conditions include the following:

- natural conditions of the site of the HP, including topographic, geotechnical, seismic, hydrological, and climatic conditions;
- the composition of the HP structures in accordance with their integrated purpose and the scheme of energy use of the river as well as its water management and water energy parameters;
- architectural appearance;
- construction conditions;
- operating conditions and reliability of the structures;
- environmental protection.

These conditions for each HP have their own specifics in connection with which the layout is selected as result of a technical and economic comparison of the variants most appropriate in the given natural conditions, with the selection of the most rational types and designs of structures for each variant, their location, and construction technology. The development of layout variants is based on materials from surveys, studies, forecasts of changes in existing conditions, analysis of existing experience in the design, and construction and operation of HPs under similar conditions as well as current trends in dam design and the achievements of scientific and technological progress. At the same time, the choice of types of structures, construction technology, and layout solutions is inextricably linked. It should be noted that the technical and economic indicators of the designed HP largely depend on the rational choice of the type and technology of erecting a concrete dam, the cost of which in many high-pressure HPs is more than 50% of the total cost of the main structures.

When choosing layout solutions, the possibility of using concrete dams to passing construction and operating discharges to supply water to the units of a HEP is considered.

In connection with a wide variety of possible layout solutions for HPs with concrete dams, the following are the main provisions and principles for the development of layouts.

Topographic and geological conditions. The topographic conditions of the sites of the HP determine the length of the head front of structures, the conditions for the location of spillways, HEP buildings and other main structures of the HP, and the passage of construction discharges as well as the placement of secondary and temporary structures, including construction base and automobile roads.

The geological conditions of the structures' foundation and banks' seismic conditions must meet the requirements of reliable and safe operation of head and other structures of the HP.

Topographic, geological, and seismic conditions to a large extent determine the choice of the type of dam and the layout of the HP structures. Massive gravity dams can be designed in various topographic and engineering-geological conditions in narrow and wide sites and arch and arch-gravity dams – under appropriate geological conditions in relatively narrow sites mainly with respect to the width of the site at the level of the dam crest to the height of the dam less than five.

Hydrological conditions. The hydrological conditions depend on the discharges passed through the HP including the maximum flood discharge as well as discharges during the construction period. The maximum designed flood discharges are for large HP:

- Krasnoyarsk – 29.8 thousand m^3/s;
- Sayano-Shushensk – 24.4 thousand m^3/s;
- Bureya – 33.1 thousand m^3/s in Russia;
- Son La (Figure 2.10) – 60.08 thousand m^3/s in Vietnam;
- Itaipu – 72 thousand m^3/s in Brazil-Paraguay (see Figure 1.43);
- Tukuri – 100 thousand m^3/s (Figure 2.1) in Brazil.

The observed maximum discharge in Yangtze river in the area of the Three Gorge HEP in China reached 110 thousand m^3/s in 1860 and 1870 [24]. At the same time, the flow capacity that must be extinguished when the maximum design floods are passed can reach more than 100 million kW [184,194].

The climate especially in areas with a sharply continental climate with a harsh winter has an impact on the choice of type and design of structures and the conditions for construction and work.

Figure 2.1 Tucuri HEP (Brazil) with a capacity of 8.3 million kW and a dam 108.0 m high.

The composition of the structures of the HP, its water supply, and water energy parameters. Under modern conditions as a rule, the integrated use of HPs is envisaged: in the interests of energy, water supply, irrigation, water transport, and other sectors of the economy and for flood protection, taking into account environmental protection requirements. In this regard, HPs along with dams may include HEP structures or SPP, locks and ship elevators, water intakes, and fish passages. In this case, it is necessary to take into account the development and deployment of production forces and industrial facilities in the construction area in accordance with state and regional programs of socioeconomic development.

In some cases, the layout solutions provide for the possibility of developing the HP in the future with increasing head, expanding HEP and construction shipping structures.

The main purpose of many HPs is the energy use of river flow. Such hydraulic plants are called a hydropower HP or HEP. To create head in a selected section of the river, the following basic schemes of HP are used [23,65]:

• **Dam** in which the head on the HEP is created due to the head water of the river level with the formation of a reservoir, which is also used to regulate the flow (daily, seasonal, and many years) and ensure the required HEP regime. Under this scheme, the structures of HEP are part of the HP and the conditions for the location of the HEP building in many cases determine its layout. The dam scheme is used in flat and mountainous conditions. For example, Dnepr HEP was built in Ukraine (see Figure 1.8). Krasnoyarsk (see Figures 1.21 and 2.4), the Bratsk (see Figures 1.22 and 2.5), Sayano-Shushensk (see Figures 1.28 and 2.6), Chirkei (see Figure 1.22), and Bureya (see Figure 2.7) in Russia, Toktogul (see Figure 1.23) in Kyrgyzstan, Three Gorges (see Figure 1.45) in China, Grande Dixence (see Figure 1.19) in Switzerland, Hoover (see Figure 1.11) and Dworshak in the USA, Itaipu (see Figures 1.43 and 2.9) in Brazil-Paraguay, and Son La (see Figure 2.10) in Vietnam (a joint project of Russia and Vietnam).

- **Derivation** in which the head on the HEP is formed mainly due to derivation which is carried out in most cases in the form of head tunnels and water conduits. A dam with a small reservoir, which can also be used for daily or weekly regulation is being constructed to take water to the HEP.
- **Mixed dam derivation** in which the head on the HEP is formed due to the dam and derivation. For example, at the Inguri HEP (Georgia) with a capacity of 1.3 million kW, the maximum head at the HEP equal to 404 m is created by an arch dam 272 m high (see Figures 1.29 and 1.30) and head derivation 16 km long.

Derivation and dam derivation schemes are mainly used in mountain conditions.

Depending on the head, hydraulic plant and HEP are divided into the following:

- low head – up to 20 m;
- medium head – from 20 to 50 m;
- high head – more than 50 m.

The layout, types, and constructive decisions of the structures depend on the main water supply and water energy parameters: levels of the upstream (NHL and SRL – normal and forced levels, respectively, and DVL – dead volume level) and downstream, which regulates the volume of the reservoir, water discharges, heads, discharges, and installed capacity of HEP. At the same time, the main parameters of the HP should be linked to the parameters above and below the HEP located in the cascade of HEP on the river.

General HP plan is created on the basis of the layout of the main structures and should be linked to the district layout, the prospect of development of the district, including industrial facilities, settlements, and transport highways. At the same time, she recommends using the HP as a bridge for transport and reliable communication with external roads. All structures of the HP should be connected by automobile roads and accessible for repair work in normal operation and in emergency situations.

The architectural appearance of the HP is developed taking into account aesthetic requirements as a single volumetric and spatial composition of the structure's complex linked to the surrounding landscape. The proportions and scale of the structures and the rhythm, color, and texture of their surface should be subject to a single composite design contributing to the aesthetic perception of the HP in harmony with the surrounding landscape (Figure 2.2).

Teshkov gravity dam – the largest HS in Czechoslovakia – was built in 1920 after a great flood. The height of the dam is 41 m, the length along the crest is 220 m, and the width varies from 7 m (in the middle) to 37 m (at the bases). The oval reservoir is filled with pure drinking water, which is to a lot of fish. Swimming there is strictly forbidden, but for a fee you can buy a ticket and fish-read. The importance and beauty of the dam is evidenced by the fact that it was recognized as an object of cultural heritage in 2010 and a technical monument of national importance in 1964.

The best HP forming a single whole with the environment form unique ensembles that organically fit into the natural landscape, harmoniously combined with the functional purpose of the hydraulic plants.

Many HPs with concrete dams have become famous monuments of industrial architecture, which embodied the outstanding technical achievements of their time, for example, Hoover in the USA (see Figure 1.11), Dnepr HEP in Ukraine (see Figure 1.8),

Figure 2.2 Teshkov gravity dam 41 m high (Dvur Kralove nad Labem, Czech Republic, 1920).

Krasnoyarsk HP in Russia (see Figure 1.21), Roseland in France (see Figure 1.38), Daniel Johnson (Manicoagan-5) in Canada (see Figure 1.39), Itaipu in Brazil-Paraguay (see Figure 1.43), etc.

Construction conditions. When choosing the layout of the HP structures, the placement of temporary structures, the construction and economic conditions of the district, and existing communications are taken into account. A significant impact on the layout is provided by the passing of construction discharges throughout the construction period, the phased commissioning of structures in stages, the conditions for the production of works using effective methods of construction and high productivity equipment, and reduction in construction time.

The wider the river and the wider the range of discharges changes and the longer the construction period, the more difficult it is to solve the scheme for passing construction discharges and transferring them to the corresponding spillways as the dam grows and queues are introduced. The utilization of construction spillways in a concrete dam is maximally used to pass construction discharges, including through unfinished deep operational spillways and turbine conduits. An important task is to ensure the reliability of the structures under construction during the entire construction.

In many cases, the construction of dams is on a critical path, determining the total construction period and the deadline for commissioning the first and subsequent stages. Ensuring favorable conditions for their construction becomes a significant factor in reducing construction time.

Important indicators determined by the construction conditions and affecting the choice of layout are the volume of work, concrete casting technology, the total cost, and construction period, as well as the cost and timing of the commissioning of the first and subsequent stages (start-up complexes).

In the context of long periods of construction of large HP, the construction and commissioning of structures in stages with a dam and a HEP building not fully erected allows for regulation of river flows and generation of electric energy at HEP during ongoing construction. It ensures a reduction in the payback period of investments and an increase in overall efficiency. Therefore, during the construction of almost all large HP, the introduction of stages is provided. At the Sayano-Shushensk HEP, the first units were commissioned for 5 years of construction at an intermediate head with volumes completed, which amounted to 30% of the total volumes.

At the world's largest HEP Three Gorges (China), the construction of which began in December 1994, the first concrete was laid in 1997, the first aggregates were introduced and navigation was opened on a permanent shipping lock in 2003, and the capacity of HEP amounted to 9.8 million kW in 2005. In 2009, construction was completed with the commissioning of all units with a capacity increase of up to 18.2 million kW.

Operation conditions. The layout of the HP structures should ensure their reliable operation and safety during the entire design period of operation of the HP and favorable operating conditions including the period of temporary operation during commissioning of the structures by stages, as well as the conditions for repair and restoration work. HP structures must fulfill their functions in accordance with their integrated purpose during both temporary and permanent operation. It is necessary to ensure favorable hydraulic regimes in the upstream and downstream during flow passage and especially during the most intense periods of passage of flood discharges and ice. Therefore, the layout solutions of HP, which include structures of classes I and II and with appropriate justification and class III, are tested on spatial hydraulic models, which study the conditions for supplying water in the upstream to the water spillway structures, HEP intake, flow spreading, and channel erosion in the downstream during the operational and construction periods. It is necessary to develop such regimes for passing discharges that ensure reliable operation of structures taking into account deformations of the channel in the downstream including the stability of structures. Under conditions of normal operation of a HEP, it is necessary to exclude the increase in water levels in the downstream caused by the formation of bars and the decrease in levels in the downstream associated with channel depletion below the minimum permissible value under the conditions of deepening of the working turbine wheel.

In addition, in high-head HP, especially in relatively narrow gorges, it is necessary to ensure reliable operation of structures in the event of landslides of large masses of soil of bank slopes in the reservoir with the formation of a wave similar to the disaster on the Vaiont arch dam, as well as in the case of landslides in the downstream overlapping the riverbed with the formation of backwater, which can cause flooding of the HEP building.

The Vaiont arch dam (Italy) for its time was the highest in the world and innovative in design (see Figure 1.31). The dam has a perimeter joint, which in the central part is 50 m above its bottom. There are also three subhorizontal joints that are designed to improve the dam in the arched direction. During the construction process, a greater disturbance of the rock mass was revealed within the limits of the support of the dam than expected, and therefore, it was decided to strengthen the abutment at the down face of the dam almost over the entire height using prestressed anchors in combination with concrete pylons. A total of 180 active anchors were installed with a pulling force of 1 MN each.

Geological surveys in the area of the future reservoir showed that there is a risk of landslide phenomena on the left bank of the river near the dam after filling the reservoir, but none of the experts including *L. Muller* could have suggested that it could happen so quickly (in 15–30 seconds) [115].

On the evening of October 9, 1963, an incredibly fast landslide suddenly arose which encompassed huge masses of rocks with a volume of 300 million m^3. During the landslide, the rocks slid by almost 500 m, blocked the 100-m gorge, and captured the opposite slope at a distance of 260 m. The landslide displaced water from the reservoir, which overlapped the dam, while the wave height at the right bank was 135 m and 110 m at the left. Having collapsed in the downstream, the water caused flood in the valley of the river Piave, which almost completely destroyed the villages of Longarone, Figaro, and others. About 2,000 people died.

The change in the speed and nature of the landslide occurred without preliminary signs. Even on the last day before the disaster, the speed of landslide movements was 20–30 cm/day, which was several million times less than the speed of a landslide on October 9. The catastrophe happened so suddenly that the brigade assigned to monitor the movement of the landslide and carry out ongoing work not only did not have time to warn their families living in Longarone, but also died.

The research work that preceded the crash included model tests, the purpose of which was to determine the duration of the creep of the left-bank slope. On the model, the slope slides for 1 minute in the form of two parts and the second part slides during the passage of the wave arising from the slide of the first part. The model showed that in this case, there should be no overflow over the dam. A potential landslide was blocking the reservoir, and a bypass tunnel was passed to preserve the river flow on the right bank.

It would seem that the designers provided for everything:

• the landslide is inevitable, but it did not form a wave of overflow through the crest of the dam;
• the landslide body divided the reservoir into two parts, and a bypass tunnel was passed to continue the operation of the HEP.

But contrary to model studies as well as the analysis of numerous landslide phenomena in various regions of the globe [115], the whole mass slid almost simultaneously; the sliding time was about 40 seconds (the effective destroy time was about 15 seconds, which was recorded by seismographs).

The arch dam resplendent withstood multiple overloads, thereby confirming the reliability of arch-type dams. Only a small part of the crest at the left bank was damaged (Figure 2.3).

Figure 2.3 Vaiont arch dam before (a) and after (b) the disaster.

Thanks to the previously completed reinforcement with prestressed anchors, the bank abutment also withstood a significant increase in load; however, new cracks were discovered in both abutments of the dam. After the disaster, 92% of the installed anchors turned out to be fully operational despite external damage.

A well-known specialist in soil mechanics, *K. Terzagi*, on this occasion noted: "This is a disaster, the consequences of which were impossible to predict and which increased our knowledge."

Environmental protection. The layout of the HP structures must ensure compliance with environmental requirements during construction and operation during the development of pits and changes in water flow regimes in the upstream and downstream.

It is necessary to investigate the possibility of disturbing the stability of slopes, flooding the territory in the area of the HP, erosion of the riverbed and its reformation, changes in water quality, disturbance of spawning grounds and fish migration conditions, landscape changes, and the impact on the living conditions of the population. Therefore, layout decisions should include the necessary measures to eliminate or minimize negative environmental consequences including the following:

- bank engineering protection, slope strengthening, drainage device in the upstream and downstream in the zone of influence of the HP;
- fastening the bed in the downstream to prevent dangerous erosion and its reformation;
- preservation of the general landscape, minimization of quarry areas, dumps, and land reclamation disturbed during the construction period, linking the architectural and planning decisions of the HP with the landscape;
- ensuring normal sanitary and environmental conditions in the area of placement and influence of the HP.

Feasibility comparison of layout variants. The choice of the optimal layout of the HP structures is made as result of a technical and economic comparison and analysis of the economic efficiency of variants based on the following key indicators:

- water supply and water energy indicators;
- volume of work and the total cost of construction;
- total duration of the construction;
- terms of input and cost of the first and subsequent stages;
- operating conditions of HP;
- environmental conditions.

To achieve high efficiency, it is necessary to reduce both the total cost and duration of construction as well as the cost and timing of commissioning the first and subsequent stages subject to the requirements of operation and environmental protection.

In practice, when analyzing the effectiveness of layout variants, the method of financial (absolute) efficiency is usually used based on the ratio of revenues and expenses for the estimated life of the HP and taking into account the timing of stages and the distribution of discounted investments by year of construction.

When evaluating the efficiency, the discount coefficient (norm) is the main set norm, and when calculating hydraulic plant from HEP or SPP, $E = 0.1$ is usually taken.

The main indicators characterizing the effectiveness of investments include [90,109] the following:

- discounted revenue (*B*);
- discounted expense (*C*);
- total return on revenue (*R* – result cost ratio)

$$R = B/C \geq 1 \tag{2.1}$$

- internal rate of return – the FIRR (financial internal rate of return) is determined by the value of the discount coefficient at which the discounted revenue is equal to the discounted expense; more effective is the option with the highest FIRR;
- integrated net return value NPV (net present value) is determined by the difference in all discounted revenue and expenses

$$B - C \geq 0 \tag{2.2}$$

Preferred is the variant with higher financial performance.

2.2 Hydraulic projects in wide sites

The site parameters of the HP rendering have a significant impact on the layout of structures. Depending on the width of the site, HPs are subdivided into HPs in wide sites (usually at $L/h > 5$) in lowland and piedmont conditions and HPs in narrow sites mainly in mountainous conditions.

As a rule, HPs in wide sites are constructed with gravity dams, in which spillways are arranged to pass construction and operating discharges and water inlets and water conduits for water supply to the HEP buildings are located.

The location of the dam in the site is determined by the composition and layout of the HEP structures. The axis of the dam in the plan is usually made in a direct line, thereby achieving the most economical solution. The curvilinear or broken lines of the dam axis are less common and can be caused by the peculiarities of the geological and topographic conditions of the site, the features of the location of the structures and their layout, and the conditions for passing flood discharges. The spillway dams of the Dnepr HEP in Ukraine (see Figure 1.8) and Jordano in Brazil with a height of 95 m (see Figure 7.57) were made with a curved outline of the axis.

In the dam scheme, the main structures of HPs in wide sites include the following:

- spillway, deaf, and station dams;
- dam buildings of HEP;
- intakes for water supply and irrigation;
- locks or ship elevators.

The layout of the structures of such HP is significantly affected by the conditions for passing construction discharges, especially on much water rivers and high head HP.

In wide sites, the following layouts are mainly used, which differ in the relative position in the site of the station dam, HEP building, and spillway dams:

- station dam with HEP building and a spillway dam in the bedding part of the site (bedding layout);
- station dam with the HEP building in the bedding part of the site and spillway on the bank;
- spillway dam in the bedding part of the site and a station dam with HEP building on the bank.

At the same time, shipping structures (locks and ship elevators) are located within the bank area.

The bed layout is often used in wide sites in conditions of high-water rivers in the construction of medium-head and high-head HP, e.g., Dnepr HEP with a maximum head of 38.7 m in Ukraine, Bukhtarma – 67 m, Bratsk – 106 m, Krasnoyarsk – 101 m, Ust-Ilim – 88 m, Sayano-Shushensk – 217 m, and Bureya – 122 m in Russia, Three Gorges – 181 m in China, and Salto Caxias – 67 m in Brazil. With this arrangement, usually the spillway and station dam are adjacent to each other, and a separate wall is constructed in the downstream.

The construction scheme of the main HP is determined by natural and constructing conditions. In many cases, their construction is carried out in two stages. With this variant, the construction of structures and passing construction discharges provides the following:

- arrangement of cofferdams separating part of the bed within which a spillway dam with bottom construction and deep spill openings is erected in the pit of the first stage and the passage of construction discharges during this period carried out through a cramped river bed;
- after the remaining part of the river bed is covered with cofferdams in the pit of the second stage, the remaining structures are erected, and construction discharges are passed through the construction and bottom openings after their closure through deep spillways in the body of the spillway dam or unfinished spillways and then through spillway openings.

The structure of the Krasnoyarsk HP complex (Figure 2.4) includes the following:

- a massive gravity dam with a height of 124 m and a length along the ridge of 1,065 m, including deaf, spillway, and station dams;
- dam building HEP with 12 units of the total installed with a capacity of 6 million kW;
- inclined ship elevator located in the bank of the site.

A spillway dam with seven spillway openings with a span of 25 m each ensures a flood discharge of 14.6 thousand m³/s at the SRL.

The structure of the Bratsk HP complex (Figure 2.5) includes a gravity dam with expanded joints and a dam HEP building with an installed capacity of 4.5 million kW, while there is no ship hoist, since until the completion of the construction of the entire cascade of HP on the Angara River, there is no transit shipping.

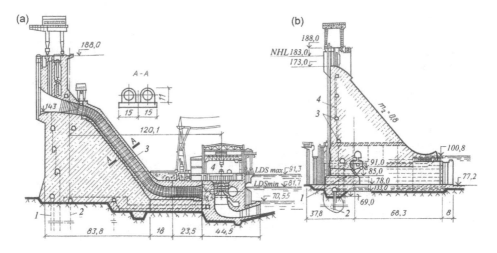

Figure 2.4 **Krasnoyarsk HEP: (a) section along the station section; (b) section in the deaf section; I– HEP building, 2 – dam, 3 – bottom spillway, 4 – cementation curtain, 5 – drainage, and 6 – galleries.**

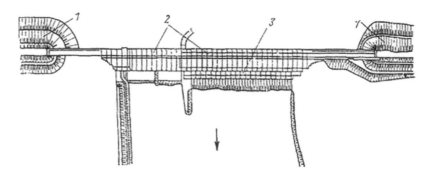

Figure 2.5 **Plan of constructions of the Bratsk HEP: I – right- and left-bank soil dams, 2 – spillway and station concrete dams, and 3 – HEP building.**

At the Sayano-Shushensk HP complex (Figure 2.6), within the arch-gravity dam 242 m high and 1,066 m long, a spillway dam is located, which provides a flood discharge of 13.6 thousand m^3/s at the SRL and a station dam with dam building HEP with 10 units with a total capacity of 6.4 million kW. In connection with the destruction of a water well that took place when floods pass, an additional coastal spillway was constructed [139].

The structure of the Bureya HP complex (Figure 2.7) includes the following:

- gravity massive dam 140 m high and 765 m long along the crest, including deaf, spillway, and station dams;
- dam HEP building with units with an installed capacity of 2 million kW.

A spillway dam with eight spillways with a span of 12 m and a springboard provides a flood pass with a flow rate of 13.1 thousand m^3/s at SRL [40,219].

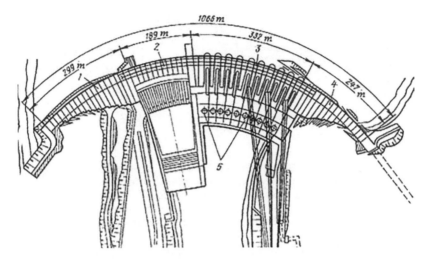

Figure 2.6 Plan of the Sayano-Shushensk HEP: I and 4 – right-bank and left-bank deaf dams; 2 – spillway dam; 3 – station dam; and 5 – HEP building.

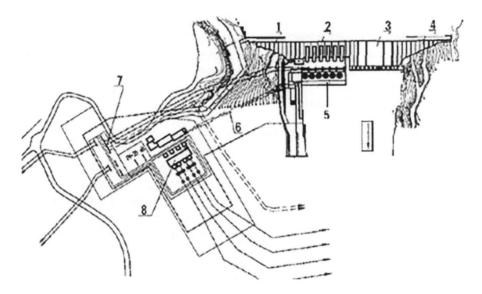

Figure 2.7 Plan of the Bureya HEP: I and 4 – deaf dams; 2 – station dam; 3 – spillway dam; 5 – HEP building; 6 – cable tunnel; and 7 and 8 – switchgear 220 and 500 kV.

During the construction of the Three Gorges hydroelectric complex (Figure 2.8a) [74,184] within the right-bank channel and partly bank part, a construction outlet channel was made through which after the main bed was blocked, one of the largest river in the world Yangtze carried passing construction discharges as well as shipping.

The parameters of the outlet channel were designed to pass a maximum flood discharge of 77 thousand m³/s with a security of 1%. A spillway dam, a left-bank station dam with a HEP building, and a deaf dam were erected in a pit fenced with cofferdams.

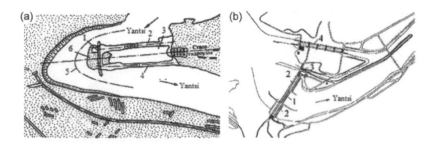

Figure 2.8 HEP Three Gorges: (a) construction scheme of discharges passing: I – longitudinal cofferdam, 2 – upstream cofferdam, 3 – pit, 4 – downstream cofferdam, 5 – axis of the construction channel, 6 – existing island and (b) structure plan: I – spillway dam, 2 – left-bank and right-bank station dams and the HEP building, 3 – ship elevator, and 4 – two-thread lock.

After the canal was blocked in the pit, a right-bank station dam with HEP was erected, and construction discharges were passed through the construction openings of the spillway dam. The structure of the Three Gorges HP (Figure 2.8b) [74,184] includes the following:

- two station dams with water inlets, pressure pipelines with a diameter of 12.5 m, and dam HEP buildings with 26 units with a total installed capacity of 18.2 million kW;
- spillway dam located between station dams;
- deaf dams;
- bank lock and ship elevator.

The length of the head front along the crest is 2,335 m, and the highest height of the dam is 181 m (see Figure 1.45). A spillway dam with a length of 483 m divided into sections of 21 m includes 22 surface spillway spans 18 m wide and 23 deep spillways with a size of 7×9 m, as well as 22 temporary construction openings 6×8.5 m in size, blocked at the last stage of construction. In the lower part of the spillway, a springboard is made.

At the maximum design flood, the HP structures provide a flow pass of 102.5 thousand m^3/s at SRL. Station dams are divided into sections with turbine conduits 24.5 m wide and 13.2 m wide fixed sections.

At the dam foundation, cementation and drainage curtains were made both from the side of the upstream and downstream faces.

The layout with the station dam with the HEP building or a spillway dam in the bed part is used in relatively wide sites or in connection with the peculiarities of topographic and geological conditions. When the station and spillway dams and the HEP building are located in the bed, the construction of the main structures can be carried out in one stage. In this variant, the following principal schemes of their construction and passing of construction costs are provided:

- device in the bank part of site of the construction channel;
- blocking the channel with cofferdams and erecting in the pit of the bed part of the spillway dam with bottom construction and deep openings or a station dam with a HEP building and a part dam with bottom construction and deep openings;

- construction discharges are passed through the construction canal and on the bank part of the site in a separate pit HEP structures or a spillway dam are erected;
- after blocking the construction channel with cofferdams, a deaf or spillway dam with deep openings is erected in it, and the passage of construction discharges is carried out through the bottom construction and after their closure through the deep openings of the spillway dam and then through the spillway openings of the dam.

Under this scheme, the Itaipu HP was built in Brazil-Paraguay and Son La in Vietnam. The structures of Itaipu HP include (Figure 2.9, see Figure 1.43) the following:

- a massive buttress dam located in the bed 196 m high with a dam HEP building with an installed capacity of 12.6 million kW (after reconstruction, 14 million kW);
- a bank spillway which provides a flood discharge of 62.2 thousand m³/s at SRL; the position of the spillway and its design were determined on the basis of hydraulic studies based on the prevention of dangerous bed erosion and the formation of deposits causing backwater in the outlet channel of the HEP [71].

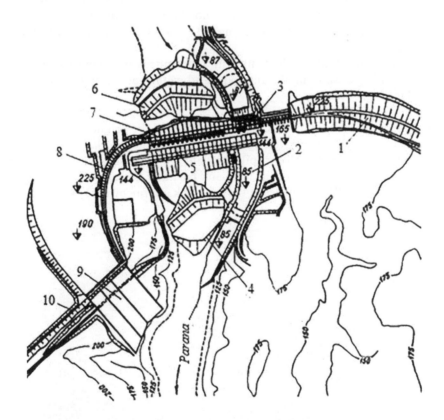

Figure 2.9 Plan of the Itaipu HEP: 1 – left-bank dam from soil materials; 2 – channel for the passage of construction discharges; 3 – time spillway; 4 – downstream cofferdam; 5 – HEP building; 6 – upstream cofferdam; 7 – concrete dam; 8 – right-bank concrete dam; 9 – spillway; and 10 – right-bank dam from soil materials.

The structure of the Son La HP (Figure 2.10) in Vietnam [43,44] includes the following:

- a gravity dam with a crest length of 870 m and a height of 140 m (with a bed width of 360 m);
- station dam located in the bed with the dam HEP building with six units with an installed capacity of 2.4 million kW;
- dam with bottom construction and deep openings;
- bank spillway.

The deep and bank spillways are designed for the passage of the design flood discharge at an SRL of 38.4 thousand m^3/s.

Construction discharges were passed through the construction canal, made off the right bank and designed for a flood discharge of 12.7 thousand m^3/s with 5% guarantee and after its closure through bottom and deep spillways.

A similar arrangement was adopted for the Salto Caxias HP (Brazil). The structures of the HP include [259] the following:

- bed spillway dam of 290 m length with 15 spillways with 16.5 m span each and segment gates 20 m high designed for a flood discharge of 49.6 thousand m^3/s;
- spillway dam with 15 bottom openings 10 × 4.5 m for passing construction costs;
- deaf dam;
- HEP structures built on the bank with four units with a capacity of 1.24 million kW including a water intake and four head metal water conduits with a diameter of 11 m (Figure 2.11).

The total front of the gravity RCC dam is 1.1 km, and the height is 67 m.

Dam construction was carried out in two stages with overlap by right-bank part of the bed and construction of a spillway dam and dam section with bottom openings followed by overlapping of the left-bank part and passage of discharges through bottom openings. HEP structures in the left-bank abutment were built independently in a separate pit.

Figure 2.10 **Son La HEP (Vietnam). View from the left bank.**

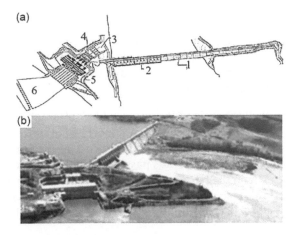

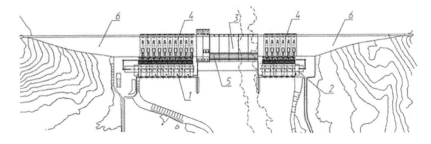

Figure 2.11 Salto Caxias HP: (a) plan: 1 – deaf dam, 2 – spillway, 3 – inlet channel, 4 – water inlet, 5 – HEP building, 6 – outlet channel and (b) HP panorama with an unfinished bed dam 200 m long.

The structures of the "Great Dam of the Ethiopian Renaissance" HEP under construction in Ethiopia with an installed capacity of 6 million kW and a gravity RCC dam 1.78 km long and 175 m high include (Figure 2.12) the following:

- bed spillway dam with a crest at the level of NHL;
- dam with bottom construction and deep spillways;
- right bank and left bank station dams with dam HEP buildings;
- deaf dams;
- spillway on the left bank.

All spillways are designed to pass the maximum flood discharge of 13 thousand m³/s with taking into account their transformation by the reservoir at the SRL. Construction discharges were passed through the bed river after bed closure in the dry seasons through the bottom construction and deep spillways and in the rainy seasons through the top of an unfinished dam 200 m long.

Figure 2.12 Plan of HEP "The Great Dam of the Ethiopian Renaissance": 1 and 2 – right- and left-bank HEP buildings, 3 – bed spillway dam, 4 – station dams, 5 – spillway dam with deep and bottom (construction) spillways, and 6 – deaf dams.

The structures of the "Iron Gate" HP built jointly by the former Yugoslavia and Romania in 1971 with a dam of total length of 1.28 km include the following:

- spillway dam in the bed part 60 m high;
- two HEP buildings with a capacity of 1.02 million kW;
- two locks in the left- and right-bank parts of the site;
- bank sections of earthen dams.

At the first stage in the foundation pits separated by cofferdams in the left- and right-bank parts of the site HEP buildings, locks and part of the spillway dam were erected. In the second stage after blocking the remaining part of the bed and another part of the spillway dam was erected, water was passed through the erected structures.

It is of interest to expand the HEP under the conditions of an operating medium-head Dnepr HEP and to build a second building (Figure 2.13). A HP consisting of a spillway dam 62 m high, a right-bank station dam with a dam HEP building, and a lock was built in 1935.

Subsequently, as a result of the regulation of the Dnepr runoff by the upstream Kremenchug reservoir, the discharge flood rate decreased from 40 to 29.5 thousand m³/s, which allowed the left-bank flooded spans of the spillway dam to be converted into HEP water intake openings and laid along reinforced concrete pipelines to aggregates of the HEP-2 building. A second shipping lock was also built. As a result of the construction of the HEP-2 building, the installed capacity of the Dnepr HEP increased from 0.65 to 1.54 million kW [62].

2.3 Hydraulic project in narrow sites

In narrow sites, HPs with gravity, arch-gravity, and arched dams are being constructed, in which in many cases spillways are arranged to pass construction and operating discharges, and water inlets and water conduits are located for supplying water to the HEP buildings.

In the construction of HP, different schemes for skipping construction costs are applied with the arrangement of bottom construction spillways, coastal construction canals, and construction tunnels. In conditions of high-pressure HP, construction

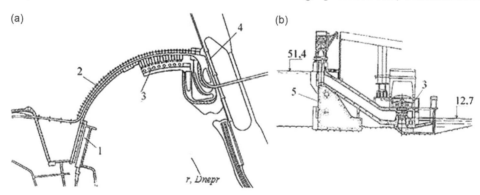

Figure 2.13 Extension of the Dnepr HEP (Ukraine): (a) plan; (b) cross-section along HEP-2; 1 – HEP-1 building, 2 – spillway dam, 3 – HEP-2 building, 4 – lock, and 5 – left-bank part of the spillway dam on which the intake of HEP-2 is arranged.

tunnels are used in many cases. Often, construction tunnels are subsequently used as spillways and HEP conduits.

During construction of HP in narrow sites, the scheme of passing construction discharges during the low season through construction tunnels with a reduced section or bottom construction spillways. During the flood period also was widely used through unfinished parts of the dam for example Kureya HEP in Russia, Yeywa HEP (Figure 2.18) in Myanmar (Burma), Hyong Dien in Vietnam (Figure 2.19).

When the dam scheme in narrow sites depends on the width of the site and the parameters of the structures, the following layouts are advisable:

•   all structures (spillway, water intake, and water conduits of the HEP) are located in the dam, the dam building of the HEP adjoins its downstream face;
•   spillway located on the bank, the HEP intake is in the dam, and the dam HEP building adjoins discharges;
•   spillway and usually the HEP intake located in the dam, and water conduits and the HEP building (underground or open) in the downstream.

As a part of HPs with a dam-derivation scheme, the spillway is usually located in the dam, and the HEP intake is on the bank.

Layouts with the placement of all structures within the dam are performed with the device:

•   separate spillway and HEP facilities in the bed part of the dam, mainly at relatively low HEP;
•   a spillway combined with a HEP station building in the bed part;
•   spillway in the bed section of the site and HEP facilities on the bank;
•   HEP facilities in the bed section of the site and spillways in the bank sections of the dam.

Kurpsai HP is located in the bed part of the site with a ratio $L/h = 3.2$. The structure of HEP plants includes (Figure 2.14):

•   gravity dam, 113 m high;
•   tunnel spillway;
•   a deep spillway providing, together with a HEP, a flood discharge of 3.88 thousand m^3/s;
•   water intake;
•   turbine conduits;
•   dam HEP building with an installed capacity of 8 million kW.

Construction discharges were passed through the construction tunnel and after its closure through deep spillways.

A similar arrangement was adopted for Iznahar HP (Spain): a gravity dam about 120 m high with a curved axis located in the bed part of the site with the ratio $L/h = 3.5$ (Figure 2.15) and a spillway located on the bank [187].

The Revelstoke HP (Canada) in the site with a ratio $L/h = 2.7$ includes gravity dam 175 m high (Figure 2.16) and the HEP building [193].

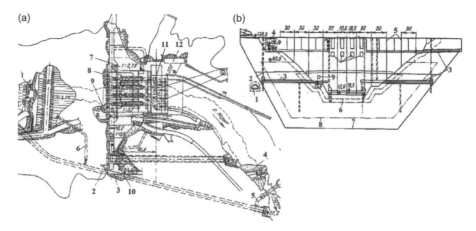

Figure 2.14 Kurpsai HP (Kyrgyzstan): (a) plan: 1 – upstream cofferdam, 2 – construction tunnel, 3 – concrete plug, 4 – downstream cofferdam, 5 and 6 – adits, 7 – dam, 8 – water intake, 9 – depth spillway, 10 – surface spillway, 11 – HEP building, 12 – control building and (b) longitudinal section along the dam: 1 – construction tunnel, 2 – cementing curtain, 3 – cementation adits, 4 – surface spillway, 5 – intersectional joints, 6 – border reinforcing and conjugating cementation, 7 – border of the cementation curtain, 8 – border of the deep drainage, and 9 – deep spillway.

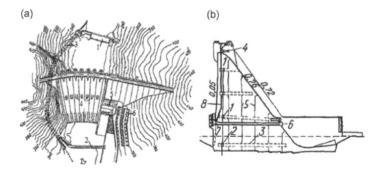

Figure 2.15 Iznahar HP: (a) plan: 1 and 2 – coffer dams, 3 – construction tunnel, 4 – spillways, 5 – HEP, 6 – transformers and (b) cross section along the dam: 1 – viewing (drainage) galleries, 2 – longitudinal gallery in dam bottom, 3 – cross drainage gallery in dam bottom, 4 – spillway, 5 – construction joints, 6 – gate of deep spillway, 7 – spillway intake, and 8 – aeration pipes.

As part of the Bhakra HP in India, a 225-m high gravity dam with a ratio $L/h = 2.3$ was constructed with surface and two deep spillways in the bed river part and two HEP buildings with a total capacity of 1.1 million kW located symmetrically on both sides from the spillway.

The total throughput of all spillway structures is 11.2 thousand m^3/s with a unit discharge of 108 m^2/s. During the construction period, discharges were passed through two construction tunnels (Figure 2.17, see Figure 1.20) [132].

Figure 2.16 **Revelstoke HP (Canada) panorama.**

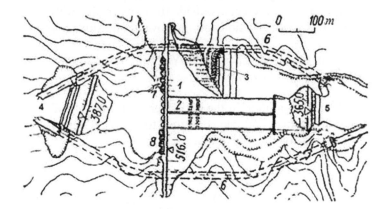

Figure 2.17 **Plan of the Bhakra HP (India): 1 – dam, 2 – spillway; 3 – left bank HEP building; 4 and 5 – upstream and downstream cofferdams; 6 – construction tunnels; 7 and 8 – water inlets of the left bank and right bank HEP.**

The structures of the Yeywa HP (former Burma, now Myanmar) under construction in the site with the ratio $L/h = 5.15$ include (Figure 2.18) the following:

- a gravity RCC dam, 134 m high, in the bed part with a surface automatic spillway of 197 m length;
- water intake;
- turbine conduits;
- the dam HEP building with four units with a total capacity of 788 MW in the left-bank part of the site;
- bank tunnel spillway on the right bank.

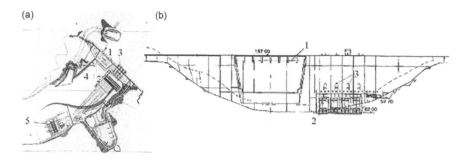

Figure 2.18 Yeywa HP (former Burma, now Myanmar): (a) plan: I − spillway, 2 − HEP building, 3 − water HEP intake, 4 − construction tunnels, and 5 − switchgear and (b) view from the downstream: I − spillway, 2 − HEP building, and 3 − head pipelines.

The flood during the construction period was carried out through two construction tunnels with a diameter of 10 m and later as the dam was erected in the wet season, also through the unfinished right-bank section of the dam. After one tunnel was transformed into a deep spillway, the second was clogged [270].

The HEP water intake was erected outside the dam profile, which provided the possibility of its advanced construction and improved conditions for concreting the dam from RCC.

The structures of the Hyong Dien HP in Vietnam in the site with the ratio $L/h = 3.1$ include (Figure 2.19) the following:

- a gravity dam, 82.5 m high, in the bed river part with a surface spillway of four openings with a span of 13 m each designed to allow a flow of 7.7 thousand m³/s for SRL and a spring-toe for rejection of the jet;

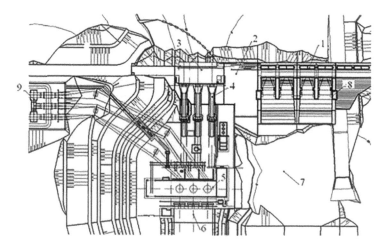

Figure 2.19 Plan of Hyong Dieng HP (Vietnam): I − spillway dam, 2 − deaf dam, 3 − water intake of HEP, 4 − head water conduits, 5 − HEP building, 6 − outlet channel of HEP, 7 − outlet channel of a spillway dam, 8 − construction channel, and 9 − outdoor switchgear.

- HEP intake in the right-bank section of the dam;
- steel-reinforced concrete head conduits of 74 m length;
- HEP building with three units with a capacity of 81 MW.

Construction discharges were passed through the construction tray on the left bank site and during floods also through the unfinished section of the bed river dam. The outlet channel of the HEP is separated from the operational spillway and the erosion pit zone by a rock whole.

The structures of the Miel-1 HP in Chile in the site with the ratio $L/h = 1.83$ include (Figure 2.20) the following:

- gravity dam, 188 m high, with a surface unregulated spillway in the bed river part;
- bank tunnel spillway with deep water intake;
- water intake;
- head tunnels;
- underground HEP building.

Passing of construction discharges carried out through the construction tunnel [227].

The structure of the Longtan HP (China) in the site with the ratio $L/h = 3.5$ includes the following:

- a gravity dam 216.5 m high in the bed river part of which was made the spillway designed for a flood discharge at SRL of 27.6 thousand m³/s with seven spillway spans of 15 m each blocked by gates 20 m high and two bottom openings 5 × 7 m also used to pass construction discharges and flush sediment;
- bank water intake with nine tunnels supplying water to the underground HEP building with a capacity of 5.4 million kW;
- two bank ship elevators with an inlet channel in the downstream connecting them to the bed river (Figure 2.21) [194,231,264].

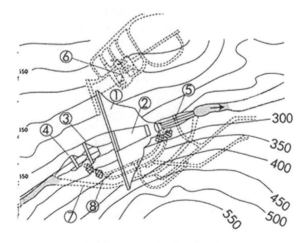

Figure 2.20 Plan of Miel-1 HP (Chile): 1 – RCC dam, 2 – surface spillway, 3 and 4 – upstream cofferdams, 5 – downstream cofferdam, 6 – underground HEP building, 7 – construction tunnel, and 8 – tunnel spillway with deep water intake.

Figure 2.21 Longtan HEP (China): 1– spillway dam, 2 – water intake, 3 – underground HEP building, 4 – outdoor switchgear; 5 – outlet HEP tunnels, 6 – construction tunnels, and 7 – ship elevator.

The concrete structures of the Kowsar HP (Iran) built in a narrow canyon include (Figure 2.22) the following[1]:

*   gravity dam 144 m high with a surface step spillway on downstream face;
*   water intake for water supply;
*   bottom outlet.

A feature of the dam erection was the construction of a metal bridge from which the construction of the dam began, through the lower part of the canyon under the bridge during the construction period flood discharges (1,000–5,000 m^3/s) passing. After the main part of the dam was erected, the lower part of RCC was erected during the inter-floods. At the same time, construction discharges were passed through a construction tunnel designed for a flow of 60 m^3/s [104,154,233].

Figure 2.22 Kowsar Dam (Iran): (a) the beginning of construction, (b) laying concrete into the narrow part of the canyon, and (c) view from the downstream after completion of construction.

The structures of the Toktogul HP located in an area with high seismicity (up to 10 points) in a narrow canyon with a ratio L/h = 1.36 (Figure 2.23, see Figure 1.23) include the following:

- spillway combined with a HEP building;
- gravity dam 215 m high;
- deep spillway to discharge flood and irrigation discharges;
- surface flood spillway;
- dam HEP building with four units with a total installed capacity of 1.2 million kW;
- water intake with turbine conduits.

In the conditions of a narrow canyon the HEP building was constructed with the units in two rows on the ceiling of which a spillway was arranged.

During the construction period, discharges were passed through the construction tunnel and after its closure through a temporary spillway of the first stage.

In HPs with arch dams and dam HEP buildings located in the bed river part, spillways are arranged in the bed and bank parts of the dam or in the abutments.

At the Castelo do Bode HP with an arch-gravity dam 115 m high at L/h = 3.5, the dam HEP building with a capacity of 137 MW is located in the bed river part. Deep spillway adjacent to HEP building designed for a flow pass of 4 thousand m^3/s was constructed curved in plan, which provided favorable conditions for dissipating the flow energy (Figure 2.24) discharged into the middle of the bed and preventing the erosion of the left bank [130].

As part of the Hitoshuze HP (Japan), three surface spillways were constructed in the arch dam: one in the bed river part and two in the bank parts of the dam. When water is discharged through two bank spillways, energy is dissipated due to the impact of jets (see Figure 8.42). On some HP with arch-gravity and thick arch dams and HEP

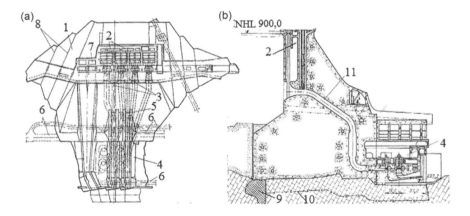

Figure 2.23 Toktogul HP: (a) plan of structures and (b) cross section: I – dam, 2 – water intake of HEP, 3 – bottom spillways, 4 – HEP building with double-row arrangement of units, 5 – chamber of gates of bottom spillways, 6 – transport tunnel, 7 – spillway I stage, 8 – transverse joints, 9 – cementing curtain, 10 – border protection cementation, and 11 – turbine conduit.

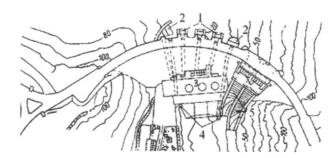

Figure 2.24 Plan of the Castelo do Bode HP: I – deep HEP inlets, 2 – bottom spillways, 3 – HEP building, 4 – outlet part of spillways, and 5 – surface spillway.

in the bed river, spillways of the "ski jump" type are located on both sides of the HEP building, for example, on a thick arch dam, Le Chastang in France 85 m high and arch dam San Rapel in Chile, 112 m high [65].

On an HP with an arch dam and spillway in the bed river part, the dam HEP building can be located within the bank parts of the dam under the spillway; the HEP building can also be moved to the downstream or executed underground [130]. The structures of the Mossyrock HP in the USA in the site with a ratio L/h = 2.15 include an arch dam 184 m high with a spillway located in the bed river part and designed for a flow pass of 7.8 thousand m^3/s. The spillway consists of four spans 13 × 15.2 with segment gates. Water using a sock-springboard is thrown into a deep stilling pool. During the construction period, two construction tunnels provided a discharge pass of 1.38 thousand m^3/s. The HEP building with a capacity of 0.45 million kW is located in the right-bank abutment dam (Figure 2.25, see Figure 8.60) [231].

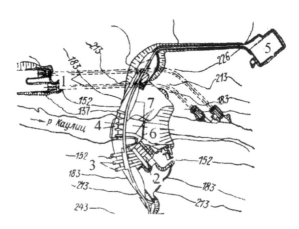

Figure 2.25 Mossyrock HP construction plan (USA): I – construction tunnels, 2 – HEP building, 3 – water intake of HEP, 4 – spillway, 5 – outdoor switchgear, 6 – stilling pool, and 7 – side walls of stilling pool.

An arch-gravity dam 139 m high with a surface spillway in the bed designed for a flow pass of 2 thousand m³/s and a dam HEP building of capacity 147 MW were erected at the Salime HP in Spain in a site with ratio $L/h = 1.8$ located under the spillway. Water from the suction pipes of the HEP units enters a headless tunnel with a diameter of 7 m through which it is discharged to a downstream behind apron wall. Transformers and switchgear are located along the left-bank wall of the water well (Figure 2.26) [187].

At the Ova Spin HP in Switzerland, in the site with ratio $L/h = 2.9$ and a thin arch dam 73 m high, a spillway plate resting on the frame structure under which the HEP building is located adjoins a spillway on the dam crest (Figure 2.27) [65,204].

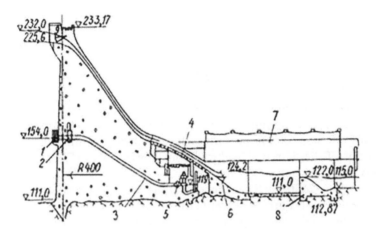

Figure 2.26 Salime HP (Spain). Cross section of dam: 1 – HEP intake, 2 – disk gate of turbine water conduit, 3 – turbine water conduit, 4 – HEP building, 5 – ball gate, 6 – splitters, 7 – outdoor switchgear, and 8 – apron wall.

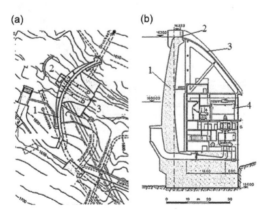

Figure 2.27 Ova Spin HP (Switzerland): (a) plan; (b) cross section: 1 – arch dam, 2 – spillway spans of the dam, 3 – spillway plate, and 4 – HEP building.

The structures of the Aldeadavila HP in Spain in the site with ratio $L/h = 1.8$ include (see Figure 8.5) the following:

- an arch-gravity dam 139.5 m high with a surface spillway with eight spans of 14 m each with segment gates and providing a flow pass of 10 thousand m^3/s at unit discharge of 89 m^2/s;
- underground HEP building with a capacity of 1.12 million kW with water intake, inlet, and outlet tunnels;
- tunnel spillway.

At the Morrow Point HP in the USA in the site with ratio $L/h = 1.58$, the following were built:

- arch dam 143 m high with deep spillways in the bed river part with a throughput of 1.16 thousand m^3/s;
- water well;
- underground HEP building located in the left-bank abutment.

An arch dam 104 m high was constructed at the Boundary HP in the USA in the site with ratio $L/h = 2.2$. Deep spillways are arranged in the bed river part with a carrying capacity of 7.2 thousand m^3/s and two surface spillways, symmetrically located in the bank sections of the dam with a throughput of 3,000 m^3/s (Figure 2.28); HEP is 0.9 million kW [195].

In the arch dam Contra in Switzerland with site $L/h = 1.73$ ratio and a height of 220 m, two surface spillways are arranged symmetrically located in the bank sections of the dam. Spillways are designed for a flow pass of 1 thousand m^3/s each; with the help of springboard

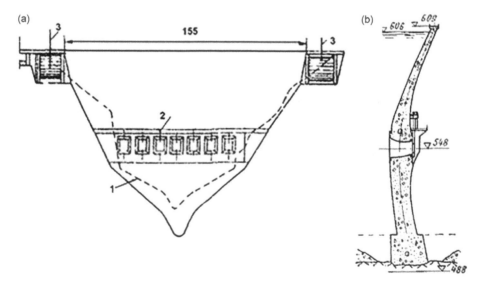

Figure 2.28 Dam Boundary (USA). View from the downstream: I – surface spillways, 2 – deep spillways, and 3 – gallery.

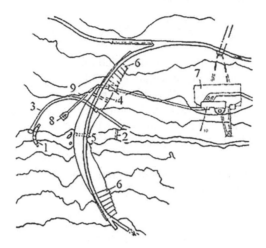

Figure 2.29 Contra HP plan: 1 – upstream cofferdam, 2 – downstream cofferdam, 3 – construction tunnel, 4 and 5 – deep gutters, 6 – surface spillways, 7 – switchgear, 8 – HEP inlet, 9 – head HEP tunnel, and 10 – HEP building.

socks the flow is thrown almost 200 m into the bed [204]. The HP also includes an underground HEP building with a capacity of 105 MW (Figure 2.29, see Figure 8.45).

The layout with the bank spillway in site ratio $L/h = 1.45$ was made at the Chirkei HP in Russia, which included (Figure 2.30, see Figure 1.32) the following:

* arch dam 233 m high with a water HEP intake placed on the upstream face of the dam;
* turbine conduits;

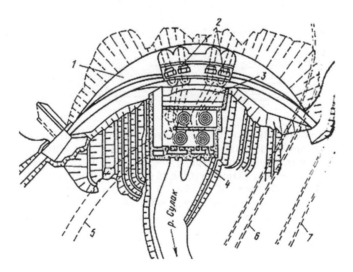

Figure 2.30 Plan of the main structures of the Chirkei HEP: 1 – dam, 2 – HEP inlets, 3 – head pipelines, 4 – HEP building, 5 – transport tunnel, 6 – construction tunnel, and 7 – tunnel spillway.

- the dam HEP building in the bed with four units with a total installed capacity of 1 million kW with their placement in two rows;
- bank spillway of the tunnel type combined with the construction tunnel.

Passing of construction discharges was carried out through the construction tunnel.

A HEP building with 16 units of capacity of 2.1 million kW was erected at the Hoover HP in the USA in the site with $L/h = 1.7$ ratio behind an arch-gravity dam 222 m high; HEP inlets and spillways are made on banks (Figure 2.31, see Figure 1.11).

With a dam-derivation scheme, the Inguri HP (Georgia) includes (Figure 2.32, see Figures 1.29 and 1.30) the following:

- arch dam 271.5 m high with surface and deep spillways designed for a flood discharge of 2.5 thousand m^3/s;
- bank HEP intake.

Water is supplied to the underground HEP building located in Abkhazia using a tunnel 17 km long,

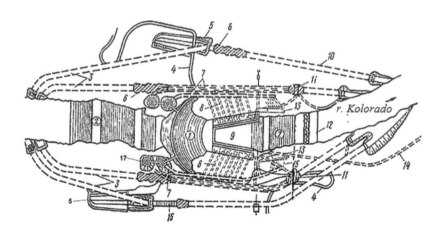

Figure 2.31 Hoover HP plan (USA): 1 – dam, 2 – cofferdam, 3 – construction spillway tunnels with diameter 5.25 m, 4 – road, 5 – spillway with lateral branch water, 6 – concrete plug, 7 – pipelines with diameter 9.15 m, 8 –pipelines with a diameter of 3.96 m, 9 – HEP building, 10 – tunnel spillway with a diameter of 15.25 m, 11 – outlets in the tunnel plug, 12 – banquet, 13 – spillway in the canyon wall, 14 – transport tunnel to HPP building, 15 – adits, 16 – time culvert opening (four pairs), and 17 – tower water inlets.

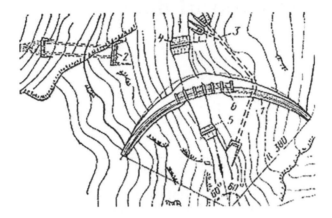

Figure 2.32 The structure plan of the Inguri HP: 1 – arch dam, 2 – HEP water intake, 3 – construction tunnel, 4 and 5 – cofferdams, and 6 – spillways.

Note

1 Project completed by IIGH.

Chapter 3

Rock foundation of dams

"Therefore everyone who hears these words of mine and puts them into practice is like a wise man who built his house on the rock.

The rain came down, the streams rose, and the winds blew and beat against that house; yet it did not fall, because it had its foundation on the rock.

But everyone who hears these words of mine and does not put them into practice is like a foolish man who built his house on sand.

The rain came down, the streams rose, and the winds blew and beat against that house, and it fell with a great crash."

Bible, Matthew 7, pp. 24–26
Gospel of Luke 6, pp. 48–49

It has always been generally accepted that rock is the most reliable material as a foundation for the construction of temples and structures, many of which have stood for centuries. However, it must be admitted that even today, our knowledge about the true properties of rock masses is limited and often not correctly interpreted.

One of the reasons that triggered the emergence of rock mechanics may be some difficulty in finding a common language between geologists and scientists, designers, and builders. A geologist due to his academic education in the bosom of natural sciences studies and describes the qualitative characteristics of rocks, and a civil engineer who has received physical and mathematical education is primarily interested in the quantitative characteristics of rock masses. The synthesis of these two approaches forms the basis of rock mechanics. The geomechanic or specialist in the field of rock mass mechanics is called upon to serve as the missing link between the geologist and the design engineer and the civil engineer, helping them to come to a mutual understanding of the problem.

The rock massif as a rule consists of rocks of various geological origin with varying degrees of disturbance and water saturation and is almost always dissected by cracks including tectonic ones.

In addition, the history of rock formation and the constant movements of the earth's crust cause natural stresses in rock masses, which significantly affect their mechanical characteristics.

One of the most active factors affecting the behavior of the rock foundation of the dam is the reservoir water, which creates a load on the foundation, at the same time penetrating into the cracks and weighing it.

3.1 Rock foundations and their role in the work of the hydraulic structure

3.1.1 Rock foundation of dams

The foundations are an integral part of the structures and as a rule are considered as a whole when designing. The concept of "foundation" includes both enormous volumes of natural soil receiving forces from structures, coastal abutments of arch and arch-gravity dams, and banks of reservoirs, as well as natural and artificial slopes in the area of construction and operation of structures. This of course includes massifs containing underground structures (tunnels, mine workings, underground HEP, and underground storage facilities).

The loads transmitted by HSs (dams, spillways, locks, pressure tunnels, and underground buildings of the HEP) can reach very high levels and are accompanied by large deformations of the foundations. The main task of engineers is to ensure the strength and stability of foundations and HSs, as well as ensuring the water tightness of the underground circuit to minimize water filtration from the created reservoirs.

Given that the foundations are natural formations with their characteristic heterogeneity and anisotropy, engineers cannot always give them the necessary properties, which poses serious challenges to ensure their reliability. In addition, it should be borne in mind that in the process of construction and filling of the reservoir and during the operation of the construction, the foundations undergo certain changes. The situation is complicated by the presence of the initial SSS due to both the weight of the overlying rocks, the processes of the formation of the earth's crust and modern tectonic movements [191].

Therefore, it is obvious that in order to study the foundations, detailed geotechnical studies are necessary, including the study of geostructural, geomorphological, and hydrological conditions, modern geological and seismic tectonic processes in the construction area, as well as changes as a result of the construction and operation of the structure. Based on these studies and based on the nature of the work of the structure and the assigned engineering tasks, appropriate models of the foundations are created on which designers and builders build their structures.

3.1.2 Engineering-geological characteristics

From an engineering-geological point of view, the strength, deformation, and filtration properties of rocks are usually divided into the following types [255]:

- rock;
- half rock;
- nonrock or soil.

These types are distinguished in accordance with the main classes of rocks, reflecting the features of structural bonds between mineral particles and aggregates. In engineering practice, rock masses and soils are usually distinguished. By definition of *N.A. Tsytovich* [173], soils are called all loose rocks of the weathering crust of the lithosphere, both incoherent (loose) and cohesive (clay), in which the bond strength is many times less than the strength of mineral particles. Rocks include solids in which the rigid bonds of mineral aggregates and grains have the strength of the same order as the strength of the particles themselves. SS 25 based on the negative experience 100 "Soils Classification" provides for the division of soils into rock and half rock (consisting of one or more minerals and having rigid structural bonds), as well as dispersed (with mechanical and water-colloidal bonds).

Rock formations are considered solids with a uniaxial compression strength equal to or greater than 5 MPa (50 kg/cm^2). Despite the conventionality of such a separation, it is justified in terms of mechanical properties and behavior under load of soil and rock foundations as well as a significant difference in research methods and methods of work.

If the soil foundations are sufficiently well studied in soil mechanics and the relationships between their composition, structure, condition, and mechanical properties, rock masses are complex fractured inhomogeneous and anisotropic mediums in which the size of the blocks is comparable with the dimensions of the supports of the structure or the diameter of the tunnel. This introduces great uncertainty in the conditions of load transfer from the structure and in the idea of the strength and deformation characteristics of the rock mass. For such foundations, scale effect is inevitable.

Rock masses as a rule consist of solids of various geological origins with varying degrees of fracture and disturbance and are almost always dissected by tectonic cracks and faults. In addition, the movements of the earth's crust that have taken place cause the presence of natural stresses of different magnitude and direction, which significantly affect the mechanical properties of rock masses. Therefore, when using the laws of mechanics to study the behavior of a rock mass, it is necessary to remember that an engineer is dealing with a discontinuous, inhomogeneous, and anisotropic medium, the characteristics of which must be studied in each case in order to clearly determine the limits of applicability of the used theoretical premises. It is obvious that it is *necessary to clearly separate the concepts of "rock (solid)" and "rock mass"* so as not to fall into the mistake made by engineers who identify the characteristics of rock (solid) and rock mass.

Considering the rock massifs as a mechanically specific medium, the following main factors should distinguished due to their composition, structure, and condition:

- fracturing and blocking;
- heterogeneity of the structure and properties;
- anisotropy of strength, deformation, and filtration properties;
- natural SSS.

For the correct use of the concepts of "homogeneity" and "heterogeneity", and "isotropy" and "anisotropy", which are of great importance for the description of rock masses, it is necessary to clarify these concepts.

A homogeneous (homogeneous) technical or geomechanical body is called so when all its parts have the same structure and therefore at any point have the same physical properties. Otherwise it is called heterogeneous. Of course it can talk about homogeneity also with respect to individual properties, and then the same body can be simultaneously inhomogeneous, if it is considered, for example its compressive strength and homogeneous, if it is proceed from another property (e.g., thermal conductivity).

A section of a body is called *isotropic* if its material in all directions in space exhibits the same properties with equal intensity, and these properties can change from point to point or from section to section, if the body is not uniform. Bodies with perfect isotropy are very rare in nature. Many rocks exhibit very strong anisotropy of strength especially when shearing. These are first of all shales and rocks cracking only in a certain direction. Therefore, isotropy is most often determined in relation to any one property, especially in those cases when solving practical problems does not require a very clear definition of boundaries and when it can talk about statistical isotropy or quasi-isotropy.

Homogeneity and heterogeneity and isotropy and anisotropy are not just properties of materials; these are signs characterizing the distribution and repeatability of the directions of certain properties and features. Homogeneous bodies or parts of the body need not be isotropic; inhomogeneous bodies on the contrary almost never are isotropic, although it is logical to assume the presence of such a case. The boundaries established between homogeneous and inhomogeneous bodies also depend on the goals of technical or scientific tasks [48,51].

3.1.3 Rock mass models

The study and analysis of any phenomenon in nature regardless of its complexity and multifactorial can be carried out on the model of this phenomenon, which in the process of cognition can be refined and improved, remaining, however, only a model to one degree or another adequate to the phenomenon under consideration. Improving a model does not necessarily mean complicating it. On the contrary, to study the influence of certain factors, it is often methodologically expedient to simplify them, for example, dividing a complex model into a number of simpler specialized models.

In engineering calculations of both structures and their foundations, simplified and schematized models are always used that reflect the most important and essential features of the objects under consideration. This requires the consistent development of a number of models of the studied massif, methodologically linked with each other and providing a solution to the main problem – a quantitative analysis of the phenomena and processes occurring in the studied rock mass under the construction and operation of the structure.

The general statement of the complex of engineering studies of rock masses can be represented as a combination of the following problems:

1. The study of the rock massif as an object of natural education, the identification of specific features that may be significant for the solved engineering problem, and the representation of a real object in the form of an *engineering-geological model* reflecting the features of its structure, composition, and condition.
2. Establishing patterns of rock behavior under loads and under various influences, determining on this basis the characteristics of the physico-mechanical properties of rocks of individual zones and elements of real massifs, and developing a *geomechanical model* of the massif.
3. Determination of stresses, displacements, and deformations and assessment of the strength and stability of rock massifs by the methods of mechanics of a deformable body using a *mathematical or physical model*.

A *geomechanical model* should be understood as a spatial or planar diagram of a rock massif, on which, for various sections and elements of the massif in general, the strength, deformation, and filtration properties of the rocks are determined in accordance with the proposed mechanism of the process or phenomenon under study and the method for studying them. The geomechanical model is built on the basis of the engineering-geological model to solve specific engineering problems.

A *mathematical model* or calculation scheme is built on the basis of a geomechanical model for the quantitative analysis of processes occurring in an array. Therefore, such a scheme should include the use of certain analysis methods. If there is a problem of assessing the stability of rock blocks and the methods of limiting equilibrium will be used, then for the transition from a geomechanical model to a mathematical model, enough information that contains the strength and filtration schemes of the massif is enough. A much larger amount of information is needed for calculating the SSS of massif using FEM or discrete elements. Each of the analysis methods used requires its own calculation scheme.

The *physical model* is really a physical and mechanical model of elastic or "equivalent" materials and designed to study the behavior of the "construction-foundation" complex under load. This model is also built on basis of a geomechanical model using the information that is necessary according to the conditions of the model material and the research methods used. However, due to the rapid development in recent years of numerical methods of calculation and software, physical modeling has lost its original meaning.

3.1.4 Basic structural features

When using rock masses as the foundations of engineering structures and first of all high dams, it must be borne in mind that studies of rock foundations of dams cannot be limited by the accumulation of data on the strength and deformation properties of the massif, as well as the identification of the main crack systems. In addition to the general picture of fracturing, attention should be paid to the presence of tectonic cracks and faults, which are signs of history left by nature and indications of the forces acting on the massif and the deformations caused by these forces. A very important element of studies of the rock foundation of dams should be the establishment of the orientation and position of the lines of tectonic discontinuities with respect to the dam. The location of structures in close proximity to the tectonic fault should avoided, as the possibility of differentiated movement of blocks separated by a gap can cause the destruction of structures.

The designers encountered problems of strengthening tectonic faults at many large sites, such as the Inguri dam (Georgia) in the right-bank abutment of which there is a tectonic fault closed by a concrete lattice (see Section 4.3.1). In the site of the Rogun Dam on the Vakhsh river (Tajikistan), there is a large tectonic fault filled with salt, which made it possible to solve the problem of either choosing the type of dam or saline protection formation from erosion by filtration flow when choosing a dam from soil materials.

The structural features of the rock foundations are closely related to physical and geological phenomena. Landslides and landslips of slopes of valleys composed of rocks are largely due to the distribution and inclination of cracks in the massifs, as well as the presence of clay material in the cracks. In this case, the stability of the structure must be ensured with the help of an appropriate seal of cracks.

In addition, the regime of tectonic movements is closely associated with seismicity, which requires careful study of seismic conditions (seismic microzoning) and geo-structural forms of massifs. This very important circumstance is usually not given due attention, but it is very important in mountainous regions that differ in tectonic activity. In calculating the stability of concrete dams erected in tectonically active areas, modern tectonic movements must be taken into account. They can differ in a positive and negative sign and reach several millimeters and even several centimeters per year.

The structural features of the structure of the valleys' banks folded by rock massifs, especially in deep canyons (gorges), are directly related to their stability. Therefore, the orientation and angles of incidence of cracks and tectonic discontinuities in the massif, their mutual intersection, and the position of the planes with respect to the surfaces of the sides completely determine their stability. Necessary engineering solutions resulting from this are fixing the banks of the valley, the depth of insertion of the dam into the abutments, the conditions of rock work, and others.

During the design and construction of the Toktogul dam in the deep canyon of the Naryn river (Kyrgyzstan), insertion of the dam into the banks could lead to an avalanche-like collapse of the surface blocks cut off from the massif by cracks of on-board rebuff. As a result, it was necessary to develop a new type of wedge-shaped dam, which was built almost without cutting into the banks (see Section 7.4).

Another very important factor is the uniformity of the deformability of the rock foundation of the dam, which is determined not only by the presence of discontinuous disturbances, but also by the uneven deformability of various sections of the foundation and can cause stress concentrations in the dam body.

The presence of karst is of great importance for HC on carbonate rocks, which occupy more than 40% of the territory of our planet. Firstly, karst is widespread in carbonate rocks, and at the same time due to the relatively high strength and resistance of these rocks to weathering agents, relief forms in river valleys appear in their distribution areas, which are favorable for creating dams and other terrestrial and underground HC. Secondly, limestones and dolomites have very valuable building properties – they are well preserved and do not erode in high and steep slopes, are stable in underground excavations, and are relatively easy to develop.

Karst changes the permeability strong and other geotechnical properties of rocks. For many years, attempts to carry out HC on carbonated karst rocks were unsuccessful, and as a result, there was a strong opinion that it was impossible to create HS in karst development areas. Karst was considered as a clear contraindication for the construction of dams and the creation of reservoirs and the construction of canals, tunnels, and other structures that delay or let water through. Such an assessment was based on the negative experience of HCs on karst rocks, when there seemed to be insoluble problems of water retention in artificial reservoirs.

When constructing concrete dams on karst limestone, the main danger lies not in the presence of karst cavities themselves (which can be concreted or cemented) but in not to miss them when studying the foundation. Undetected karst cavities, even with a continuous cementation curtain, can lead to unpleasant consequences. Therefore, the study of karst must be strictly combined with the forms of geological structures, i.e., with the nature of folding, the nature of tectonic discontinuities and fracturing.

A good example is the construction of the Kirdzhali dam in Bulgaria [41,101] (for more details, see Section 4.3.3).

The development of the technology of consolidation and compaction of rocks in adjoining and at the foundation of structures using cementation and bituminization methods, as well as the development of methods for constructing concrete supporting structures contributed to a change in outlook on the possibility of building dams on karst rocks. An example of the successful application of these methods was the construction of a number of dams of a HEPs cascade on the river Tennessee (USA) and others. In USSR, the first objects of HCs on karst rocks were Volkhov and Syzran HEP.

During HC on karst formations, taking into account the danger posed by undetected karst cavities, an additional survey of the foundation is required during the construction period after the excavation has been completed.

Currently, the accumulated experience of a special engineering-geological study of the patterns of karst development allows us to make reliable, often very bold decisions in the construction of dams in highly karst areas. In this case, protection against dangerous manifestations of karst is achieved by selecting within the massif of sites least affected by karst, as well as by creating complex antifilter curtains, shielding, or artificial clogging.

Water filtration in rock massifs occurs along cracks and not throughout the massif as in soils; therefore, the calculation of the filtration flow parameters cannot be determined by the equations of classical hydromechanics, which makes it very difficult to solve hydrogeological problems and struggle with filtration at the foundation of structures.

Hydrogeological factors play a significant role in the construction of high dams on rock foundations. It should be noted that if the rocks in the mountain masses cut by the river valleys are not aqueduct due to their natural drainage, then when filling the reservoir, cracks are filled with water. The hydrostatic pressure of the water weighs the massifs, causing them to rise, which in turn leads to uneven vertical displacements of adjacent dams and vertical deformations of high dams erected on them. Unfortunately, this phenomenon has not yet been taken into account, despite many examples of the appearance of horizontal cracks on the upstream face of high concrete dams.

A detailed discussion of possible solutions to these problems and methods used to strengthen rock foundations of hydraulic structures used in engineering practice is given in Chapter 4.

3.2 Physical characteristics of rock mass

3.2.1 Fissured rock mass

Fracturing refers to the totality of cracks in a rock mass.

Cracks as a rule are grouped into systems with their preferred azimuths and angles of incidence, although in rare cases there can be many unsystematic cracks without pronounced azimuths and angles of incidence. Such a randomly oriented arrangement of cracks is often observed with rapid cooling of igneous rocks.

The *azimuth of incidence of a crack* is the angle β made up by the direction of incidence of the crack with a direction to the North (counted clockwise). Sometimes along with the incidence azimuth, the term "strike" of the crack is used, which is the "trace" of the surface of the crack on a horizontal surface. The extension of the crack is normal to its incidence azimuth.

The *angle of incidence of the crack* is the angle α formed by the surface of the crack with the horizon. A set of parallel or quasi-parallel cracks forms a *system of cracks*.

The characteristics and orientation of the cracks systems are critical to the behavior of the rock mass and for assessment of its stability and strength.

In order to be able to carry out such an analysis, it is necessary to study in detail all the crack systems in the massif and their individual characteristics, which are as follows:

- azimuth and angle of incidence of the plane of cracks of the system;
- relative position of the systems of cracks, which determines the structure of the rock mass and the shape of the rock blocks;
- length of the cracks (continuous or intermittent);
- average and maximum length of cracks;
- density of cracks in the system;
- magnitude of crack opening;
- the presence or absence of aggregate and its characteristics;
- characteristics of the surface roughness of the cracks.

In polar coordinates, the plane of the crack is displayed by a point having the ordinates α and β; Figure 3.1 shows point M, which displays the plane of the crack with an incidence angle of $\alpha = 70°$ and an azimuth of the incidence of $\beta = 240°$. From the measured incidence angles and azimuths of the incidence of crack, a statistically generalized idea of the fracturing of the rock mass can be compiled.

The construction of fracture diagrams is based on determining the density of points, which is usually carried out by moving a unit area on the circle diagram (palette). The number of points that fall within the circle, which is recorded in the center, is calculated, and then lines of equal density of points are drawn from these numbers (Figure 3.2a).

The places of thickening of the density of points are determined by the systems of cracks, and the azimuths and angles of incidence of the corresponding systems of cracks are determined by the maximums of the thickening (Figure 3.2b).

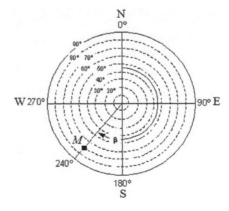

Figure 3.1 **The principle of constructing fracture diagrams.**

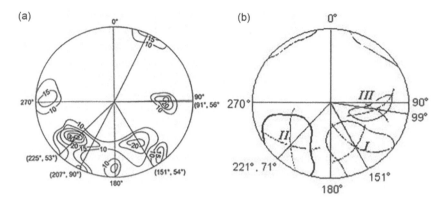

Figure 3.2 Fracture diagrams (Hoa Binh HP, Vietnam, adit No. 5).

This method of identifying crack systems and determining their parameters is convenient in the presence of clearly defined crack systems, but even in this case, it does not provide information on the reliability of the quantitative indicators obtained and their possible changes (for example, on the variances of azimuths and cracks in the given system). With less pronounced crack systems, the application of this method becomes difficult, and, naturally, the use of the apparatus of mathematical statistics begs to not only suggest but also prove the existence of crack systems and determine the values of their parameters.

Figure 3.2b shows the same fracture diagram constructed by the probabilistic method [47,51] with the determination of the centers and dispersion ellipses of three crack systems identified by the proposed probabilistic method using 600 crack points recorded in adit No. 5 of the Hoa Binh HP (Vietnam).

3.2.2 Deformability of rock mass

Deformability is one of the main characteristics of rock masses that determine the behavior of a structure built on them (or in them). It took a lot of time to understand that, despite the fact that the rock mass is composed of practically elastic rocks, it does not behave as linearly elastic but often even as a plastic body due to movements that occur along the contacts of cracks in it.

3.2.2.1 Deformability under uniaxial loading

The main parameters of the deformation of the material under load are longitudinal ε_y (along the axis of the applied load) and transverse ε_x (in the direction normal to the axis of application of the load) relative strains, transverse strain coefficient v (Poisson's ratio):

$$v = \frac{\varepsilon_x}{\varepsilon_y} \tag{3.1}$$

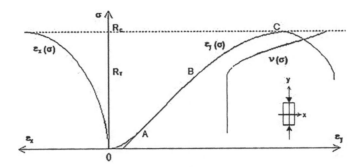

Figure 3.3 Diagrams of longitudinal (ε_y) and transverse (ε_x) strains and Poisson's ratio ν of a specimen under uniaxial loading.

and the modulus of deformation (elasticity) of the material:

$$E = \frac{\sigma}{\varepsilon_y} \tag{3.2}$$

Studies show that polycrystalline materials undergo a series of characteristic stages during deformation (Figure 3.3):

1. compaction of the material associated with the closure of microcracks (section OA);
2. elastic work of the material (section AB);
3. beginning of the intensive process of microcrack formation (the beginning of the phase of the plastic behavior of the material, point B);
4. formation of macrocracks (phase of destruction, point C).

The behavior of the rock material at different stages of its deformation is considered in more detail. In addition to the diagrams of longitudinal (ε_y) and transverse (ε_x) relative strains, a diagram of changes in the transverse strain coefficient (ν) under loading of the sample is also constructed.

The *zone of compaction* or compression of microcracks is characterized by a nonlinear dependence of ε_y in the practical absence of transverse strains ε_x.

In the *zone of elastic work,* the rock behaves almost like an elastic and homogeneous material with a constant value of the transverse strain coefficient ($\nu = $ const).

At a certain level of stress (R_τ), the values of the transverse strain coefficient begin to increase sharply, indicating the onset of the process of crack formation, leading to an increase in the volume of the sample. Material from a quasi-homogeneous state passes into a microfractured state with a disturbed internal structure. This is a *zone of plastic deformation.*

When the stress reaches the strength value R_c, the material is destroyed, and at this moment, the values of the transverse strain coefficient reach the limit value $\nu = 0.5$ for a homogeneous material. This is *the material destruction zone,* which continues at stresses already lower than its limit strength. The behavior of the material in this transcendental region of deformation can be registered only on the so-called "hard" presses or during the rapid registration of deformations during the test.

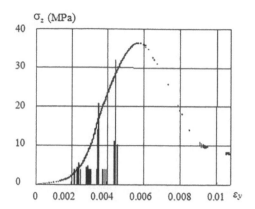

Figure 3.4 Test chart of the sample in the prelimit and trans-limit areas. *The vertical lines represent bursts of acoustic emission.*

Table 3.1 Rock classification by deformability

Degree of deformability	Modulus of deformation of massif E, 10^3 MPa (10^3 kg/cm^2)
Very slightly deformable	Over 20 (200)
Weakly deformed	From 10 (100) to 20 (200)
Moderately deformable	From 5 (50) to 10 (100)
Highly deformable	From 2 (50) to 5 (50)
Extremely deformable	Less than 2 (20)

At the Institute of Engineering of the National Autonomous University of Mexico (IENAUM), a setup was created that allows recording up to 10,000 instrument readings per second [50,53]. This setup was used to study the deformability of polycrystalline materials in bulk stress state. Figure 3.4 shows a diagram of a uniaxial test of a sample in a continuous reading mode (1,000 measurements per second) with simultaneous fixation of acoustic emission.

Examination of this diagram indicates that the destruction (microcracking) of the sample begins almost from the moment it begins to load, as evidenced by "bursts" of acoustic emission. The first large crack occurred at a stress of approximately 50% of the strength of the sample, and then the loading process continued.

In addition, the presented diagram clearly demonstrates that the deformation of the sample continues after its destruction (in trans-limit areas) but with a much higher speed.

The classification of rock massifs by deformability is given in Table 3.1 [11].

3.2.2.2 Deformability under triaxial loading

The rock mass during its loading as a rule is deprived of the possibility of free lateral expansion and, therefore, with an increase in one of the acting principal stresses, the other two also increase, and this increase in the "lateral" principal stresses continues even after the "destruction" of the material. An increase in lateral compression leads to an increase in the bearing capacity of the array.

To "automatically" reproduce this process of increasing "lateral" stresses when the first main stress increases, in the Laboratory of Rock Mechanics of JSC Institute Hydroproject (LRMIH), a special installation was developed. The main advantage of installation is that after the stress tensor reaches its ultimate state, it continues to "slide" over the surface of strength, thus making it possible to obtain not one point on the surface of strength, which is usually the end of traditional research, but surface strength plot [49].

A similar installation, which allows testing cubic samples measuring 15×15×15 cm, was then manufactured at IENAUM. The installation consists of four rigid steel plates, interconnected by steel elastic rods ("elastic elements"), reproducing the rigidity of the surrounding rock mass.

With these rods, when they are pretensioned, any combination of "lateral" main stresses can be created. Measurement of lateral loads was carried out by load cells glued to the rods. A photograph of the complete assembly is shown in Figure 3.5.

Having reached the ultimate strength, the stress tensor begins to "slide" along the surface of strength, making it possible to obtain a plot of the surface of strength with the corresponding diagram of deformation of the sample in the "beyond" region (Figure 3.6).

Subsequently, a new installation was created at IENAUM, which made it possible to simultaneously apply three independent main voltages to the sample, the values of which were controlled by a special program on a computer. This installation consisted of three mechanically independent systems of applying loads and measuring strains on a cubic sample 15×15×15 cm in size [50]. Each of these systems consisted of a hydraulic jack and two rigid metal plates interconnected by two rigid rods with spherical self-aligning joints (Figure 3.7).

A diagram of the system for applying load and strain measurements is shown in Figure 3.8.

Side jacks made it possible to create stresses of up to 22 MPa, while a vertical jack made it possible to apply up to 130 MPa to a cubic sample measuring 15×1×15 cm.

The displacements of the faces of the sample were measured by linear transducers with an accuracy of 2×10^{-3} mm. All data on loads and strains were recorded by a

Figure 3.5 Installation of volumetric loading of samples with elastic elements.

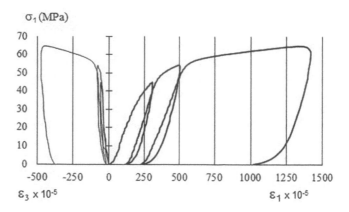

Figure 3.6 Diagram of sample deformation before and after its destruction.

Figure 3.7 Installation of volumetric testing of samples with independent application of all three main stresses.

computer with a speed of up to 1,000 measurements per second, which made it possible to measure the behavior of the sample in time.

At the same time, the acoustic emission emitted by the sample upon destruction was measured using a piezoelectric sensor. A photograph of the sample after the test is shown in Figure 3.9.

3.2.2.3 Field deformability study of mass

Given that the rock mass is composed of rock blocks of various sizes and fractures, its deformability depends on the area of application of the load. This effect is called

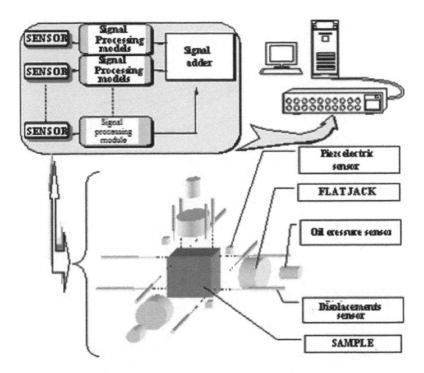

Figure 3.8 Load application and strain measurement systems.

Figure 3.9 Gypsum sample after volumetric stress test.

the "scale effect." Studies of the deformability of rock foundations, performed by seismic-acoustic methods, showed that with an increase in the volume of the loaded massif, the decrease in its elastic modulus did not occur smoothly but abruptly when changing from one structural size to another. Figure 3.10 shows the scale curves of the average or modal velocities of elastic waves (characterizing the elastic modulus) depending on the volume of the studied rock mass for various rocks [53,144].

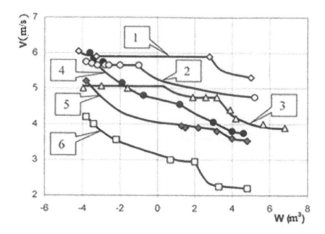

Figure 3.10 Scale curves of average velocities of longitudinal elastic waves (V) depending on the volume of the studied mass (W) for the rock foundations of various HPs: 1 – granites (Dnestr), 2 – porphyry basalts (Talnakh), 3 – siltstones (Arkhyz), 4 – gneisses, crystalline schists (Kirdzhali), 5 – limestones (Toktogul), and 6 – limestones (Inguri).

It should also be borne in mind that the determination of the foundation deformation moduli is carried out even before the construction of the dam. During the construction process, the rock foundation is decompressed and broken due to excavation of the pit for the construction, explosion, weathering, and construction work. As a result, after completion of the dam construction and filling of the reservoir, the deformability characteristics of the foundation may turn out to be significantly different from those in the design.

In addition, both the deformation moduli and the transverse expansion coefficients of the massif are usually anisotropic, i.e., they vary in space depending on the orientation of the applied loads at each point. Figure 3.11 shows the experimental diagrams of the propagation of stresses in a layered block medium depending on the direction of application of the load [213].

If the deformability of the foundation along the length and width of the dam foundation is uniform, then these are the most favorable conditions for the work of the dam, which do not require removal of the rock to replace it with concrete. As calculations showed [51], a uniform change in the foundation deformation modulus has little effect on the pattern of stress distribution both in the dam body and in the foundation. There are many examples of exploited gravity and buttress dams built on very deformable foundations. For example, the Rappbode Dam (Germany) is 105 m high built on the foundation with a deformation modulus of 800 MPa, and the Bayna Basht dam (Yugoslavia), buttress dam 90 m high, is built on shales and sandstones with a deformation modulus of 400 MPa [51]. At the same time, in dams built on hard rock foundations, cracks are often observed at the contact of the dam with the base. This was the case at the dam of the Bratsk HEP (height 125 m), built on very strong and hard basalts with a deformation modulus of 25,000 MPa, where a gap in the cementation curtain and a local restoration of uplift were observed.

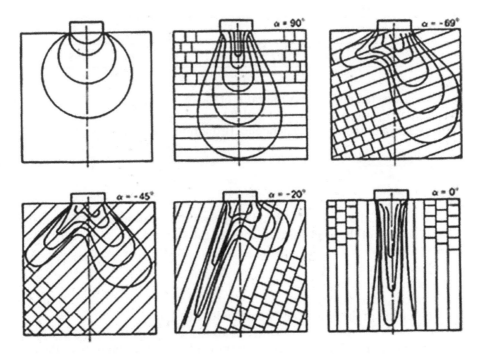

Figure 3.11 Evolution of the diagram of maximum compressive stresses in a layered block base when the angle α changes between the direction of the load and the bedding.

In practice, several methods are used to determine the deformation characteristics of rock massifs in the field; these are seismic-acoustic and geomechanical research methods (stamps, cameras, dilatometers).

Of the geomechanical methods for determining the modulus of deformation of rock masses, the stamp method is most widely used.

This method consists in applying a load to the surface of a rock massif by means of a stamp and measuring the displacements (settlement) of the massif surface under and near the stamp. When interpreting the test results, the rock mass is approximated by an elastic and isotropic medium, for which the Boussinesq solution for the elastic half-space is valid:

$$E = KP \frac{(1-v^2)}{w_o\, B} \tag{3.3}$$

where
    K – shape factor of the stamp;
    P – force acting on the stamp;
    v – Poisson's ratio of the mass;
    w_o – settlement under the center of the stamp;
    B – characteristic stamp size.

Table 3.2 K factors

B/A	I	1.2	1.6	1.8	2	3	4	5
K	0.87	0.94	1.07	1.313	1.18	1.40	1.55	1.68

When using a rectangular stamp, the value of the shape coefficient K can be determined depending on the ratio of its sides A and B according to Table 3.2 [51]:

When using a round stamp, dependence (Equation 3.3) can be represented as follows:

$$E = \frac{KpR(1-v^2)}{w} \tag{3.4}$$

where

K – coefficient, the value of which depends on the location of the measuring point of the settlement of the base (for the center of the stamp $K = 2$ and for the edge of the stamp $K = 4/\pi$);

p – pressure under the stamp;

R – radius of the stamp;

v – Poisson's ratio of the base;

w – settlement of the base at any point under the stamp.

This dependence is valid only with uniform loading of the stamp over the area, i.e., for flexible stamp. When using hard dies in the equation, it is necessary to introduce a correction taking into account the influence of this stiffness, because otherwise, the obtained values of the modulus of elasticity or deformation will be overestimated.

Figure 3.12 shows the definition of the foundation deformation modulus under the Metlac bridge support in Mexico by the method of a round stamp with a diameter of 117 cm. For the possibility of measuring precipitation in the center of the stamp, a hole was provided in it. Settlements were measured by dial gauges in the center of the stamp, on its faces, and on the free surface of the rock to allow the construction of a bowl of deflection and its comparison with the theoretical outline.

To build a theoretical outline of the bow of the surface deflection of a perfectly elastic medium under a round stamp of radius R, the expression [25] can be used:

$$w = \frac{\left[4pR(1-v^2)\right]}{\pi E} \int_{0}^{\pi/2} \sqrt{[1-(r/R)^2 \sin^2 \phi]} \, d\phi \tag{3.5}$$

where r – the distance from the considered surface point to the axis of the stamp, provided that r is less than R.

To facilitate the use of this dependence, Table 3.3 shows the values of the elliptic integral $S = \int_{0}^{\pi/2} \sqrt{\left[1-(r/R)^2 \sin^2 \phi\right]} \, d\phi$.

To build the shape of the bow of the deflection outside the limits of the stamp, i.e., when $r \geq R$, in a first approximation, we can use the dependence:

Figure 3.12 Type of installation for determining the modulus of deformation of a rock mass using the round stamp method.

Table 3.3 Values of the elliptic integral

r/R	0	0.1	0.2	0.3	0.4	0.5	0.6	0.7	0.8	0.9	1
S	1.571	1.567	1.555	1.535	1.506	1.467	1.417	1.355	1.278	1.171	1

$$w = \frac{4pR^2(1-v^2)}{\pi E r}$$

(3.6)

Figure 3.13 shows the results of a test conducted to determine the modulus of conglomerate deformation at the foundation of the support Metlac road bridge on the Mexico City – Veracruz freeway. As can be seen from the diagram, the load was applied for several loading and unloading cycles, and the results of the fourth and fifth cycles lay inside the hysteresis loop of the third cycle, which indicates the practical stabilization of the process of conglomerate deformation in the elastic zone. With the adopted Poisson's ratio $v = 0.3$, the value of the elastic modulus $E = 48,300 \text{ kg/cm}^2$ was obtained.

To check the correctness of the obtained modulus value, it was necessary to verify that the obtained sediment values correspond to the theoretical outline of the deflection bowl for the selected value of the elastic modulus. Figure 3.14 shows such a comparison.

A quite acceptable correspondence between the calculated and experimental data allows us to consider the obtained value of the elastic modulus correct.

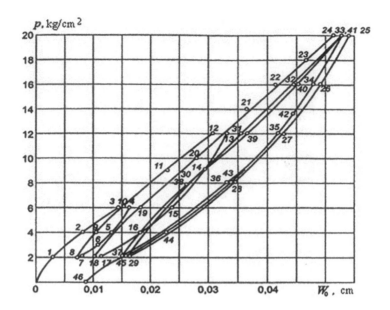

Figure 3.13 The results of the study of the modulus of elasticity of the conglomerate.

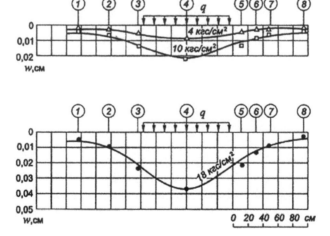

Figure 3.14 Comparison of the experimentally obtained values of the settlements of the surface of the base with the theoretical outline of the bowl of deflection according to the obtained modulus of elasticity of the mass (1–8 months of measuring the sediment).

As an express method for determining the deformation characteristics of a rock mass, the *dilatometer or pressiometer method* is often used.

A dilatometer is a cylindrical device lowered into the well, which allows transmitting uniform (or sector) internal pressure to the well walls, simultaneously measuring the deformation of the well diameter in different directions. As a rule, portable dilatometers are used, which are alternately lowered into a number of wells, determining in each of them the deformation characteristics of the massif at various depths. This makes it possible to quickly and without large costs get an idea of the quality of the rock foundation for a large number of dispersed wells.

Seismoacoustic methods are the only ones that allow us to assess the deformation homogeneity of rock masses on a scale comparable to the size of the structure. Seismoacoustic methods based on the use of elastic waves of a wide range of frequencies (from 30–50 Hz to 50 and more kHz) make it possible to determine the elastic and deformation properties of almost any given volume of rocks. Currently, the most widely used are as follows:

- ground-based seismic studies by the correlation method of refracted waves on longitudinal and nonlongitudinal profiles;
- seismic and acoustic measurements by the method of refracted waves according to the method of longitudinal profiling in mining;
- seismic and acoustic studies using the method of multipoint transmission;
- seismic, acoustic, and ultrasonic studies in wells and holes.

The use of seismic-acoustic research methods is based on the fact that the propagation velocities of longitudinal and transverse waves directly depend on the density (ρ) and deformability (E_d) of the medium. Two infinite volumetric elastic waves propagate in an infinite homogeneous and isotropic medium: longitudinal P and transverse S. Having determined the propagation velocity of longitudinal V_p and transverse V_s of elastic waves excited in the massif by the vibration source in the section under consideration, the so-called dynamic elastic modulus of the massif E_∂ and the coefficient Poisson v by dependencies are as follows:

$$E_\partial = \rho V_s^2 \frac{\left(3V_p^2 - 4V_s^2\right)}{\left(V_p^2 - V_s^2\right)} \tag{3.7}$$

$$E_\partial = \rho V_p^2 \frac{(1+v)(1-2v)}{(1-v)} \tag{3.8}$$

$$E_\partial = 2\rho V_s^2 (1+V) \tag{3.9}$$

$$v = \frac{\left(V_p^2 - 2V_s^2\right)}{2\left(V_p^2 - V_s^2\right)} \tag{3.10}$$

The elastic moduli calculated from these dependencies, as a rule, exceed the moduli obtained under static loading of the rock mass. This is due to the viscosity characteristic of real rock massifs, which manifests itself depending on the magnitude and duration of the applied load [147]. When seismic waves propagate and when the arising stresses act for a very short time, rock masses behave as perfectly elastic bodies.

Nevertheless, a comprehensive study of the rock mass by various methods allows us to establish a correlation between the dynamic and static moduli of elasticity (or deformation). It can suggest the following dependence for recording such a correlation:

$$E_{cm} = E_{\partial}\left[1 - e^{-aE_{\partial}^2}\right]$$

(3.11)

where: a is the coefficient determined experimentally for each type of rock.

The great advantage of seismic acoustic methods is their simplicity, low cost, and the ability to "sound" large volumes of rock mass in different directions. It should be noted, however, that the accuracy of determining the elastic moduli and Poisson's ratios by these methods strongly depends on the accuracy of determining the velocities, which are included in all the dependences squared [147].

3.2.2.4 Moduli of deformation obtained by different methods

It is obvious that various methods for determining the deformation moduli of rock masses will give different results first all due to the presence of a scale factor, since in each of the determination methods, different volumes of the array are involved in the core.

For example, a number of methods were used to determine the deformation modulus of the foundation of the arch dam of Malpasset (France), and all of them gave different values:

According to the results of laboratory tests:

• On cylindrical rock samples d = 145 mm	E = 58,000 MPa
According to the results of field research	
• Dilatometers in wells d = 76 mm	E = 32,000 MPa
• Dilatometers in wells d = 165 mm	E = 22,000 MPa
• On stamp d = 280 mm	E = 14,000 MPa
• Geophysical methods	E = 38,000 MPa
From calculated abutment offsets	
Dams in nature	E = 22,500 MPa

The problem with determining the geomechanical characteristics of the rock massif arose during the resumption of the construction of the Rogun HP after a 14-year break, during which their significant changes could occur. All these years, underground workings were in a flooded state, which, of course, had an impact on the geomechanical characteristics of the rock mass, which also serves as the foundation for the dam. It was necessary to carry out new comprehensive studies of the deformability and bearing capacity of the strata of unevenly inter-bedded different-grained sandstones and siltstones.

Specialists of the Center for the Service of Geodynamic Observations in the Energy Industry (CSGOEI) conducted a set of pressiometric, geophysical, and computational studies using mathematical models to assess the strength and deformation characteristics of a rock mass containing the underground workings of the Rogun HEP.

Table 3.4 Summary of strain modulus values obtained by various methods

Method for determination		Value of the module E (MPa)			
		Siltstones		Siltstones	
		Unchanged mass	Weakened zone	Unchanged mass	Weakened zone
Hoek-Brown rating assessment		15,000	-	6,000	-
Geophysical methods and seismic exploration		11,000[a] / 8,400	8,300[a] / 6,250	8,100[a] / 6,100	5,800[a] / 4,300
Pressiometry		-	4,600[a]	-	3,500
Designed methods	Flat model	9,000	6,000	5,500	4,000
	Volume model	7,500	5,000	4,000	2,670

a The numerator indicates the magnitude of the module for the second loading cycle E_{II}, and the denominator indicates the absolute (total) strain modulus E_Σ.

The calculations were carried out in an elastic-plastic formulation using the FEM with reproduction of the NSS of the massif, decompression zones around the recesses, fixing the workings with concrete, passive and active anchors, and the actual sequence of driving and fixing the workings.

In the process of calculations, the mathematical model was calibrated based on a comparison of the results of calculations with the data of field measurements of the convergence of the walls of the machine hall, after which the predicted calculations of the SSS system were completed during the completion of underground structures to the operational state [249].

Analyzing the results of determining the deformation modulus by various methods, a summary of which is given in Table 3.4, one cannot but note their significant difference in general. At the same time, there is good convergence of the results of determining the deformation moduli for sandstones and siltstones according to calculation methods and seismic measurements (E_Σ values) and for siltstones also according to pressiometers (for weakened zones) and rating estimates (for an undisturbed massif).

A summary of the obtained values of the deformation moduli is presented in Figure 3.15.

Summarizing the above results, it is possible to draw the following conclusions:

• The most reliable averaged values of the deformation moduli of sandstones and siltstones of the rock mass of the underground engine hall of the Rogun HEP were obtained by calculations on mathematical models based on the values of the convergence of the underground mine walls measured in nature;

• The obtained coincidence of the module values calculated on the mathematical model from the displacements of the walls of the hall (occurred over a long period) with the value E_Σ determined from the geophysical data (characterizing the total deformations in contrast to the E_{II} module) made it possible to conclude that one of the effective methods of real estimation modulus of deformation of rock masses is the calculation method based on measurements of natural deformations of rock masses.

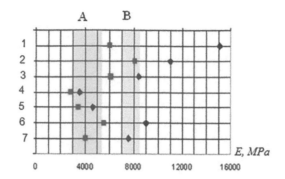

Figure 3.15 Summary of deformation modulus values obtained by various methods: A – sandstones and B – siltstones; 1 – rating, 2 – geophysics D_u, 3 – geophysics D_Σ, 4 – pressiometer E_u, 5 – pressiometer $E_{\parallel}$, 6 – flat model, and 7 – volume model.

Table 3.5 Strength classification of rocks

Variety of soils	Limit strength on uniaxial compression R_c (MPa)
Very durable	>120
Durable	120–50
Medium durable	50–15
Low durable	15–5
Reduced durable	5–3
Low durable	3–1
Very low durable	<1

3.2.3 Strength of rock

3.2.3.1 Compression strength under uniaxial loading

In any field of practical use of applied mechanics, it is always necessary to know the limit strength of the material used. This strength determines the maximum load that the material can withstand without breaking.

Determining uniaxial compressive strength (R_c) represents the first step in evaluating the strength of a material. Compressive strength is one of the most important characteristics for judging the strength of a rock material in a specimen. According to SS 25100-95 limit strength of uniaxial compression of R_c in a water-saturated state, rocks are divided according to Table 3.5.

There are a number of factors, both external and internal, affecting the determined value of compressive strength.

External factors:

• friction at the contacts of the plates of the press with the ends of the sample;
• geometry of the sample (shape, size, and their ratio);
• load application speed;
• temperature and humidity at the time of the test.

Internal factors:

*   type of rock;
*   connectivity, texture, mineralogy, humidity, and porosity;
*   grain size;
*   fracture of the sample;
*   orientation of crystals and weakening planes.

The larger the sample size, the more defects and cracks it may contain, which leads to a decrease in the determined strength value. This is the so-called "scale factor".

The determination of uniaxial compression strength is carried out, as a rule, on cylindrical (sometimes on prismatic or cubic) samples in accordance with the recommendations developed by the International Society for the Rock Mechanics (ISRM) [267].

The main requirements of these recommendations are as follows:

*   the test sample should be a straight cylinder with a ratio of length to diameter from 2.5 to 3 with a diameter of at least 54 mm, while the diameter of the sample should be 10 times larger than the maximum rock grain;
*   the ends of the sample should be parallel to each other and perpendicular to the longitudinal axis of the sample; the deviation from the plane at the ends should not exceed 0.02 mm, and the deviation from the perpendicularity of the axis should not exceed 0.001 radian (about 3.5 minutes) or 0.05 mm per 50 mm of length;
*   the side surfaces of the sample should be smooth and free from flaws, and the maximum deviation should not exceed 0.3 mm over the entire height of the sample;
*   the application of the load should be carried out without interruption at a constant speed, so that the destruction occurs after 5–15 minutes after the start of loading; the loading speed should preferably be in the range from 0.5 to 1 MPa/s;
*   the use of gaskets between the ends of the sample and the plates of the press is not allowed;
*   load registration should be accurate to 1%;
*   for each determination, it is advisable to test at least five samples.

These recommendations were taken as the basis by most laboratories, including in our country, although deviations are allowed in some cases, in particular, regarding the use of thick cardboard gaskets for testing short samples, which reduces the effect of shear stresses at the ends of the sample.

To obtain the strength value (R_c), the maximum experimental load (P) is divided by the cross-sectional area of the sample (F):

$$R_c = \frac{P}{F} \tag{3.12}$$

3.2.3.2 Tension strength under uniaxial loading

The tensile strength of rocks can be determined both by direct tensile tests and by the splitting method, the so-called "Brazilian method", when the specimen is cracked in the press by its diameter (Figure 3.16).

Figure 3.16 Determination of the tensile strength of a cylindrical rock specimen by the "Brazilian method" (splitting).

In this test, more reliable and uniform results are obtained than in direct tensile tests, since in this case, the effects of microcracks are excluded, which can firstly reduce the cross section of the sample and secondly cause an eccentricity of the application of tensile load.

The tensile strength during cracking of the sample is determined by the dependence

$$R_p = \frac{2P}{\pi DL} \tag{3.13}$$

where
D – diameter of the sample;
L – length of the generatrix of the cylinder.

Detailed studies conducted in France by *Turenc* and *Denis* [51] showed that only the Brazilian method provides a uniform field of tensile stresses in the sample and makes it possible to obtain reliable strength values.

3.2.3.3 Strength in triaxial stress state

In the vast majority of cases, rock foundations are in a volumetric stress state, and it is required to determine the strength of the material at any combination of acting stresses.

This problem is especially difficult when there are dealing with fragile polycrystalline materials such as rock and concrete. The destruction of such materials has been the subject of numerous theoretical and experimental studies of recent years. They continue to this day, given the complexity of the problem [221].

The process of brittle fracture is associated with the formation of micro- and then macrocracks and with the deformation of the material over the entire range of application of the load. The very concept of material strength in a complex volumetric stress state needs to be clarified since the process of microdestruction of the material begins practically when the stresses reach 0.4–0.5 R_c and continue passing from one level of microdestruction to another. The presence of lateral compression of the material sample in the array does not allow it to "fall apart" into parts (as occurs with a uniaxial or plane stress state), and it continues to withstand an ever-increasing load with a significant increase in deformations. Therefore, it can define the concept of the strength of a brittle polycrystalline material in bulk stress state as follows:

> The strength of the polycrystalline material in the volumetric stress state is the combination of three principal stresses at which a sharp increase in sample deformation occurs or that combination of three principal stresses at which the material ceases to meet design requirements.

To determine the strength, a number of empirical criteria were proposed, each of which describes the strength within certain limits of the operating stresses and for certain materials.

The stress state at the time of failure can be represented by three principal stresses $\sigma_1, \sigma_2, \sigma_3$, which in the coordinate system of the principal stresses represent a point. Combining all such fracture points obtains the fracture surface in the coordinates of the main stresses:

$$f(\sigma_1, \sigma_2, \sigma_3) = 0 \tag{3.14}$$

In the future, assume that the compression is positive and $\sigma_1 \geq \sigma_2 \geq \sigma_3$. It can be assumed that the surface strength will not exist under hydrostatic compression $(\sigma_1 = \sigma_2 = \sigma_3)$, when fracture is considered impossible. In addition, the condition of the hierarchy of principal stresses $\sigma_1 \geq \sigma_2 \geq \sigma_3$ requires that the existence of a strength surface be considered only where this condition is satisfied. The strength surface intersects the axes σ_1 and σ_3 at points corresponding to the uniaxial compressive and tensile strengths.

3.2.3.4 Strength criteria

To assess the behavior of the material in a complex stress state, a number of theories, methods, and criteria were proposed. All of them proceed from different assumptions; some researchers consider normal or tangential stresses to be the main reason for the onset of the limit state (fracture of the material), others proceed from limit strains, and others continue to consider various fracture criteria for brittle fracture and slide fracture [116].

Mohr criterion proposed by Otto Mohr in 1900 [239] was most widely used in engineering practice.

This criterion is based on the premise that the strength of the material at a point depends only on the maximum and minimum principal stresses and does not depend on the average principal stress. For this analysis, the stress state is represented graphically in the form of Mohr stress circles.

If the maximum and minimum principal stresses during the destruction of material samples are experimentally determined, then a family of Mohr circles can be constructed, and their envelope will represent the strength surface of this material (Figure 3.17).

A generalization of the results of numerous tests of rocks shows that a parabola of the second degree is acceptable for describing the limit state of rocks (Figure 3.18a):

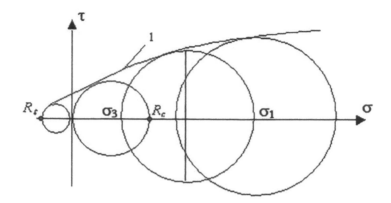

Figure 3.17 Mohr circle diagram: I – envelope.

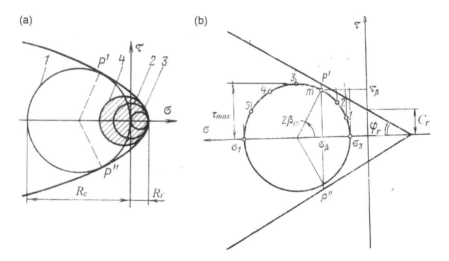

Figure 3.18 Envelope of the limit circles Mohr: (a) parabolic and (b) straight.

$$\sigma = -a\tau^2 + R_t \tag{3.15}$$

$$\text{where}: a = \left(\sqrt{1+n} + \sqrt{n}\right)^2, \; n = \frac{R_t}{R_c} \tag{3.16}$$

R_t and R_c – uniaxial tensile and compression strength actually; stress signs are accepted according to the theory of elasticity: "+" – tension and "–" – compression.

When processing a few experimental data, the envelope is often approximated by a straight line (Figure 3.18b):

$$\tau = -\sigma tg\varphi + C \tag{3.17}$$

where: φ – angle of internal friction and C – cohesion.

The strengths R_t and R_c are associated with φ and C dependence [101]:

$$R = 2C \times tg\left(45° \pm \frac{\varphi}{2}\right) \tag{3.18}$$

where: the "+" sign corresponds to R_c, and the "–" sign corresponds to R_t.

According to Mohr, failure (transition beyond the yield strength) occurs when the stress circle characterizing the stress state at the point touches the envelope of Mohr limit circles $\tau = \tau(\sigma)$. The points of tangency of the stress circle of the envelope P' and P'' determine the orientation of two sites along which fracture develops in the form of a slide.

The safety factor for slide along the site is written as

$$\theta = \frac{-\sigma\, tg\varphi + C}{\tau} \tag{3.19}$$

For an isotropic material, there is a site at which θ takes the smallest value with a parabolic envelope (Figure 3.18a):

$$\theta_m = \frac{\sqrt{R_t - \sigma_1} + \sqrt{R_t - \sigma_3}}{(\sigma_1 - \sigma_3)\sqrt{a}} \tag{3.20}$$

with a straight envelope (Figure 3.18b):

$$\theta_m = \frac{2}{(\sigma_1 - \sigma_3)}\sqrt{(-\sigma_1 tg\varphi + C)(-\sigma_3 tg\varphi + C)} \tag{3.21}$$

It is often believed that sites with maximum tangential stresses τ_m are potential slide sites; however, as follows from Figure 3.19, this is not so.

The safety factor for a given site is determined by the formula (3.19), in which σ and τ are the normal and tangential stresses on the site.

Hoek and Brown criterion proposed in 1980 is most often used among the many criteria available in the technical literature. In accordance with its latest version [210], this criterion can be represented by the following dependence:

$$\sigma_1' = \sigma_3' + R_c\left(m_b \frac{\sigma_3'}{R_c} + s\right)^a \tag{3.22}$$

where

σ_1 – maximum principal stress at the time of failure;

σ_3 – minimum principal stress at the time of failure;

R_c – the uniaxial compression strength of the rock sample;

m_b and s – coefficients depending on the characteristics of the massif and the degree of its fracture (for a monolithic rock $s = 1$);

It is accepted that compression is positive and $\sigma_1 \geq \sigma_2 \geq \sigma_3$.

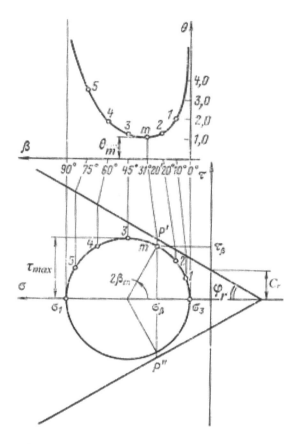

Figure 3.19 Scheme for determining the safety factor θ_m at a point according to Mohr.

The following dependencies are used usually to calculate these coefficients:

$$m_b = m_i \exp\left(\frac{GSI - 100}{28}\right) \tag{3.23}$$

$$s = \exp\left(\frac{GSI - 100}{9}\right) \tag{3.24}$$

$$a = \frac{1}{2} + \frac{1}{6}\left(e^{-GSI/15} - e^{-20/3}\right) \tag{3.25}$$

where
GSI – Geological Strength Index of the rock mass, which is a generalized characteristic of its quality;

D – factor accounting for rock damage resulting from explosions or unloading of the massif; if studies are conducted on laboratory samples, then the following are accepted: GSI = 100 and $D = 0$, which implies minimal sample disturbance;

m_i – coefficient, the value of which for various types of rocks is determined on on the basis of laboratory tests of rock samples or according to Table 3.6 [210].

Table 3.6 The parameter m_i for rock groups

Type of rocks	Group	Texture of rocks			
		Coarse-grained	Medium-grained	Fine-grained	Very fine-grained
Sedimentary	Clastic	Conglomerate (22)	Sandstone 19	Alevrolit 9	Argylllit 4
	Organic	← Chalk 7 →			
		← Coal (8–21) →			
	Carbonate	Breccia (20)	Organogenic Lime-stone (10)	Crystalline Limestone 8	-
	Chemical	-	Gypsum 16	Anhydride 13	-
Metamorphic	Nonshaved	Marble 9	Horn (19)	Quartzite 24	-
	Weakly Shattered	Migmatite (30)	Amphibolite 25–31	Milonite (6)	-
	Shaleed	Gneiss 33	Clay Shale 4–8	Phyllite (10)	Crystal shale 9
Erupted	Bright Dark colored	Granite 33 Granodiorite (30) Diorit 28 Gabbro 27 Norit 22	Dolerite (19)	Riolite (16) Dacite (17) Andesite 19 Basalt 17)	Obsidian (19)
	Volcanogeni-clastic	Agglomerate (20)	Breccia (18)	Tuf (15)	-

The tensile strength of the rock mass is determined by the following dependence [210]:

$$\sigma_{tm} = -\frac{s R_c}{m_b} \tag{3.26}$$

In parentheses are the expected values.

The equivalent envelope of Mohr circles can be represented as follows:

$$Y = \log A + BX \tag{3.27}$$

$$Y = \log \frac{\tau}{R_c}; \quad X = \log\left(\frac{\sigma_n - \sigma_{tm}}{R_c}\right) \tag{3.28}$$

For $GSI < 25$, when $s = 0$,

$$\frac{\partial \sigma_1'}{\partial \sigma_3'} = 1 + am_b^a \left(\frac{\sigma_3'}{R_c}\right)^{a-1} \tag{3.29}$$

Gaziev criterion was proposed in 1984 [49] as a result of studies conducted at the LRMIH. This criterion was initially developed for the conditions of "classical" volumetric loading, when the second and third principal stresses are equal to each other.

Further research was continued at IENAUM, as a result of which a universal criterion was formulated for any combination of principal stresses [214].

Taking into account the considerable difficulty in obtaining a purely theoretical criterion for the strength of brittle polycrystalline materials in a complex stressed state, the most promising way to solve this problem is to develop a phenomenological criterion based on the most important factors determining the destruction of such materials [215,241].

Such a criterion should meet the following conditions:

1. describe the real model of fracture of brittle material;
2. evaluate the strength of the material at any combination of principal stresses;
3. meet the boundary conditions:
    - when $\sigma_2 = \sigma_3 = 0$, the first principal stress σ_1 should be equal to the uniaxial compression strength $\sigma_1 = R_c$,
    - when $\sigma_1 = \sigma_2 = 0$, the third principal stress σ_3 should be equal to the uniaxial tensile strength $\sigma_3 = -R_t$,
4. condition $\sigma_1 \geq \sigma_2 \geq \sigma_3$ must be satisfied;
5. should be simple for practical use and include a limited number of empirical coefficients requiring additional experimental determination.

Under the assumption that the main parameters determining the strength of brittle polycrystalline materials are:

- first invariant of the stress tensor:

$$J_1 = \sigma_1 + \sigma_2 + \sigma_3; \tag{3.30}$$

- second invariant of the stress deviator:

$$J_2 = (\sigma_1 - \sigma_2)^2 + (\sigma_2 - \sigma_3)^2 + (\sigma_1 - \sigma_3)^2; \tag{3.31}$$

- axial compression strength of the material R_c;
- tensile strength of the material R_t (as a parameter of strength) is always positive and
- ratio between the smaller principal stresses σ_2 and σ_3.

The following strength criterion was proposed [53,214]:

$$\frac{\sigma_* + m}{1 + m} = \left(\frac{\tau_* - m}{1 - m} \right)^n \tag{3.32}$$

where

$$\sigma_* = \frac{\sigma_1 + \sigma_2 + \sigma_3}{R_c} \tag{3.33}$$

$$\tau_* = \sqrt{\frac{(\sigma_1 - \sigma_2)^2 + (\sigma_2 - \sigma_3)^2 + (\sigma_1 - \sigma_3)^2}{2 R_c^2}} \tag{3.34}$$

$$m = R_t / R_c \tag{3.35}$$

The value of the exponent n in a first approximation can be taken equal to $n = 1.15$–1.30 (determined by the selection in each case).

The criterion dependence (Equation 3.32) describes the strength surface in the space of principal stresses (Figure 3.20).

Hierarchy of principal stresses ($\sigma_1 \geq \sigma_2 \geq \sigma_3$) requires considering only a part of this surface, where the indicated condition is satisfied: this is a sector with a central angle in space of 60° formed by the planes $\sigma_1 = \sigma_2$ and $\sigma_2 = \sigma_3$ and limited by the strength surface (Figure 3.20), the longitudinal generators of which are parabolas of varying degrees.

The intersection of this surface with the plane $\sigma_2 = \sigma_3$ describes the strength of polycrystalline materials with a "classical" bulk loading of the sample, when the second and third principal stresses are equal to each other ($\sigma_2 = \sigma_3$). In this case, the criterion can be written as follows:

$$\frac{\sigma_1 + 2\sigma_3 + R_t}{R_c + R_t} = \left(\frac{\sigma_1 - \sigma_3 - R_t}{R_c - R_t} \right)^{1,3} \tag{3.36}$$

In this form, this criterion was proposed by E. Gaziev and A. Morozov in 1984 [49].

In modern understanding, the criteria of Mohr, Hoeck-Brown, and Gaziev are not theories of strength, since they lack a connection between deformations and stresses during loading; these criteria serve to assess the level of acting stresses in relation to the strength of the rock mass.

The energy of rock destruction is directly related to its strength. As a matter of fact, this position is laid down in the idea of assessing the strength of a material by the energy of its shape.

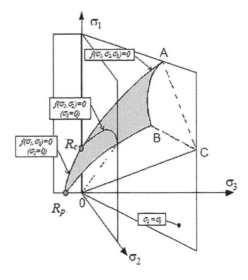

Figure 3.20 Strength surface according to Gaziev criterion.

To analyze the work performed by external forces, it is advisable to use the proposed N.I. Bezukhov [25] with the expressions of "generalized stress" or "stress intensity":

$$\tau_i = \frac{1}{\sqrt{2}}\sqrt{(\sigma_1-\sigma_2)^2+(\sigma_2-\sigma_3)^2+(\sigma_1-\sigma_3)^2} \tag{3.37}$$

and "strain intensity":

$$\varepsilon_i = \frac{1}{\sqrt{2}}\sqrt{(\varepsilon_1-\varepsilon_2)^2+(\varepsilon_2-\varepsilon_3)^2+(\varepsilon_1-\varepsilon_3)^2} \tag{3.38}$$

These two parameters are directly proportional to the square root of the second invariant of the stress tensor and strain tensor.

At the time of fracture, the stress intensity is $\tau_i = (\tau)_{cr}$. In the case of a uniaxial test, $(\tau_i)_{cr}$ is equal to the uniaxial compression strength $(\tau_i)_{cr} = R_c$, and under the "classical" triaxial loading (at $\sigma_2 = \sigma_3$), it is equal to the maximum shear strength at the moment of fracture onset $(\tau_i)_{cr} = (\sigma_1 - \sigma_3)$.

The work of external forming forces at the moment of destruction in the elementary volume of the sample in this case can be determined by the following dependence:

$$E_{cr} = \int_0^{\varepsilon_{cr}} \tau_i(\varepsilon_i)\,d\varepsilon_i \tag{3.39}$$

Under volume loading, the moment of the onset of failure or the limiting value of the stress intensity at the time of the onset of failure can determined by the proposed criteria (Equation 3.32).

The diagram of deformation of a rock sample before destruction is presented in Figure 3.21.

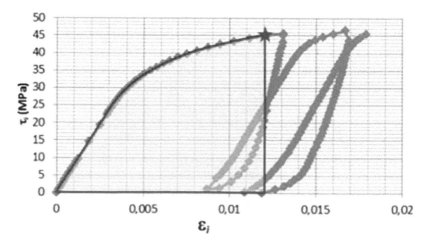

Figure 3.21 The diagram of the deformation of the sample until the moment of destruction.

Table 3.7 Uniaxial loading test results

Icon on Figure 3.22	Material	R_c (MPa)	ε_{lcr}	E_{cr} (kJ/m³)	Link
1	Diabase, breccia	130–153	0.0143–0.0175	849–1,140	A
2	Diabase with quartz	87–128	0.0121–0.0141	474–788	
3	Diabase	60–76	0.0102–0.0123	336–432	
4	Metadiabases	90–95	0.0108–0.0121	416–528	
5	Metadiabases	53–58	0.0084–0.0105	217–265	
6	Metadiabases	30–35	0.0070–0.0094	157–187	
7	Quartz slates	40–45	0.0038–0.0089	191–282	
8	Chlorites, slates	17–24	0.0071–0.0089	89–183	
9	Slates, breccias	18–28	0.0115–0.0133	59–153	
S	Sandstone	143	0.0084	601	B
T	Tuff	44	0.0074	165	
DC	Diabase (Coggins)	341	0.00738	1,531	C
B	Basalt	223	0.00761	1,092	
C_{21}	Concrete	51.7	0.0087	390	D
G	Gypsum	12	0.0029	19	

Table 3.8 Volumetric test result stress

Icon on Figure 3.22	Material	Strength (MPa)	Stresses of destruction (MPa)	τ_{cr} (MPa)	ε_{icr}	E_{cr} (kJ/m³)
C_2	Concrete	$R_c = 45$ $R_t = 2.9$	$\sigma_{1cr} = 63.3$ $\sigma_{2cr} = \sigma_{3cr} = 2.9$	43.35	0.00612	180
C_3	Cement solution	$R_c = 39.4$ $R_t = 2.7$	$\sigma_{1cr} = 46.1$ $\sigma_{2cr} = \sigma_{3cr} = 1.0$	45.10	0.01216	340
C_4	Concrete	$R_c = 45$ $R_t = 3.1$	$\sigma_{1cr} = 63.44$ $\sigma_{2cr} = \sigma_{3cr} = 2.97$	60.47	0.00942	415
C_5	Cement solution	$R_c = 39.4$ $R_t = 2.7$	$\sigma_{1cr} = 61$ $\sigma_{2cr} = 4.65$ $\sigma_{3cr} = 3.13$	57.14	0.01230	381

The corresponding value of the specific energy of destruction will be equal to the area of the hatched (by pink color) zone in the diagram (Table 3.7). The diagram of the dependence of the strength of the material from energy necessary for its destruction presented in Figure 3.22.

A. Pininska, J., Lukaszewski, P., 1991. The relationships between post-failure state and compression strength of Sudetic fractured rocks. *Bulletin of the International Association of Engineering Geology*, [50].
B. Kawamoto, T., Saito, T., 1991. The behavior of rock-like materials in some controlled strain states. *7-th International Congress on Rock Mechanics*, Aachen (Germany), vol. 1, pp. 161–166.
C. Miller, R.P., 1965. Engineering classification and index properties for intact rock. Thesis doctoral, University of Illinois, Urbana.
D. Gaziev E. Rupture energy evaluation for brittle materials, *International Journal of Solids and Structures*, vol. 38, 2001, pp. 7681–7690.

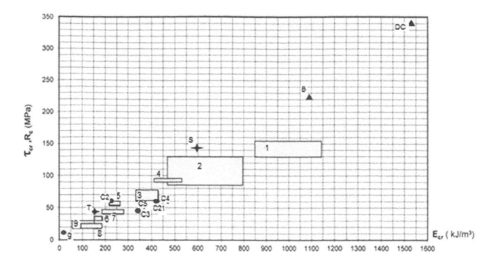

Figure 3.22 The diagram of the dependence of the strength of the material from energy necessary for its destruction. Decryption conditional icons are given in Tables 3.7 and 3.8.

3.2.4 Slide strength of rock mass

The slide strength of a rock is one of its most important characteristics, which is often of greater interest than the compressive and tensile strength.

Typically, the slide strength of a rock is defined by a linear relationship:

$$[\tau] = c + f\sigma,\tag{3.40}$$

where the parameters c and f do not carry a specific physical meaning but are only mathematical parameters of a straight line, replacing the curvilinear dependence $\tau - \sigma$ over a certain stresses range.

In addition to the indicated parameters c and f, which determine the strength according to the "monolithic" rock or along the crack with "links", it should mention the angle of internal friction φ or the friction coefficient $\mu = tg\varphi$ which is the residual strength along the already formed surface of the crack. This friction coefficient cannot be identified with the parameter f, if only because they correspond to different states of the rock in the slide zone, and their identification can lead to serious errors (f is equal to μ only if the residual strength is reached).

3.2.4.1 Slide strength along rock on fissure

When it comes to the slide strength of rock masses, then, as a rule, this refers to slide strength along rock cracks or weakening surfaces.

The surface of the walls of cracks is very rarely absolutely smooth and almost always has one or another roughness. Therefore, the process of slide along a crack is associated with overcoming or chipping of these irregularities and depends on the

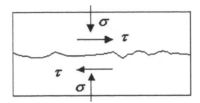

Figure 3.23 Displacement on closed rock fissure.

geometry of these irregularities, the strength of the rock material of the walls, and the magnitude of the applied normal and slide loads (Figure 3.23).

In order to be able to move, the upper part of the rock mass above the crack should rise while cleaving or crushing roughness, i.e., any deformation or destruction of the material in the crack will be associated with its *dilatancy*, that is, with volume expansion. If the material does not have the ability to expand in volume, then it cannot be destroyed.

Rock fissures dilatancy is a very important property that plays a decisive role in the stability of rock masses and their anchoring. A fissure limited in its opening capabilities will cause additional normal stresses and, therefore, withstand higher slide loads. It should be noted that dilatancy manifests itself most of all in closed rock fissures, while in open or filled fractures, their compression can be observed.

The main parameter determining the dilatancy of a fissure during slide is its roughness and the history of displacements that occurred in the past.

If we assume that the rock is weightless (normal stresses on the contact are equal to zero), then the dilatancy of the fissure will be determined by the displacement of the upper part relative to the lower one with a rise by i_o, which is the initial average roughness.

In the presence of a normal stress σ, some of the steepest irregularities will chip or crease, depending on the crushing strength R^* of the crack walls. As a result, the real dilatancy angle of the crack i will be smaller and determined by the initial angle i_o, normal stress σ, and the material strength of the bumps on crushing R^* (Figure 3.24).

An analysis of the available results (unfortunately not numerous) allows us to suggest the following dependence for determining the angle i [47]:

$$i = i_0\left(1 - \frac{\sigma}{R^*}\right) \tag{3.41}$$

where

i – the dilatancy angle in the process of displacement along the crack;

i_0 – initial angle of elevation in the direction of slide (initial roughness);

σ – normal compression stress in the plane of the crack;

R^* – the compressive strength of the crack wall material;

m – the index of brittleness of the rock, which with a sufficient degree of accuracy can be taken equal to 10 (Figure 3.25).

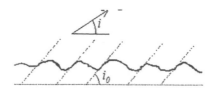

Figure 3.24 Statistically average *i* and effective angles i_o of dilatancy.

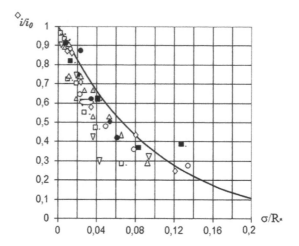

Figure 3.25 The dependence of the angle of latitude *i* in the process of displacement along the crack on the normal stress σ: ○□◊∇△◻ – experiments of Barton [188]; ■ – green shales in the alignment of the Andijan dam; ● – crack No. 750 in the site of the Toktogul dam.

The slide strength along the fissure can be written as follows:

$$[\tau] = \sigma\, tg(i + \varphi) \tag{3.42}$$

or

$$[\tau] = \sigma\, tg\, [i_o(1 - \sigma/R^*)^{10} + \phi] \tag{3.43}$$

In conclusion, it can be noted that when high normal stresses are reached, close to 30%–40% of *R**, the influence of angle *i* decreases significantly and the strength curve $[\tau] = f(\sigma)$ takes the form shown in Figure 3.26.

Thus, at high normal stresses or at low strength of the fissure walls, the slide strength along the fissure can become equal to the residual strength.

Residual strength is the slide strength with a constant slide force and with a constant normal load. This is the so-called "smooth" fissure displacement.

First it is to clarify the concept of a "smooth" fissure. In real rock masses, there are no absolutely smooth fissures, and therefore, by a smooth crack it means an idealized

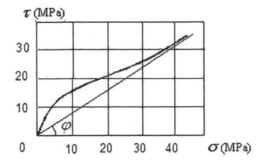

Figure 3.26 Dependence $[\tau] = f(\sigma)$ over a wide range of changes in normal stress.

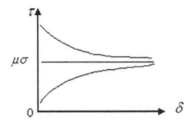

Figure 3.27 In the process of displacement along a rock fissure, it acquires its characteristic roughness corresponding to the coefficient of friction of rock $[\tau] = f(\sigma)$.

fissure, the walls of which do not have macroirregularities, and displacement along this fissure can occur under constant normal and slide loads, i.e., no matter the amount of displacement.

It should be borne in mind that for each rock, there is a "smoothness" of the fissure, i.e., its characteristic roughness. If the initial roughness in the fissure is greater than "characteristic", then these irregularities will be chipped, and the surface of the crack will be made "smooth".

If the roughness of the rubbing surfaces is less than the characteristic roughness of this rock, then the process of sliding the surface will acquire the required roughness of the "smooth" surface, at which the displacement will occur at constant normal and slide loads (Figure 3.27).

Coulson [47] studied 10 rocks of various mineralogical compositions, including basalt, granite, limestone, sandstone, gneiss, and dolomite. Samples of these rocks had various types of roughness from polished to sand blasted (Figure 3.28). Despite the different types of initial roughness, the obtained friction coefficients for each of the rocks had a difference not exceeding 0.05.

Thus, the slide strength along "smooth" fissures, or the so-called residual slide strength along a fissure, can be written as follows:

$$[\tau] = \mu\sigma \tag{3.44}$$

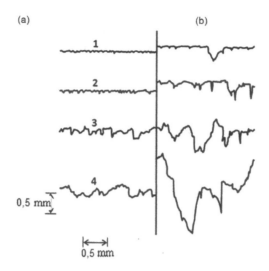

Figure 3.28 **Different initial roughness of samples of basalt (a) and limestone (b):**
1 – polirovanny, 2 – smoothed, 3 – leveled, and 4 – treated sand blasting
apparatus.

where

μ – the coefficient of friction;

σ – the normal stress in the plane of the fissure.

Sometimes together with the coefficient of friction, the concept of the angle of friction φ is used, which is determined from the coefficient of friction $\mu = tg\varphi$ or $\varphi = arctg\mu$.

Rocks rich in quartz or feldspar (sandstones, granites, etc.) have an internal friction angle of about 30° (sandstones – from 25° to 40°); carbonate rocks (limestones, dolomites, and marble) – from 32° to 36°; gneiss – from 18° to 30°; rocks containing mica – from 14° to 26°; rock consisting mainly of clay materials – from 4° to 14°; and most natural soils, consisting of clay, loam, sandy loam and sand – from 12° to 30°. The friction coefficient μ is practically in the range from 0.5 to 0.9.

In the absence of clay materials in the fissure contact, flooding practically does not affect the coefficient of friction, unless, of course, the rock material itself does not lose its strength in the presence of water. As experimental studies show, water is not a "lubricant" of the contact, and the friction coefficients for dry and wet rock surfaces practically coincide.

In abroad, the slide strength formula proposed by N. Barton [188] was widely used:

$$\tau = \sigma\ tg\left[\varphi + \zeta\lg\left(\frac{R^*}{\sigma}\right)\right] \tag{3.45}$$

where: ζ – fissure roughness coefficient, varying in the interval from 0 to 20, which can be obtained by one of three ways:

- comparing the profile of the fissure under consideration with standard profiles given in the works of N. Barton [188] and V. I. Rechitsky [137];

- statistical processing of the fissure profile;
- reverse calculation as a result of the analysis of the data of one of the experiments on the slide.

3.2.4.2 Slide strength of rock mass

Slide strength along a rock mass is associated with its destruction along a fractured or block medium, in which a slide surface is formed partly along fissures and partly along pillars of healthy rock. The large factual material accumulated to date on field and laboratory studies of slide strength shows that the fracture process begins with the appearance of a zone of tensile cracks at the base of the sliding stamp, and then a surface is formed in this disturbed zone, on which slide.

The experimental studies showed that with an increase in the slide load (T), the rise of the loaded face of the pillar begins with the appearance of a steeply falling tensile crack under the loaded face. With a further increase in the slide load, tensile cracks form under the upper face of the pillar and a compression zone under the lower face. Further displacement or rotation of the pillar occurs with the formation of a "slide" crack at the base of the pillar, which in fact is also a tensile crack, not a cleavage, and its direction practically coincides with the direction of the slide force T (Figure 3.29).

It is interesting to note that as shown by experimental studies conducted on a monolithic and block foundation models, the destruction of both models regardless of their different structure was of the same nature and came from the concentration of tensile stresses at the upper and compressive stresses at the bottom faces of the pillar (stamp).

As mentioned above, the slide strength of a rock is set conditionally by a linear relationship:

$$[\tau] = c + f\sigma, \tag{3.46}$$

where: parameters c and f do not carry a specific physical meaning but are only mathematical parameters of a straight line, replacing the curvilinear dependence $\tau - \sigma$ over a certain stress range.

3.2.4.3 Methods for determining slide strength

Under laboratory conditions, the slide strength of rock material is usually determined by the "direct slide" scheme. For this purpose, many different installations were created, including for testing large samples with a diameter of up to 80–100 cm [147].

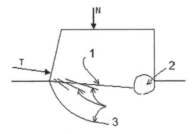

Figure 3.29 The fracture pattern at the base of the pillar during slide: 1 – slide crack, 2 – compression zone, and 3 – tensile cracks.

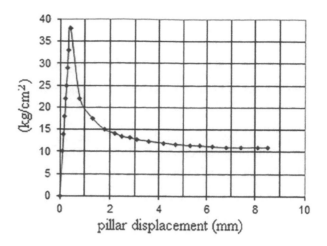

Figure 3.30 The stress-displacement diagram for a slide (shear) of gneiss at $\sigma = 15$ kg/cm^2.

The typical stress-displacement diagram when cutting a gneiss sample is shown in Figure 3.30. However, the results obtained are not characteristic enough for rock masses, and therefore, field testing methods should be recognized as the main methods for determining slide strength.

Determination of the shear strength of rock massifs is carried out as a rule in the field on rock pillars or concrete dies both by existing fissures and by the mass.

These studies are usually carried out in special underground workings or adits, the penetration of which should be carried out without the use of explosives. In the center of these adits, columns from the rock to be studied are left, which are then cut off from the roof, thereby obtaining rock pillars freed from the rock mass at the sides and from above and retaining contact only at the base. When performing operations to "free" the pillar from the surrounding massif, it is necessary to strive to preserve the slide region from unloading and decompression.

To be able to apply vertical and horizontal loads to the pillar and, if possible, evenly distribute them over the entire contact area, the pillars are put in a reinforced concrete shirt or clip. A normal load (usually vertical) is applied using hydraulic jacks, planed steel plates, and rollers between them to allow tangential displacement of the pillar. Sometimes instead of sliding the pillars, concrete stamps slide against the rough surface of the rock mass.

Slide loads are also applied using hydraulic jacks, usually at an angle to the slide surface, to eliminate the shear moment relative to the center of gravity of the shear plane. As the studies show, the considered load application scheme usually gives rather distorted results with limited experimental pillar sizes. It is likely that this loading pattern can more successfully be used in determining slide strength from a weakened contact or gap at the base of the stamp. Such studies are carried out both for pillars that do not contain a mature fissure and for pillars containing the investigated fissure or interlayer.

Naturally, with the increase in the size of the pillar, the reliability of the results obtained increases. In this regard, for particularly important structures, studies are often carried out on pillars or stamps, the slide area of which is measured by several square meters or dozens of them.

So during the construction of the dam of the Bratsk HEP on the river Angara, a pillar of rock with a size of 7×7 m in plan and 5.5–6 m in height [70] was tested. To determine the slide characteristics of the rock base along the fissures, this block was cut directly in the dam site. The normal load on the shear surface consisted of the dead weight of the rock above this surface and the vertical component of the inclined slide force, which was created by flat jacks concreted into the side trench. Using these jacks, a force of 15,000 tons could be applied at an angle of $10°47'$ to the horizon. The slide of the pillar along the fissure occurred at a slide force of 3,350 tons. This load was applied in steps with maintaining at each step for more than 2 hours until the deformations were completely stabilized. The total test duration was approximately 30 hours.

In 1964–1966 in the study of the rocky base under the concrete dam of the Krasnoyarsk HEP, a slide of two pillars was carried out [70]. These samples, $8\times12\times7$ m in size, were carefully cut down in the weakest fractured granites near the lower edge of the dam, and the upper part of the cells was dressed in a reinforced concrete shirt to a height of 6 m. The vertical load, amounting to 7,000 tons, was created by loading the pillars with massive concrete blocks, and the horizontal load of about 11,000 tons was created by flat jacks resting against the lower edge of the dam, which had been partially erected by that time.

Similar tests were carried out on the construction of the Juupia HP on the river Paraná in Brazil, where the size of the rock pillar was $5.5\times5.5\times4.6$ m [218].

To determine the slide strength from a rock fissure filled with clay material at the base of the Vouglan arch dam in France, a rock pillar was measured in the underground excavation with a size of 2.2×2 m, located directly on the indicated clay layer. The whole was dressed in a reinforced concrete shirt, to which normal and slide loads were applied using two groups of flat jacks. To eliminate friction between the jacks which created a normal load and the reinforced concrete pillar of the pillar, two neoprene plates were installed alternating with teflon films. The test was carried out at three consecutive normal loads: 1–3 MPa. Slide rate was limited to 0.04 mm/hour [191].

Analyzing the process of destruction of the rock material during the shift of targets, the following stages can be noted:

1. With an increase in the shear stress, elastic or quasi-elastic shear deformations first occur. Vertical deformations under the upper face of the pillar with an increase in shear force are initially equal to zero and then gradually begin to increase, indicating the expansion of the studied slide zone.
2. Then the process of crack formation begins, and the tensile cracks are oriented at a certain angle to the future slide surface. With a further increase in slide force, these cracks open, and secondary cracks appear. A slide zone is being formed. Vertical deformations on the lower edge of the pillar initially indicate a compression of the slide zone, and then, with a certain amount of slide, close to destructive (Figure 3.30), they also begin to increase, indicating that the expansion process has begun.

3. When the shear stress reaches the maximum value of τ_{lim}, a cleavage occurs along the formed slide surface. The force in the jacks creating a slide load drops rapidly, and when the experiment continues, after reaching a certain amount of displacement along the slide plane, the slide force stabilizes at a certain level corresponding to the residual slide strength or friction coefficient.

The fact that the fracture process in the slide zone begins even before the shear stress reaches its maximum value is also evidenced by the nature of the vertical displacements in Figure 3.31, which shows a diagram of the vertical displacements during the slide of the rock in the site of the Alto Rabagao Dam in Portugal at $\sigma = 1$ kg/cm^2. As can be seen from these graphs, for a certain amount of slide, the vertical displacements on the lower edge of the sliding stamp change sign, indicating a begun increase in volume in the slide zone. Similar results were obtained when conducting unique slide tests at the Krasnoyarsk HEP [70,197]. Muller observed the same phenomenon when conducting research in Japan [240].

An increase in the volume of rock material in the slide zone indicates the onset of fracture, i.e., about the material reaching its ultimate strength. It follows that preventing the possibility of expansion of the rock mass in the slide zone using anchors can significantly increase its slide strength.

After the formation of a continuous sliding surface, the upper part separated from the lower one gains more freedom in its displacement, as a result of which the horizontal displacements grow much faster. The shear stress drops sharply to a value of τ_o, which characterizes the friction at the contact and maintains a constant value at a given normal load.

In cases where it is not possible to conduct studies of slide strength, the recommendations of SR should be used [11]. These recommendations are based on a summary of the wide experience gained in our country in the design and construction of HS.

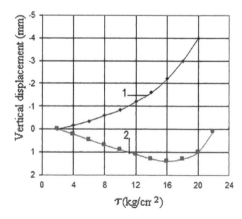

Figure 3.31 Vertical displacements during the slide of a pillar of rock in the site of the dam Alto Rabagao (Portugal). A positive vertical displacement was considered a downward displacement: I – pressure face and 2 – lower face.

3.2.5 Rock mass quality indicator

This indicator, called "Rock Quality Designation" or RQD, was proposed in 1968 by *Donald Deere* as a quantitative indicator of the quality of a rock mass [51]. It is more representative than the "core exit".

RQD is determined by the percentage of cores of length solid rock each more than 10 cm (4 in) long:

$$RQD = \frac{\text{sum of cores length solid rock more 10 cm}}{\text{total length of drilling core}} \times 100$$

D. Deer proposed to classify the quality of rock masses as follows according to the obtained RQD values (Table 3.9).

Determination of RQD is one of the most common methods for assessing the degree of fracture of rock masses by core quality.

This parameter can also be estimated from the speed of propagation of elastic waves in a rock mass during seismic exploration. Considering that the presence of cracks reduces the propagation velocity of elastic compression waves in the massif, it is possible to determine the parameter RQD from the ratio of the velocity of elastic waves in a fractured massif to the velocity in an undisturbed rock sample [252]:

$$RQD = \left(\frac{V_m}{V_c}\right)^2 \times 100 \tag{3.47}$$

where
   V_m – the speed of elastic compression waves in the fractured massif;
   V_c – the same in an undisturbed rock sample.

In Russia regulatory documents, the RQD value is used to determine the degree of fracture of rock masses [11].

3.2.6 Natural stress state

"Natural stress state" (NSS) of the massif means natural geostatic and geodynamic stresses existing in the massif prior to the start of construction of the structure.

Geostatic stresses are created by gravitational force (gravity), which acts everywhere and is the basis for the formation of the natural SSS of rock masses. The force of

Table 3.9 Correlation between RQD and rock mass quality

No.	RQD (%)	Rock quality
I	<25	Very poor
2	25–50	Bad
3	50–75	Average
4	75–90	Good
5	90–100	Very good

gravity is counteracted by the forces of elastic deformation of the rocks, which provide the equilibrium state of the massifs. At the same time, the structure and composition of the massif and its geometrical parameters (relief, thickness of individual layers, and their location relative to the horizon), as well as the physical–mechanical properties of the rocks composing the massif, have a significant influence on the formation of this state.

Geodynamic stresses are caused by tectonic as well as cosmic forces. Most researchers are inclined to believe that the excess of horizontal stresses over vertical in some parts of the globe is associated with tectonic forces. All types of tectonic movements apparently lead to the formation of an excess (relative to the geostatic) stress field, where horizontal compressive, tensile, and shear stresses can be observed.

The influence of cosmic gravitational forces on the stress state of the earth's crust is currently explained on the basis of data on the dependence of earthquakes on Moon phases. Analytical studies [114] showed that the gravitational forces caused by the Moon although extremely slowly but continuously affect the Earth's shell and displace it in both hemispheres to the equatorial region. The physical and geographical contours of the continents confirm these conclusions.

When building on a rock foundation and especially when building underground structures in a rock, it is necessary to take into account the NSS of the massif, as rock behavior at depth [260], as well as its strength and deformability, is largely determined by the magnitude and direction of the existing stress tensor [254]. The natural compression of rocks increases with depth and gives the masses such mechanical properties that are crucial for their practical use.

The first hypothesis about the distribution of stresses along the depth of the rock mass was put forward by the Swiss geologist *Albert Geim* in 1878 who drew attention to the high stresses in the rock around large transalpine tunnels, not only vertical but also horizontal. He suggested that stresses in the depths of the massif are distributed according to hydrostatic law, i.e.,

$$\sigma_h = \sigma_v = \gamma ch \tag{3.48}$$

where
  σ_h and σ_v – horizontal and vertical stresses in the mass;
  γ_c – volumetric weight of the rock;
  h – the depth of the considered point of the rock mass from the day surface.

However, engineering practice has not confirmed this hypothesis. It is more logical to assume the presence of proportionality between vertical and horizontal stresses

$$\sigma_h = k\sigma_B \tag{3.49}$$

where k – the lateral pressure coefficient, the value of which can vary over a very wide range depending on local conditions and can be either less than unity or significantly exceed it (Figure 3.32).

Horizontal stresses decrease when approaching the slope. Due to the transition from a triaxial (volumetric) stress state in the depth of the massif to two-dimensional stress state near the daytime surface of the slope, crack formation occurs in the massif.

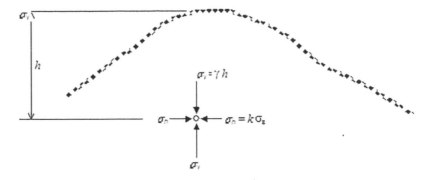

Figure 3.32 **The hypothesis of the distribution of stresses in the mountain massif.**

This phenomenon referred to as "unloading" of the massif leads to the formation of unloading cracks or cracks of on-board rebuff, as a result of which, at a distance of several tens of meters from the slope surface, the horizontal stresses perpendicular to the slope strike become equal to zero.

Horizontal stresses decrease when approaching the slope. Due to the transition from a triaxial (volumetric) stress state in the depth of the massif to two-dimensional stress state near the daytime surface of the slope, crack formation occurs in the massif. This phenomenon, referred to as "unloading" of the massif, leads to the formation of unloading cracks of on-board rebuff, as a result of which, at a distance of several tens of meters from the slope surface, the horizontal stresses perpendicular to the slope strike become equal to zero.

Interesting studies to identify the effect of loading history on stress state formation and, in particular, on the magnitude of the lateral rebound coefficient for sands and clays were conducted at the University of Illinois. In a special laboratory setup, the soil was first slowly loaded to very large vertical pressures – about 175 kg/cm^2 and then slowly unloaded. In the course of the entire experiment, measurements were made of the coefficient k. Under loading, the coefficient k was almost constant: 0.35–0.45 for sand and 0.4–0.7 for clays, depending on their physical characteristics. During unloading, it was found that the resulting horizontal stress in the soil exceeds vertical and the coefficient k quickly exceeds unity, reaching values of 2–2.5 or more when the vertical stresses tend to zero. Such a process of creating horizontal stresses can occur in sedimentary rocks.

It must be borne in mind that the formation of a stress state in a rock mass is associated not only with its own weight and loading history, but also with the history of tectonic movements.

Niels Hust [258] measured the vertical and horizontal stresses in a number of mines in Sweden and established the following:

1. directions of the main stresses in the vault and the floor of the mine coincide;
2. directions of the main stresses coincide with the directions of the maximum fracture of the rock mass and with the directions of the mountain ranges on the surface;
3. horizontal stresses exceed vertical stresses by 1.5–8 times ($k = 1.5$–8).

Hoek and *Brown* [209] analyzed the available data on measurements of natural stresses in rock masses in Australia, Canada, USA, Scandinavia, South Africa, Great Britain, France, India, and Malaysia and concluded that vertical stresses are approximately equal to stresses from the weight of overlying rocks (Figure 3.33), and horizontal stresses vary over a very wide range (Figure 3.34).

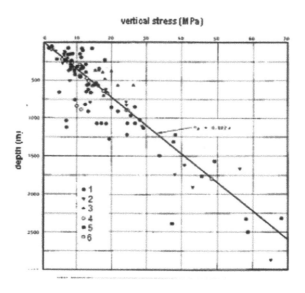

Figure 3.33 Dependence of vertical stresses on depth: 1 – Australia, 2 – USA, 3 – Canada, 4 – Scandinavia, 5 – South Africa, and 6 – other regions.

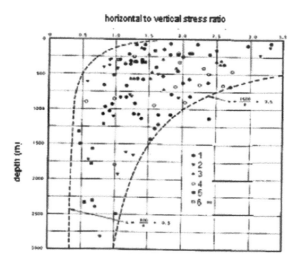

Figure 3.34 Change in the ratio between horizontal and vertical stresses with depth: 1 – Australia, 2 – USA, 3 – Canada, 4 – Scandinavia, 5 – South Africa, and 6 – other regions.

The abscissa axis in Figure 3.33 shows the ratio of the average horizontal stress to the average vertical stress at a given depth. As can be seen from the data presented, this ratio is in the following range:

$$\frac{100}{z} + 0.3 \leq \frac{\sigma_\delta}{\sigma_\delta} \leq \frac{1,500}{z} + 0.5 \tag{3.50}$$

These data also indicate that at depths of up to 500 m, horizontal stresses significantly exceed vertical ones. At depths of more than 1 km, horizontal stresses are close in magnitude to vertical stresses, which corresponds to the hypothesis of *A. Geim.*

Studies show that in rock masses, there can be no big difference in the acting stresses since at very high horizontal stresses at great depths, cracking and plastic deformation processes begin, which leads to a decrease in the difference between horizontal and vertical stresses.

A characteristic feature of the stress state of the near-surface parts of the earth's crust is that they are formed in fractured, anisotropic, and inhomogeneous environments, which are rock massifs. Consequently, the *stress field in such massifs has a discrete, inhomogeneous, and anisotropic character.* This means that the main stresses at two relatively close points of the rock mass can differ in magnitude and direction from each other. Hence, the importance of the experimental study of the stress state of rocks in general and the study of stresses at different scale levels in particular is clear. At the same time, this indicates that the study of the stress state of rock masses both theoretically and experimentally is a very difficult task.

The stress state of the rock mass formed in natural conditions is called primary (the stress state of the undisturbed mass). In general, the intensity of primary stresses increases with depth however, and at relatively small depths where modern construction and mining operations are usually carried out, stresses in rocks are significant and cannot be neglected [150].

The stress state arising as a result of surface or underground rock excavations and changing the primary (natural) stress field is called secondary (stress state of the disturbed massif). The primary stress state in sections of narrow river canyons, near large cracks and other disjunctive disturbances, has its own specificity and is close in nature to the secondary stress field.

There is also a stress state of the third kind, which appears as a result of a change in the primary and secondary state due to the influence of additional loads from the weight of high dams, water pressure in hydraulic tunnels, etc. (stress state of the load) [150].

The listed types of stress state can be explained with the example of the construction of a hydraulic tunnel. Before the tunnel is drilled along its future route, there is some primary stress field in the rock mass. During tunneling, this field changes and a zone of secondary stresses appears near the tunnel cavity. The power of this zone depends on the parameters of the primary field, the diameter and shape of the tunnel, the physico-mechanical characteristics of the rock, and also the method of penetration [150]. During the operation of the tunnel, the hydrostatic pressure of the water in the tunnel affects the rock resulting in a new tertiary stress state (stresses of the third kind).

The characteristic structural features of rock massifs – their discreteness, anisotropy, and heterogeneity – are reflected in both primary and secondary and tertiary stress fields. As a result, in real conditions, they are dealing with a complex picture of stresses from which ordinary field measurements performed in mine workings, due to their discreteness, capture only individual fragments of the stress field. In order to restore the main features of the acting stress field from these data, it is necessary to know the laws of the formation of primary stresses in a homogeneous medium, the relationship between the stress state of geostructural blocks of different scales, as well as the main laws of the appearance and distribution of secondary stresses. For non-homogeneous media, it is possible to reliably solve these problems only by combining theoretical calculations with special experimental (field and model) studies.

The stress state of rock masses to a greater extent than other characteristics of their state depends on the scale of research – the rock volume W, which determines the results of single measurements. Geophysical methods allow you to conduct research on any scale chosen by the researcher. This scale can vary from units of cubic centimeters to hundreds of cubic meters or more. This is a great advantage of geophysical methods and opens up great opportunities for studying different-scale stress fields, which is not available for other methods [150].

When comparing the results of geophysical and static methods for determining stresses, it is very important to take into account the scale of each method. For example, the results of the most common static method of unloading well ends correspond to measurement bases of no more than a few cm. Therefore, they can be compared only with data from local ultrasonic measurements in the same wells but not with data from acoustic and moreover seismic methods.

The question of the ratio of different-scale stress fields and the linking of the complex of data of static and geophysical methods despite its exceptional importance is still not sufficiently developed [251].

It can also be noted that the engineering activity of people affects the SSS of the earth's crust. It should be said that in the XX century, man spontaneously or consciously began to significantly change the state of the Earth's shells: atmosphere, hydrosphere, lithosphere, and biosphere. Engineering activities related to the pumping out of groundwater, oil, and gas, mining, and the construction of giant dams and reservoirs ultimately affect the SSS of rock masses in the uppermost layers of the earth's crust.

3.3 Water reservoir influence on rock mass and dam

The main factor of the force acting on the dam body and its rock foundation (not counting the dynamic effects and air temperature) is the external and internal water pressure. The SSS of rock masses in the sides of reservoirs and abutments of dams is significantly determined by the presence of filtering water in the cracks.

An important factor determining the effect of water on the rock foundation and on the dam is the seasonal fluctuations in the water level of the reservoir. So, for example, on the Sayano-Shushensk dam, seasonal filling of the reservoir from 500 to 540 m was carried out in 4.5 months (usually from May to September) and emptying in 7.5 months. In the reservoir of the Inguri arch dam, 271.5 m high, the water level fluctuations amount to 90–100 m.

The reservoir is the most active and dynamic component of the "dam-foundation-reservoir" system and its role is to exert a force on the deformation of both the dam and the foundation, both in the form of external hydrostatic or hydrodynamic pressure and as an internal weighing effect.

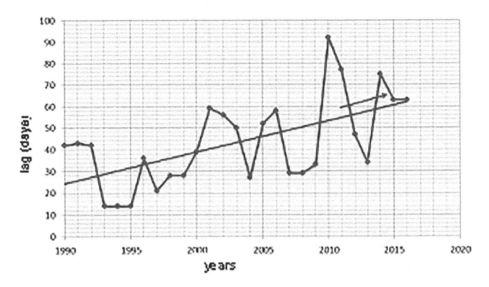

Figure 3.35 The lag of the horizontal displacements of the dam crest from the date LUS reaches its maximum value.

Both of these effects are associated with the time necessary for the deformation of the structure and its foundation, and therefore the observed deformations lag behind the changes in the water level in the reservoir. Figure 3.35 shows a diagram of the delay in the offset of the crest of the Sayano-Shushensk dam from the date the reservoir level changes.

3.3.1 Abutment deformations

When the reservoir level is raised by a high dam, large accumulated masses of water before being realized in the form of volumetric hydrodynamic forces in the rock mass act on the reservoir bed mainly as a surface load. Such a load leads as is known to deformations of the sides of the canyon, which significantly affect the SSS of dams [148].

The effect of the weight of the reservoir water on the properties and condition of the enclosing massif is most clearly manifested with a waterproof reservoir bed. In this case, the influence and the range of the additional load on the surface part of a given section of the earth's crust are fairly well traced by the results of numerical calculations of deformations and displacements of various sections of the foundation.

Figure 3.36 shows the results of measurements of the horizontal displacements of the upper part of the earth's crust in the zone of influence of the reservoir of the Inguri arch dam [145,172].

The results of these measurements indicate that when the reservoir is filled against the background of a general deflection of the central part of its bottom, there is a distinct tendency to increase peripheral areas. At the same time, the convergence of the upper parts of the gorge slope and some distension of the lower ones were noted. In the complex geological structure of the Inguri valley, the field of ongoing deformations

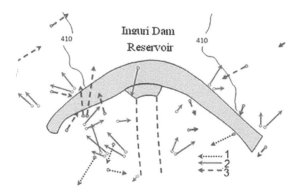

Figure 3.36 Horizontal displacements of various blocks of the rock foundation of the Inguri arch dam (Georgia) in the process of filling the reservoir: 1, 2, and 3 – inclinometers.

has a complex mosaic character, due to the deformation properties of the geological media. With depth, the region of development of the observed strains and their intensity gradually decrease.

Judging by the measurement data, the depth of the influence of the load on the weight of the accumulated water in the head part of the Inguri reservoir does not exceed 1,500–1,800 m. The noted features of the deformation of the bed of the Inguri reservoir are well confirmed by geodetic [145] and regime geophysical observations [250]. This is most pronounced during the construction of high dams, creating deep reservoirs with a large amount of accumulated water.

A large complex of studies carried out by the Laboratory of hydraulic structures of the Sayano-Shushensk arch-gravity dam clearly indicates the significant effect of fluctuations in the water level in the reservoir (annually in the range of 39–40 m) on the deformation of rock abutments of the dam. Figure 3.37 shows diagrams of the transverse (tangential) displacements of the outside sections of the dam (No. 10 and No. 55) equipped with reverse plumb lines. As can be seen in the diagrams of long-term observations, these deformations of the coast of the reservoir are directly related to seasonal fluctuations in water level.

3.3.2 Weighing and lifting the rock foundations

As water penetrates into the cracks of the rock mass, its gradual weighing occurs. For example, observations of the settlements of the Inguri arch dam and its foundation both by surface topographic reference points and by reference points installed in the drainage gallery in the rock foundation of the dam showed that precipitation from the increasing weight of the dam was recorded during the construction process. After the filling of the reservoir began, these precipitations continued under the influence of the weight of water [85,172]. However, as the water penetrated the cracks of the rock mass and created ever-increasing weighing, a rise in the foundation was noted at almost all measurement points, including the benchmark located in the downstream of the dam at a distance of 400 m.

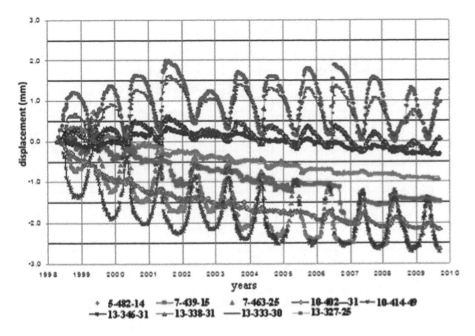

Figure 3.37 Contact displacements at the foundation of sections 5, 7, 10, and 13 of Sayano-Shushensk dam. The legend indicates section number, marks, and distance from the upstream face.

At the foundation of the Inguri dam, the first time the reservoir was filled, the time of water saturation and weighing of the massif was approximately 12 months. It is interesting to note that a rise in geodetic marks at each rise in the water level in the reservoir was also observed in the downstream of the dam at a distance of more than 1.5 km (Figure 3.38).

In 2002, specialists of CSGOEI carried out geophysical studies of the foundation of the Sayano-Shushensk HEP and stress studies by the method of hydraulic fracturing. According to more than 100 experimental results, it was found that in the studied part of the massif with increasing water level in the reservoir, the maximum horizontal stresses increased, and the vertical ones decreased, which indicated an increase in weighing forces (Figure 3.39). The higher the dam, the greater the pressure penetrating the water cracks and the higher the degree of decompression of the massif.

3.3.3 Uplift in fissures

Fluctuations in the reservoir level during its operation are accompanied by a perforation of the SSS and the filtration regime of the massif. In the surrounding geological media, various natural and technogenic geodynamic processes are activated, the development of which can lead to dangerous phenomena for the construction, such as damage to local parts of the foundation due to the growth of large cracks, the appearance of zones of anomalously high filtration, the collapse of individual blocks of the massif, etc. [151].

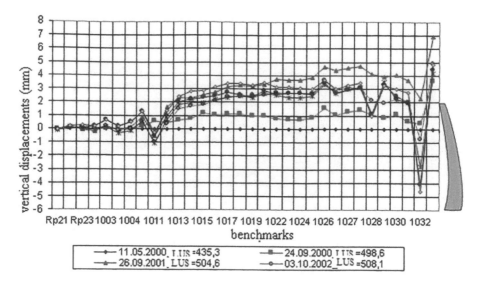

Figure 3.38 Vertical displacements of the right bank of the Inguri river in the downstream of the dam at high level water in the reservoir.

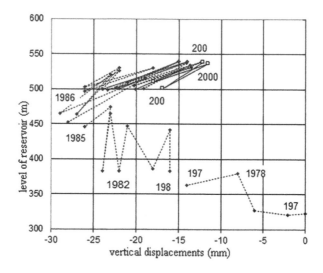

Figure 3.39 Vertical movements of the base of the central section of the Sayano-
Shushensk dam for the period from 1977 to 2001.

During rapid and deep drawdown of water reservoir in the massif zones of excessive
fracture, pressure may arise in which the processes of fracture of the massif develop.
An illustration of this can be seen in Figure 3.40, which shows a graph of the changes
in the elastic wave velocities at the foundation of the Inguri dam in the process of lowering the reservoir level.

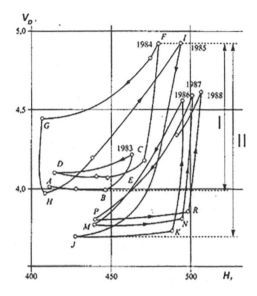

Figure 3.40 The graph of the changes in the elastic wave velocities at the base of the Inguri dam when the reservoir level changes: I – changes during normal drawdown and II – changes with rapid and deep drawdown of water reservoir.

The graph shows a sharp (30%–40%) drop in speeds in 1996 after a quick (more than 2 m/day) lowering of the reservoir, and the nature of the manifestation of this effect significantly depended on the speed and depth of lowering [250].

Fluctuations in the reservoir level during its filling and lowering also play a negative role in the formation of decompression zones characterized by reduced strength.

Filtration through rocky foundation and abutment dams is associated with various adverse effects including a decrease in the bearing capacity of rock foundations due to the creation of water pressure forces over the surfaces of cracks.

In this case, depending on the orientation of the cracks, slide forces are created on the rock blocks, or uplift is created along the potential slide surfaces, which reduces the slide resistance forces due to a decrease in the normal components of the acting forces. The compressive stresses in the area of the filtered water decrease by the value of the filtration pressure p, and the Mohr circle shifts to the left approaching the envelope of the limit circles of Mohr 1 for a dry mass (Figure 3.41).

In addition, filtering water creates volume forces S at the foundation of the dam and weighing forces W in the slopes of the reservoir (more precisely in that part of the slopes where the groundwater level changed after the creation of the reservoir). These forces cause stress in the dam and deformation of the gorge. Until recently, these force effects on dams were not taken into account, but at present, attempts to take them into account by calculation are known.

Serious problems with a high drawdown rate occurred in the reservoir of the La Amistad dam in Mexico in 1996 [51]. The dam, 90 m high, was built in 1969 and has been functioning perfectly for 27 years. The severe drought that befell the northern

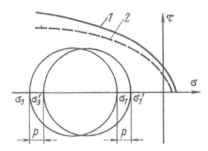

Figure 3.41 The effect of free water in cracks on the strength of a rock mass:
1 – envelope of the Mohr limit circles for a dry mass and 2 – the same
for a flooded mass.

part of Mexico for 4 years led to a significant drawdown of almost all reservoirs in the
north of the country, including the reservoir of the La Amistad dam, and in 1996, this
drawdown turned out to be especially deep and quick. In the reservoir bed composed
of horizontally layered limestones, karst voids filled with water were found.

When the reservoir was quickly lowered, the water pressure in the voids did not
have time to decrease, which led to the appearance of huge fistulas in shallow water:
water was squeezed out of the limestone layers throwing out "geysers", and then water
rushed into these voids by a waterfall. Gradually, the pressure was equalized and gap-
ing holes with a diameter of several meters remained at the bottom of the reservoir at
its edge (Figure 3.42). The sealing of these newly formed concentrated filtration paths
was carried out by stone filling with monolithic concrete.

When assigning modes of lowering and filling of a high dam reservoir, certain restrictions must be assigned.

Figure 3.42 Dam La Amistad (Mexico). Fistulas in the bottom of the reservoir resulting from a rapid lowering of water level in the reservoir.

3.3.4 Reducing of rock strength by humidification

In addition to uplift, water penetrating into cracks and pores of massif of bank abutments can cause a decrease in shear strength in formations containing clay components.

A classic and at the same time tragic example of the collapse of the reservoir's side is the Vaiont arch dam: a giant landslide with a volume of about 300 million m^3 almost completely filled the reservoir (Figure 3.43), displacing the water from the reservoir (for more details, see Section 2.1).

A similar phenomenon was observed during the first filling of the reservoir of the 200-m high arch dam of Simapan in Mexico in November–December 1993, when an unstable massif was discovered on board the reservoir directly opposite to the dam. The volume of the creeping rock mass was estimated at 14.6 million m^3. The collapse of such a huge rock mass in the reservoir posed a serious danger to the dam under construction. The calculations showed that with a further increase in the level of the reservoir, the movement of the massif should stop. So, it fortunately happened [51].

The SSS of rock masses on the sides of reservoir and abutment of high dams is significantly determined by the influence of filtering water penetrating into the cracks. The creation of a deep reservoir not only causes the loading of a local area of the earth's crust but also simultaneously changes the hydrogeological conditions in the zone of its influence. These factors have a negative effect on the properties and condition of the enclosing massif; however, they are often underestimated when designing and analyzing the operation of high dams.

Of particular importance is the influence of water on the strength and stability of rock masses during the construction of high dams. When filling a high dam reservoir, the combined effect of the following factors can have a very adverse effect on the bearing capacity of the abutment and foundation of high dams:

Figure 3.43 **The collapse of Mount Mont Tok (left bank) in the reservoir of the Vaiont arch dam.**

- loss of rock abutment weight due to the weighing effect;
- reduction of the slide strength of the massif due to crack opening and decompression of the rock mass;
- with a simultaneous loss of strength due to wetting of the existing clay interlayers - it should be borne in mind that their action occurs at the time of the growth of hydrostatic pressure on the dam and the growth of slide forces in the rock mass. If a possible seismic effect is added to these factors, the consequences can be catastrophic.

An exhortative example is the catastrophe of the arch dam Malpasset, which occurred in France on December 2, 1959.

Malpasset arch dam of double curvature was erected in 1950–1952 in the south of France on the small river Reiran. The height of the dam was 65 m, the thickness of the dam on the crest was 1.5 m, and in the lower part, it was 6.82 m (Figure 3.44). The dam was intended for irrigation and water supply.

On December 2, 1959, at 9 p.m. local time during the first filling of the reservoir, the dam suddenly collapsed. The official version of the French design bureau Coyne et Bellier on the causes of the disaster is set out below.

A peculiarity of the geological structure of the foundation composed of gneisses predetermined the formation of an unstable thrust block, the so-called "dihedron" in the left-bank abutment (almost along the entire length, from the bed to the pier).

The downstream face of the dihedron was formed by a fault filled with clay material; the top face was formed in the direction of gneiss stratification as a result of tensile deformations which usually occur under the upstream face of the dam.

The stability conditions of the left-bank abutment were due to the fact that in addition to the slide forces from the dam, it was exposed to the pressure of the filtration flow of extremely large magnitude from the upstream side. Two circumstances contributed to this. Firstly, on the left bank, the force from dam coincided with the direction of the layering and concentrated in the gneiss layer of almost constant thickness and was not distributed to the sides (Figure 3.45). Secondly, the filtration coefficient in the gneiss formation under load decreased by 100 times compared with the surrounding rock.

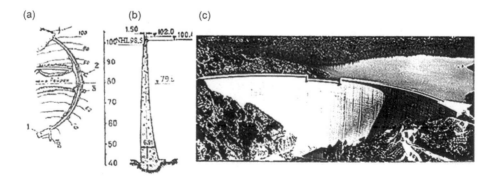

Figure 3.44 Malpasset arch dam (France): (a) plan; (b) a section along the key console; 1 – left bank and 2 – spillway; and (c) view from the downstream before the disaster.

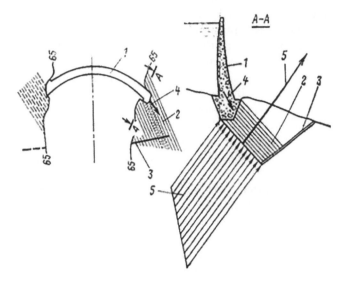

Figure 3.45 Malpasset arch dam (France): sectional cut at mark 65 m and cut along AA; 1 – dam, 2 – gneiss layers, 3 – fault, 4 – force from the dam, and 5 – hydrostatic pressure.

Thus, in the left-bank abutment densely formed "underground dam" on the continuation of the main dam.

On the upper face of the dihedron, hydrostatic pressure developed over time as a result of which the stability of the dihedron was impaired, and the dam began to move with the dihedron. This displacement discovered in July 1956 occurred with increasing speed as the water elevation increased. For the year from July 1958 to July 1959, the displacement was equal to the displacement for the previous 5 years.

During the last four days before the disaster, the headwater level was raised by 5 m and for the first time reached 100.12 m. As a result of the increased load on the dam and hydrostatic pressure on the "underground dam", the dam displacement sharply increased. Ultimately, slide strength was surpassed along the bottom face of the dihedron after which it shifted and then the destruction of the dam, which was explosive in nature and occurred within a few seconds.

On November 15, 1959, increased water filtration was discovered across the right bank about 20 m from the dam. On November 19, heavy rains began in the Malpasset region. Over an almost two-week period, 500 mm of rain fell (while in the last 24 hours before the accident, 130 mm fell). On December 2, due to the fact that the rain continued and the water level in the upstream did not reach the dam crest by only 28 cm, the dam staff requested permission from the Var Department administration to open the flood spillway gates.

Fearing flooding of the construction site of the A8 highway, which is being built 200 m downstream (the bulls of the bridge whose concrete had been poured recently) could have been damaged, the district administration banned the spillway. At 18 hours 00 minutes, permission was received to partially open the spillway gates – it was opened with a flow rate of 40 m^3/s, which is extremely small for the operational drawdown of the reservoir. At 21 hours 13 minutes, the collapse of the pressure front of the dam occurred. A water breakthrough created a wave 40 m high moving at a speed of 70 km/h. Within a few minutes, two small villages were completely destroyed – Malpasset and Boson as well as the construction site of the A8 highway. According to surviving witnesses, at the time of the accident, a heavy crack was heard from the dam side, and then doors and windows were knocked out in all houses by an air shock wave (indicating the immediate destruction of the dam; a huge wall of water worked like a piston, moving in a narrow canyon, squeezing the air in front yourself). Then a wave with a height of "only" 3 m reached Frejus (almost 10 km from the dam), flooding its entire western half, and then went into the sea. The official results of the disaster: as of January 15, 1960, 423 people are reported dead and/or missing, including 135 children.

As a result of the accident, the Malpasset dam was completely destroyed. There remained a small edge of the dam on the right bank and pier on the left bank, shifted 2 m horizontally from the initial position (Figure 3.46).

The destruction process can be divided into stages (Figure 3.47).

The first stage began with the ejection of the dihedron, which was accompanied by the lifting of the dam, with the exception of the upper part of the right-bank abutment [79,101] and then by turning the dam relative to the inclined axis near the right bank. The maximum displacement during rotation was 1.2 m.

At the second stage, the turn of the dam stopped because the left-bank pier opposed this. However, the pier could not absorb the force from the dam and slide about 2 m along the tangent to the arch, after which the arch dam having no support along the entire left-bank abutment collapsed.

The above analysis of the causes and mechanism of the catastrophe largely predetermined the basic principles of calculating the stability of arch dams (see Section 6.5.7). When assessing the reliability and safety of high dams, it is necessary to consider the integrated system "dam-foundation-reservoir" in which all elements are closely interconnected. The Banzao dam in the Chinese province of Henan was designed for a flood with a probability of 1 time per 1,000 years. However, in August 1975, a flood occurred with a probability of 1 time in 2,000 years [24]. At first, the dam in the upstream was destroyed; the breakthrough wave reached the Banzao Dam and destroyed

Figure 3.46 Remains from the destroyed arch dam Malpasset (France): A – offset pier.

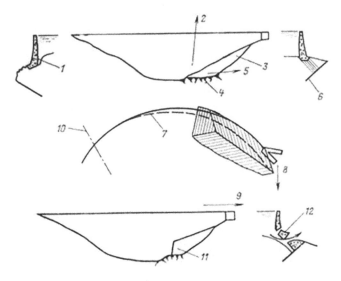

Figure 3.47 Malpasset arch dam (France). Destruction sequence diagram; I − slit, 2 − rise of the dam, 3 − dihedron, 4 − destruction of dihedral bottom, 5 − displacement, 6 − fault, 7 − turn of the dam, 8 − displacement of the dam, 9 − secondary displacement of pier, 10 − axis of rotation, II − breach, and I2 − dam destruction.

Figure 3.48 **The destruction of the Banzao Dam in China.**

it (Figure 3.48) and then another 62 dams. The official death toll was 26 thousand people, but as a result of subsequent epidemics and starvation, the total number of victims was 171 thousand.

3.4 Stability of rock mass

3.4.1 Types of unstable mass and schemes of stability analysis

The main factor determining the strength and stability of rock massifs is their natural disturbance and fissuring, which create blocks of various shapes and sizes in the massif, which determines the internal structure of the massifs.

In nature, there are so many forms and structures of unstable massifs that it is almost impossible to create design schemes for all cases. Nevertheless, it can distinguish the most common forms in engineering practice for which the corresponding mathematical models and calculation methods have been created:

1. layered rock massifs with falling layers in the direction of the slope;
2. rock masses on the polygonal displacement surface;
3. soil and rock slopes and slopes with a curved displacement surface;
4. vertical rock blocks standing on a subhorizontal displacement surface;
5. voluminous rock blocks cut off from the massif by two or three planes of displacement and separation.

In addition, there may be separate unstable blocks separated from the massif as well as overhanging blocks on steep slopes and in underground workings, which need to be fixed with anchors or retaining structures to prevent their collapse.

3.4.2 Statement of the stability analysis problem

Until recently, it was believed that the stability of rock masses and slopes can be estimated using the classical laws of continuum mechanics (for example FEM calculations). Attempts have also been made to use circular cylindrical displacement surfaces for this purpose, which are successfully used to assess the stability of soil slopes. But if it is possible to use the laws of continuum mechanics to soils whose particle sizes are substantially smaller than those considered in engineering calculations, then for rock masses whose block sizes are comparable to the dimensions of the massif, these laws should be applied with great care.

To assess the stability of rock masses, the *method of limiting equilibrium* is used. Other methods including the FEM can be used to analyze and identify the weakest zones in the rock mass, which can serve as potential displacement surfaces of the mass. But stability assessment is recommended to be performed by the method of limit equilibrium.

The use of calculation schemes in which a shifting array is considered as a single rigid body is very conditional and may lead to erroneous recommendations. This is especially true for assessing the stability of layered rock masses, in which the mutual arrangement and interaction of the blocks determine the kinematic scheme and the nature of the possible displacement.

In each specific case, the design scheme should proceed from the possible mechanism of displacement of the massif. This condition determines the requirements for engineering–geological surveys in order to determine the structure of the massif, the size and shape of its shifting blocks as well as the characteristics of slide strength on existing crack surfaces.

The main factors determining the stability of the rock mass are as follows:

1. internal structure of the massif determined by the nature of its fissuring in conjunction with the size and shape of the shifting blocks;
2. slide strength along cracks or weakened zones;

3. forces acting on the massif: dead weight of the massif, applied external forces and loads (structures, anchors), the presence of filtration flow, and seismic acceleration during an earthquake.

In calculating the stability of rock masses, the following concepts are used:

- stability coefficient and
- deficit of stability.

The stability coefficient is the ratio of resistance forces to acting forces:

$$k = \frac{R}{T} \geq 1 \text{ or } k \geq k_o \tag{3.51}$$

where
  R – the forces of resistance (passive forces);
  T – acting slide forces (active forces).

The stability deficit is the difference between the slide and holding forces:

$$S = T - R < 0 \text{ or } S = T - R < S_o \tag{3.52}$$

If in order to ensure stability the safety factor must be greater than its standard value k_o, then the stability deficit must be less than its standard value S_o.

The main advantage of the safety factor k is its dimensionness, which allows comparing the degree of stability or reliability of different massifs. But as will be shown below, in some cases, these values are conditional and it is not always possible to say that a massif with a stability coefficient of 1.5 is more stable than another massif with a stability coefficient of 1.3.

In calculating the safety factor, the external holding forces (anchor forces, key strength, etc.) must be summed with the slide strength forces in the numerator.

The stability deficit gives directly the magnitude of the force that must be applied to the massif to ensure its stability.

Based on the nature of the slope topography, the fissuring of the massif, and the possible displacement mechanism, a possible displacement surface is determined. With extended slopes when the direction of displacement coincides with the azimuth of the slope fall, it is possible to consider two-dimensional problems. In all critical cases, it is necessary to consider three-dimensional stability problems with volumetric calculation of massifs and displacement areas.

When analyzing stability, it must be in mind that the limiting state does not occur simultaneously on the entire displacement surface, and therefore, the slide resistance forces are not constant on the entire displacement surface. In addition, sliding and holding forces are not equal at all points on the displacement surface.

The stability margin determined under such conditions may turn out to be erroneous since local displacements can develop while the entire massif can be in a state of stable equilibrium.

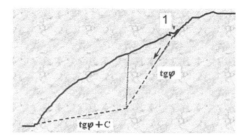

Figure 3.49 **Breakaway crack on the surface of the slope – the first sign of the process of buckling.**

For a massif located on a polygonal surface, displacements on the steeply dipping part can begin much earlier than the limiting state on the dipping part of the surface is reached. Often, you can observe on the surface of the slope the separation crack, which is the first evidence of the beginning of the process of buckling (Figure 3.49).

Taking into account that the displacement of the massif is not considered in the method of limit equilibrium, it is necessary to choose the correct design scheme that will provide the physical possibility of displacement of the massif along the selected surface.

Stability calculations should be considered as a tool for a qualitative analysis of the degree of influence of various factors on the stability of the massif. If the geology of the massif is quite simple, and the outlines of the displacement plane and the characteristics of its slide strength are known, then the obtained value of the stability coefficient can be considered quite reasonable. If the geological conditions are complex, the position and shape of the displacement surface, the characteristics of the slide strength along it, and also the position of the filtration flow are unknown, then solving the problem becomes problematic. The most reliable method for assessing the stability of a rock mass is the method of probabilistic analysis, which allows one to take into account the variances (mean square deviations) of all factors involved in the calculation and to obtain the value of the probability of rock collapse.

But carrying out a probabilistic analysis is not always possible both due to the lack of distribution curves of natural parameters and due to the mathematical complexity of such an analysis. In this case, it is possible to analyze the influence of the parameters involved in the calculation on the value of the stability coefficient and select the most influential parameters, which should be studied with the greatest care during research.

For example, if the most influential parameters when they change in real practical limits can reduce the value of the safety factor by 20%, the adoption of a safety factor of 1.3 or 1.4 may be sufficient. If variations of these parameters can change the value of the safety factor by 50%, then its value $k = 2$ may be insufficient. Hence, it is obvious that it is impossible to a priori designate and even more normalize the "necessary" value of the safety factor.

Considering the safety factor of rock mass k as a product of the set of basic design parameters p_i, such as

- tilt angles of potential displacement surfaces;
- parameters of slide strength on these surfaces;
- forces of filtration pressure on the shifting array;
- parameters of seismic effects, etc.,

the influence of the possible deviation of each of these parameters on the value of the safety factor $\Delta k/\Delta p_i$ is determined.

$$k = f\left(p_i\right), \left(i = 1, 2, \ldots\right) \tag{3.53}$$

This makes it possible to single out both the most influential parameters and secondary parameters, the determination and refinement of which do not require a lot of time and money.

A similar analysis was performed when calculating the stability of the right-bank abutment of the Naghlo dam in Afghanistan (Figure 3.50).

It was necessary to choose the most reliable mounting method as the HEP provided electricity to the capital of the country.

The entire right bank is composed mainly of small and large-point gneisses with layers of marbled limestone falling toward the bed at an angle of 65°–80°. Judging by the deformations recorded in fissure 1, a large volume of the rock mass could be involved in the movement limited by a steeply falling fissure 1 and a hollow falling zone of highly crushed rock passing through the places of formation bending.

An analysis of the influence on the slope stability of the following factors was carried out:

- rise and lowering of water level in the reservoir;
- possibility of reducing as a result of soaking the parameters of slide strength along a shallow-falling fissured zone ($\mu_{II} = \tan\varphi_{II}$ and c_{II}) as well as the parameter $\mu_I = \tan\varphi_I$ along a steeply falling crack 1;
- effect on the stability of the slope of the possible cutting of the rock in the upper part of the slope.

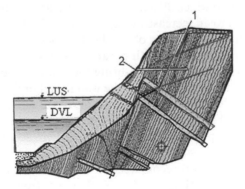

Figure 3.50 The structure of the right bank of the Naghlo dam (Afghanistan).

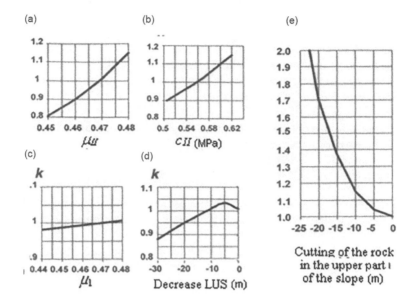

Figure 3.51 Influence of parameters on the slope stability factor: (a) μ_{II} along a shallow-falling displacement surface; (b) c_{II} on a shallow-falling displacement surface; (c) μ_I along the crack (I d) of the water level in the reservoir; (e) the depth of cut of the soil on the crest of the slope.

Diagrams illustrating the effect of all of the above parameters on the safety margin are presented in Figure 3.51.

Of all the parameters, the strength parameters along the shallow-falling zone have the strongest influence, while the decrease in the coefficient of friction along a steeply falling crack 1 has practically no effect on the value of the safety factor. It is interesting to note that a rise in the water level in the reservoir by 110 m (up to the NHL mark) caused an increase in the slope stability by almost 15%, while a decrease in the parameter μ_{II} in the shallow displacement zone by 2.3% leads to a decrease in the slope stability by the same 15%.

Therefore, it can be assumed that before the reservoir was filled, the slope was in a state close to the limit, in which all natural slopes are usually located. A rise in the reservoir level could lead to an increase in slope stability by 15%, if this rise were not accompanied by simultaneous wetting of the displacement surface, which in turn could lead to a decrease in its slider strength parameters. As was shown, a decrease in the parameter μ_{II} by only 2.3% could again bring the slope to the state of initial limiting equilibrium. The subsequent decrease in the water level in the reservoir in connection with the operation of the HEP could cause the massif to move, which was actually observed. Drawdown of the reservoir by 7–8 m led to a sharp increase in displacements, which attenuated at a depth of lowering of 5 m at which, according to the calculation, the slope has the maximum stability (Figure 3.51d).

To stabilize the slope, it was recommended to cut the soil on the crest of the slope. As the calculations showed (Figure 3.51e), such a cut strongly affects the increase in stability: taking 20 m of soil leads to an increase in the safety factor by 70%.

This example shows the role of the safety factor not as an absolute criterion but as an analysis tool as an indicator of the sensitivity of the rock mass stability to changes in certain determining factors.

3.4.3 Stability analysis by a polygonal surface

The principle of calculation is a sequential analysis of the stability of rock blocks, starting from the upper located on sections of the surface displacements with the steepest incidence angles.

As already mentioned, the limiting state almost never occurs simultaneously on the entire potential displacement surface, and the fracture process begins in the upper parts of the slope, which is manifested by the appearance of cracks and local displacements at the top.

With a polygonal displacement surface consisting of three planes (Figure 3.52), the stability analysis begins with a consideration of the stability of the upper block No. 3. If it is unstable and its displacement is possible, it will transfer the stability deficit to the downstream block No. 2:

$$S_3 = G_3\left[\left(\sin\alpha_3 + n\cos\alpha_3\right) - \mu_3\left(\cos\alpha_3 - n\sin\alpha_3\right)\right] - C_3 A_3 \tag{3.54}$$

where

G_3 – dead weight of block No. 3;

α_3 – the angle of incidence of the displacement plane;

$\mu_3 = \tan\varphi_3$ – coefficient of friction along the considered displacement plane;

C_3 – cohesion on the same displacement plane;

n – horizontal coefficient of seismic acceleration (in case of an earthquake);

A_3 – surface area of the displacement.

When calculating the long-term stability, it is possible not to take into account the adhesion forces on a steeply falling crack, where the displacement process can begin long before the mass loss of general stability, and therefore, only the friction coefficient conditionally taken as residual strength should be taken into account. Thus, when considering the upper block No. 3 of the massif, the cohesion forces along the displacement surface can be neglected.

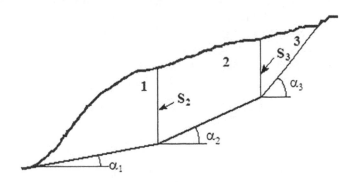

Figure 3.52 Design scheme of massif on a polygonal displacement surface.

From consideration of the stability condition of the next block No. 2, the force transmitted by it to the block No.1

$$S_2 = G_2(\sin\alpha_2 + n\cos\alpha_2) - \mu_2(\cos\alpha_2 - n\sin\alpha_2)] + \\ + S_3\left[\cos(\alpha_3 - \alpha_2) - \mu_2\sin(\alpha_3 - \alpha_2)\right] - C_2 A_2 \tag{3.55}$$

In the general case for a block with number i, when numbering blocks from bottom to top along the displacement surface, the deficit is written in the form

$$S_i = G_i\left[(\sin\alpha_i + n\cos\alpha_i) - \mu_i\cos\alpha_i - n\sin\alpha_i\right] + \\ + S_{i+1}\left[\cos(\alpha_{i+1} - \alpha_i) - \mu_{i=1}\sin(\alpha_{i+1} - \alpha_i)\right] - C_i A_i \tag{3.56}$$

Since as a rule a fractured rock mass does not have tensile strength, the blocks transmit only positive values of their stability deficit. At negative values of Si deficiency, they are taken equal to zero.

The stability margin of the entire massif excluding the section with a steep drop where, as mentioned above, the limit state is assumed can be determined by the dependence:

$$k = \frac{\mu_1\left[G_1(\cos\alpha_1 - n\sin\alpha_1) + S_2\sin(\alpha_2 - \alpha_1)\right] + C_1 A_1}{G_1(\sin\alpha_1 + n\cos\alpha_1) + S_2\cos(\alpha_2 - \alpha_1)} \tag{3.57}$$

This method of calculating the stability of rock masses based on calculating the stability deficit was adopted by JSC Hydroproject Institute as the main one [135,136] and was also recommended for use by the USSR Ministry of Transport Construction [143].

3.4.4 Accounting of filtration flow pressure

The hydrostatic pressure of the filtration flow is determined for each of the blocks under consideration taking into account the depth (ζ_1 and ζ_2) of immersion of their base below the curve of the free surface of the filtering water (Figure 3.53).

When solving the two-dimensional problem, the hydrostatic pressure forces Q1, Q2, and Q3 are determined as shown in Figure 3.53, and when considering the volume stability problem, they are calculated taking into account the corresponding areas and the shape of the block faces.

Taking these forces into account, the stability deficit for block i is written as follows:

$$S_i = G_i\left[(\sin\alpha_i + n\cos\alpha_i) - \mu_i(\cos\alpha_i - n\sin\alpha_i)\right] + \\ + (Q_{2i} - Q_{1i})(\cos\alpha_i + \mu_i\sin\alpha_i) + \mu_i Q_{3i} + \\ + S_{i+1}\left[\cos(\alpha_{i+1} - \alpha_i) - \mu_i\sin(\alpha_{i+1} - \alpha_i)\right] - c_i\omega_i \tag{3.58}$$

The safety factor can be determined by the formula

$$k = \frac{\mu_1\left[G_1(\cos\alpha_1 - n\sin\alpha_1) + S_2\sin(\alpha_2 - \alpha_1)\right] + C_1 A_1}{G_1(\sin\alpha_1 + n\cos\alpha_1) + S_2\cos(\alpha_2 - \alpha_1)} \tag{3.59}$$

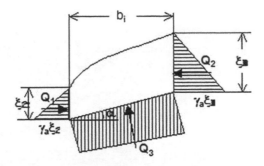

Figure 3.53 Scheme of hydrostatic pressure forces on the block face.

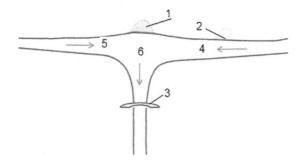

Figure 3.54 Fernando Hiriart arch dam plan (3): I and 2 – unstable massifs A and B and 4, 5, and 6 – rivers of San Juan, Tula and, Moctezuma.

3.4.5 Example of stability analysis of water reservoir bank

Fernando Hiriart's arched dam, 200 m high, was built on the river Moctezuma at the confluence of the Tula and San Juan rivers (Figure 3.54).

In the process of filling the reservoir in November 1993, unstable massifs were discovered on board of the reservoir, one of which was located in the immediate vicinity.

The established network of geodetic benchmarks made it possible to determine the magnitudes and directions of displacements. In order to continue filling the reservoir, it was necessary to evaluate the volume of the shifting massif and the possibility of its collapse, as well as the speed of collapse and the possible wave height in the reservoir. The practical lack of geological and geotechnical information about the massif made it impossible to determine either the position of the displacement surface or the parameters of slide strength along it. The data of some geological and geophysical studies, the topographic plan, and the results of measuring displacements by geodetic reference points in the landslide zone were used. The topography of the side of the reservoir with a shifting massif is presented in Figure 3.56.

To determine the position of the displacement surface and the corresponding estimate of the volume of the creeping massif, it is proceeded from the following logical premises. The supposed displacement surface should allow the massif to move

Figure 3.55 A landslide in the reservoir of the Fernando Hiriart Dam.

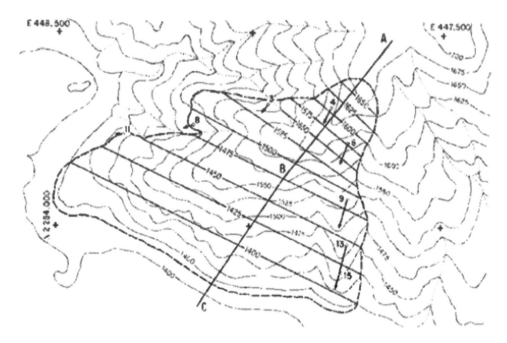

Figure 3.56 Topography of the bank with the contour of the shifting massif, displacement vectors, and horizontals of the hypothetical displacement surface.

in the direction of the displacement vectors fixed by the reference points. Therefore, the horizontals of the displacement surface should be normal to the direction of the displacement vectors of the massif and should wedge out onto the surface of the slope in places of cracks fixed on the slope, corresponding to the outcrops of the displacement surface on the day surface of the slope, as well as in the places of ravines and depressions on the topographic plan. Variants of proposed displacement surface are

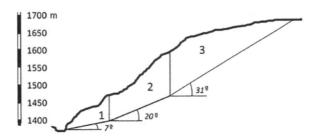

Figure 3.57 **ABC section.**

presented in the form of contour lines of this surface in Figure 3.56. The section along the ABC line is shown in Figure 3.57. To facilitate the calculation of stability, the displacement surface was replaced by a polygonal surface consisting of three planes with incidence angles of 7°, 20°, and 31°.

The volume of the creeping rock mass was $14.6 \times 106 \, m^3$, which at a mass density of $\gamma = 2.4 \, t/m^3$ gave a weight of 35×10^6 tons. The collapse of such a huge rock mass in the reservoir posed a serious danger to the dam under construction.

Given that the massif is already shifting, i.e., is in the limiting state, it could be assumed that the adhesion along the displacement surface is zero and the slide strength is determined only by the coefficient of friction, the same over the entire surface:

$$c_1 = c_2 = c_3 = 0; \mu_1 = \mu_2 = \mu_3 = \mu \tag{3.60}$$

A volumetric stability calculation was performed for various positions of the water level in the reservoir under the assumption that wetting of the displacement surface does not affect the friction coefficient. Since the filling of the reservoir was carried out slowly, only the weighing effect of water was taken into account.

This calculation made it possible to determine the value of the coefficient of friction over the displacement surface at which the slope will be in the limiting equilibrium state (at $S_1 = 0$). The corresponding diagram of the dependence of the required coefficient of friction on the position of the water level in the reservoir is presented in Figure 3.58. An examination of this diagram showed that

1. the maximum strength required to ensure the stability of the massif was obtained by raising the reservoir elevation to mark 1,475 m; it could be assumed that the displacement of the massif began when the water level in the reservoir rose to this level;
2. when the water level reached the level of 1,500–1,510 m, the resistance forces began to exceed the slide forces, and the observation data for the displacements of the benchmarks confirmed that the displacement process slowed down (Figure 3.59);
3. the stability of the massif is very sensitive to the water level in the reservoir and to the value of the coefficient of friction over the displacement surface; a change of this coefficient by a few tenths is enough for the massif to lose stability.

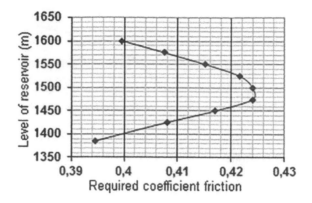

Figure 3.58 Dependence of the required coefficient of friction on the position of the water level in the reservoir.

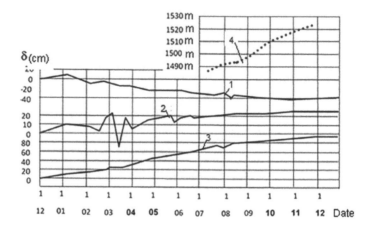

Figure 3.59 Diagrams of the displacement of benchmark No. 6 on the landslide body: 1 – vertical, 2 – to the north, 3 – to the east, and 4 – to the level of reservoir.

Based on the calculation results, it was concluded that after reaching 1,560 m in the NHL reservoir, the displacement of the massif should stop, and its collapse will not occur if during partial lowering of the reservoir, the level does not fall below 1,520 m. Subsequent events fully confirmed the validity of this conclusion that the massif shifted and stopped after filling the reservoir. This also confirmed the validity of the calculation and the hypothesis about the position and shape of the displacement surface.

3.4.6 Stability calculation on curve surfaces

If in relatively strong rock masses divided into blocks by fissure systems, collapse can occur along the planes of these fissures, then in "homogeneous" massifs composed

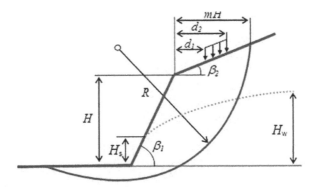

Figure 3.60 Scheme for calculating the stability of a slope on a cylindrical displacement surface [51].

of weak rock and half-rock, as well as screes and soils, the fracture surface can form along the line of least resistance.

Surveys of collapses or landslides have shown that this surface has a circular or parabolic shape. Traditionally in the analysis of stability, a circular shape was considered, which made it possible to use the equation of moments. The commonly used calculation scheme is shown in Figure 3.60.

In the above diagram:

where

H – height of the slope;
β_1 – angle of inclination of the slope;
β_2 – angle of inclination of the upper part of the slope;
m – coefficient of the distance to the exit point of the displacement surface on the crest of the slope;
H_w – groundwater level away from the slope;
H_s – level of water leach on the slope;
R – radius of the displacement surface;
d_1 and d_2 – the distances from the edge of the slope to the boundaries of the external load applied to the buildings on the crest of the slope, if any.

The entire shifting massif is divided into 10–20 vertical compartments with a width of Δx. The design scheme of the compartment is shown in Figure 3.61, where

A_i – total area of the compartment;
w_i – area of the "suspended" part of the compartment;
$G_i = \gamma_R A_i$ (weight of rock compartment per linear meter of slope width);
$F_{si} = -\gamma_w w_i$ ("weighting" force);
$F_p = \gamma w\, w_i\, sin\alpha_{wi}$ (filtration pressure force);
γ_R – volumetric weight of the rock mass (soil);
γ_w – volumetric weight of water.

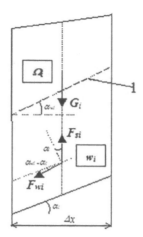

Figure 3.61 Computed design of the compartment: 1 – curve of the free flow surface.

To determine the safety factor, it is first necessary to find the position of the center of the critical circular displacement surface and its radius, which is associated with the iterative calculation process. This complicates the analysis and makes it necessary to use computer programs.

3.4.7 Stability analysis of volume rock blocks

When analyzing the stability of rock masses, the problem often arises of assessing the stability of bulk rock blocks separated from the main massif by cracks or other weakening surfaces (Figure 3.62).

Figure 3.62 Volumetric rock block lying on two fissure planes.

Figure 3.63 shows a similar collapse in April 2018 on the banks of the Vakhsh river in the area of the Rogun HP construction.

The design diagram of a volumetric block located on two adjacent crack planes is shown in Figure 3.64. The planes ABC (1) and ADC (2) were the displacement planes of the block, while the plane EFG (3) was the possible plane of separation of the block from the main massif. The block displacement occurred in the direction of the rib EA. The BCD plane is the day surface of the block, and the BCD plane conventionally represented the free surface of the filtration flow.

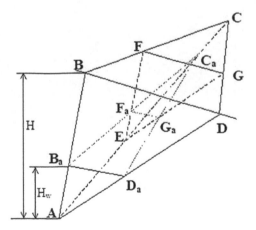

Figure 3.63 **Collapse rock mass on the bank of the Vakhsh river.**

Figure 3.64 **The design scheme of the volume unit.**

The equations for the analytical and graphical calculation of the stability of such a block are given in the monograph [51]; however, these calculations are more appropriate to perform using existing computer programs.

3.4.8 Stability of rock blocks separated by fissures of board rebuff

When constructing HEP in mountainous regions or in narrow canyons, one sometimes encounters the presence of rock blocks separated from the main massif by fissures of board rebuff and standing on subhorizontal plane of fissures (Figure 3.65).

Such a problem arose during the construction of the Patla derivation HEP in Mexico, where the machine building was located at the bottom of a deep gorge the slopes of which were replete with subvertical fissures onboard rebuff. These columnar blocks standing on subhorizontal fissures or on bedding planes under the action of atmospheric water accumulating in a vertical fissure began a slow displacement to the edge of the cliff. To prevent their collapse in the gorge, it was necessary to strengthen them on the slope.

Figure 3.66 shows a photograph of the strengthening of the rock block by monolithic concrete at a height of 80 m above the building of the HEP.

Figures 3.67 and 3.68 show the calculation scheme of the block and the diagrams of the effect on the safety factor of the stability of sediments accumulated in a vertical crack, as well as possible seismic effects, based on which the reinforcement parameters were calculated.

Figure 3.65 Column blocks separated from the massif by fissures of board rebuff (Necaxa River, Mexico).

Figure 3.66 **Strengthening a columnar rock block on a slope at an altitude of 80 m above the building of the Patla HEP.**

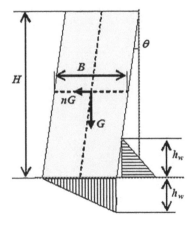

Figure 3.67 **Block design.**

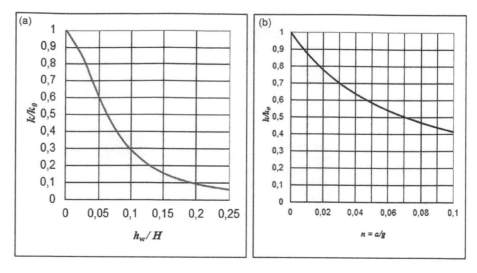

Figure 3.68 The effect of water in the crack (a) and seismic (b) on the stability of the block.

Chapter 4

Interaction of concrete dams with the foundation and improvement of foundations

"The dam must enter into marriage with foundation. However, in most events this matrimony is unequal: the dam is young, beautiful, slim and well built, while the foundation is decrepit and weak; his face dissected by wrinkles and fissures.... He gets all the hardships that he patiently endures, being fortunately smarter, than all design calculations, and he copes with difficulties. If he becomes weak, he gets injections. If, in spite of everything, he loses his internal equilibrium, then he destroyed. Then comes the end of her existence."

L. Muller (1908–1988) Austrian geologist

4.1 Work features of concrete dams with rock foundation

About the inseparable bond of a concrete dam and a rock foundation forming a single static system is better than *L. Mueller* cannot be said. And if the design and parameters of the dam and the properties and quality of the concrete are selected during the design process and are carried out during its construction, then the foundation with all its features, diverse and contradictory and favorable and unfavorable, is a predetermined material formed over the course of a long geological development.

In history of the construction of concrete dams, catastrophic accidents and severe damage to concrete dams on rock foundations are known, most of which are associated with insufficient bearing capacity of the foundation.

The rock foundations represent a natural geological medium, the physical and mechanical properties of which such as strength, deformability, water permeability, heterogeneity, and anisotropy are associated with the features of the constituent rocks, the nature of their disturbance, and NSS due to modern tectonic movements of the earth's crust [54]. Therefore, the data of engineering–geological surveys and geomechanical studies are of great importance as a result of which the geological structure and the physico-mechanical properties of the rocks and the rock foundation as a whole are determined (see Chapter 3).

In the process of construction and operation, the NSS of the rock foundation and abutments changes significantly. These changes are caused first of all by loads from the structures and water of the reservoir and the reformation of the filtration regime in the banks as well as by unloading during the development of the pit under the dam, filling cracks when performing cementation in the foundation, and the dynamic effect during blasting.

A change in the filtration regime is accompanied by a decrease in the strength and deformation properties of rock and the parameters of slide resistance along fissures.

Especially, these processes develop intensively during the first filling of the reservoir and cyclic fluctuations of the water level in it. These changes apply to the active area of the foundation and determine the features of the interaction of the dam with the geological medium [72]. Therefore, it is important to determine the boundaries of the active area interaction of the foundation with the structure, conduct comprehensive detailed engineering–geological surveys and studies in them, and design the structures taking into account their interaction with the active area.

As a result of the study and research of the active area, a comprehensive engineering-geological model is created in which the composition, structure, stress state and physico-mechanical properties of the rock foundation are reproduced and regularities of rock behavior under loads and effects are established. On the basis of the geotechnical model, a mathematical model of the "concrete dam-rock foundation-reservoir" system is developed, which reproduces the geological structure of the foundation, dam geometry and the sequence of its construction, including pit excavation and filling of the reservoir, and physico-mechanical characteristics of concrete and rock foundation. On the mathematical model, all the necessary studies of the SSS and the filtration regime of the system and adjustments of the project are carried out if necessary.

In addition, based on the study of natural and technogenic geodynamic processes in the region, their influence on the geological medium is estimated including the assessment of modern tectonic movements and the level of seismic hazard [28,90,97,98]. When monitoring during the construction, filling of the reservoir, and operation of structures, changes in the state of the active area are taken into account.

In the project process, despite the desire to maximally approximate the design scheme of the "dam-foundation-reservoir" system to the actual one, certain assumptions have to be made, which causes some conventionality of the design model and may lead to a mismatch between the design forecast and the actual state of the dam and the process of its construction and operation. A number of these assumptions are confirmed or refuted during the monitoring process [38]; first of all, this concerns the engineering-geological medium. If necessary, additional surveys and studies are carried out to clarify the engineering-geological model.

Significant adjustments were made during the construction of dams: Kirdzhali arch-gravity, 77 m high (Bulgaria) [41], gravity dams of the Bureya and Boguchan HEP in Russia [44], and many others. So, at the completion of the construction of the Kariba arch dam, 128 m high (Zimbabwe), it was revealed that the quality of the rock mass in the right-bank abutment is significantly lower than that provided in the project according to surveys. Therefore, it was necessary to carry out the mining method of concrete buttresses for transmission dam load on a healthy rock [233].

The NSS should be taken into account in the calculations of the "dam-foundation-reservoir" system for arch and spacer gravity dams for which stresses and deformations at the abutments are crucial [52,94,165].

The main load determining the SSS of the contact area and the foundation is the pressure of the water on the "dam-foundation-reservoir" system and the filtration volumetric forces in the foundation and in the bank abutments.

During the initial filling of the reservoir and subsequent cyclic fluctuations in the upstream level during operation, the rocks are saturated with water, which leads to a change in the filtration regime, SSS, and the physico-mechanical properties of the foundation. A danger to the structure can be caused by a rapid rise in the level and especially during the initial filling of reservoirs of high dams, when the rapid loading of the reservoir bed can lead to an increase of displacements of rock foundation, abutments, and accordingly the dam.

Often, there is a lack of synchronism in the change in the level of the upstream and the displacements of the dam and the drainage discharges in the foundation (for example, at the dams of the Krasnoyarsk and Sayano-Shushensk HEPs). This is associated with the propagation of the filtration flow in the depth of the fissured foundation and with a change in filtration forces in space and time [166].

The nature of the changes in the rock foundations depending on the type of dam and features when filling the reservoir can differ significantly. With a poorly permeable foundation filling, the reservoir creates a significant additional load on the surface zone of the Earth's crust, which causes the predominant development of processes of compaction and consolidation of rocks. The dams of the Bratsk and Ust-Ilim HEP are located on a layer of poorly permeable diabases underlain by more permeable rocks, and the diabases act as a natural apron, which forms large vertical filtration forces. Relatively large settlements of these dams are associated with 75 and 60 mm, respectively.

With substantially water-permeable rocks of the foundation, the process of its decompression under the influence of weighing forces and filtration forces in fissures receives predominant development [148,151].

When filling and cyclically changing the level of the reservoir, water penetrates into the fissures of the rock mass saturating it; the movement of water along the fissures causes a wedging effect expanding them and the formation of new cracks. There is decompression and a decrease in the strength of the rocks and an increase in their deformability and permeability; moreover, these unfavorable processes during long-term operation can develop gradually causing aging of the foundation and ultimately destruction [113].

Suffusion processes have an additional physicochemical effect: the dissolution of minerals (gypsum, anhydride, etc.) and similar processes in the aggregate of fissures (clay, etc.) and their soaking and removal.

During quick and deep drawdown of the reservoir, overpressure zones are formed in the fissures of rock mass, which can lead to cracking formation especially in the conditions of permeable mass and a change in the SSS and the filtration regime. These factors should be considered when designing and operating and limiting the speed of drawdown and raising reservoir levels.

The most critical place in the "dam-foundation-reservoir" system is the zone of contact between the dam and the foundation.

Domestic (Bratsk and Ust-Ilim gravity dams, Sayano-Shushensk arch-gravity dam, and Inguri arch dam) and foreign (arch dams Kelnbrein and Schlegeis in Austria, gravity dams Fontana and Dworshak in the USA, Daniel Johnson multi-arch dam in Canada, etc.) experience shows that there were problems with crack formation in the rock foundation under the upstream face of the dam.

The decompaction of the rock foundation under the upstream face of concrete dams is largely due to the fact that, as a rule, constructions for connection of the dam with the foundation have not been specifically developed [211]. Strengthening cementation and an antifiltration curtain – these traditional measures in most cases only contributed to the appearance of tensile deformations and cracking under the upstream face. Under the conditions of cyclic force effects (fluctuations in the water level in the reservoir and seasonal fluctuations in air temperature) on the rock foundation, progressive local destruction is possible, which leads to increased piezometric heads and filtration and ultimately a change in the static scheme of the "dam-foundation" system work.

The cracking zone deep in the foundation under the upstream face of the concrete dam is about 0.3 of dam height, and the active zone of the foundation for high dams may exceed their height. The experience of operating high dams with large reservoirs shows that the dimensions and work conditions of the active zone are determined by the engineering-geological features of foundation and filtration conditions, the height and design features of the dam, its connection with the foundation, loads, and effects on the "dam-foundation-reservoir" system. The reliability of determining the SSS of the "dam-foundation" system largely depends on the reliable determination of the dimensions of the active zone. When designing the Sayano-Shushensk dam, the dimensions of the active zone were limited by a depth of 50 m; but as shown by regime geophysical studies and observations of foundation deformations during dam operation, the dimensions of the designed active area were understated several times.

Insufficient consideration of the size and properties of this zone and the NSS of the rock foundation was one of the reasons for the "abnormal" behavior of the dam, the decompaction of the rock under the upstream face of the dam, and the increased permeability of the foundation of the Sayano-Shushensk and several other dams.

The analysis of field observations of uplift in the bed part of the foundation of high dams shows that the observed values usually exceed the design values, while the work of the drainage curtain is more effective than the design. In bank abutments where volumetric filtration takes place, the uplift also exceeds the design value [68,96].

There is a hypothesis [166,167] that in the reservoir area of large HPs with high dams, during operation, the crust under the load from the weight of the reservoir's water (for example, the weight of the Sayano-Shushensk reservoir is 30 billion tons) bends, and the bowl of deflection may spread away from the reservoir for tens kilometers. This process of gradual fading has been going on for decades. So, during the period of operation of the Sayano-Shushensk HEP, the settlement of the reservoir bed in the area of the dam site was about 30 cm. The maximum settlement of the reservoir bed of the Mid HP on the river Colorado in the USA with a reservoir volume of 35 km^3 and a foundation of shale and granite-gneisses for 28 years of operation amounted to 20 cm.

At the Sayano-Shushensk dam, a long-term decrease in the length of the chords of arches is also observed, which indicates the process of rapprochement of bank abutments in the dam site [133]. This process has a significant impact on the dam with increasing arch stresses with the redistribution of loads perceived by the foundation and bank mass. Similar phenomena were observed during the operation of the Chirkey and Inguri dam (see Section 3.3.1). The influence and development of settlement bed reservoirs and bank abutments of large dams should be analyzed during construction and operation.

The behavior of the contact zone is also affected by temperature effects during seasonal changes in external temperatures in severe climatic conditions [68,175,177,186]. On the Ust-Ilim, Sayano-Shushensk, and Bratsk dams, the value of irreversible openings of the contact joint under the influence of annual fluctuations in outdoor temperatures was about 20%. Deformation of the contact joint is cyclical in nature with maximum disclosure in the winter. The danger of these cyclic fluctuations is that they contribute to a gradual increase in time of the length and opening of the contact. Over the 10 years (from 1976 to 1986), on the Ust-Ilim dam, the absolute value of contact opening increased almost 1.5 times [96].

The SSS of the contact zone can vary significantly depending on the temperature of concrete regulation during the construction of the dam, the conditions for the staged construction and monolithic joints, and loading the dam in the process of filling the reservoir; this may increase the zone of tensile deformations.

Zones of tensile deformations and correspondingly decompaction were observed at the foundation of the extruded sections of the Bratsk and Ust-Ilim dams at an intermediate level of water in the reservoir (80% of the NHL) as well as the Sayano-Shushensk dam with incomplete head. In sections of the Bratsk and Sayano-Shushensk dams, an unfavorable stress distribution in the contact zone was formed when the reservoirs of the first stage were filled. Hydrostatic pressure was perceived by the extruded profiles of the dam in the absence of the last pillar, which later turned out to be underloaded.

The decompaction of the foundation in the station sections of the dam in the absence of a machine room can be caused by negative temperatures in the winter as well as a certain weakening of the rock foundation at the downstream face of the dam due to explosions during the development of the HEP pit.

Some compression with an improvement in the SSS of the contact zone under the upstream face can be obtained in a concrete dam with pillar cutting with a certain sequence of concreting and monolithic pillars [98,102].

Built-in conduits of bottom and deep spillways (permanent and temporary) and turbine pipelines of HEP affect the SSS of the contact zone. According to computational studies, the weakening of the profile of the Ust-Ilim dam due to the built-in water conduit led to an increase in the zone of tensile stresses under the upstream face by an average of 20% [75].

The ratio of the deformation modulus of concrete E_c and the rock foundation E_r and the inhomogeneous deformability of the rock foundation are the most important parameters that affect the formation of tensile deformation zone and the opening of the contact joint as well as the stress concentration at the downstream face.

According to field and computational studies, the greater the area of decompaction in the foundation under the upstream face of the dam, the higher the deformation modulus of the rock mass relative to concrete [102,175]. With an increase in the foundation deformation modulus, the magnitude of the compressive stresses in the foundation also increases at the downstream face of the dam. On average, for gravity dams, the ratio E_r/E_c in the range of 0.3–0.5 can be considered acceptable.

The location of compliant rocks under the upstream face is favorable and more rigid under the downstream face; the location of rigid rocks under the upstream face is unfavorable in which the zone of tensile deformations under the upstream face and the concentration of compressive stresses under the downstream face increase.

The ratio of the deformation characteristics of concrete and the foundation significantly affects the dynamic characteristics of the "dam-foundation" system and the distribution of dynamic stresses in the contact zone. An increase of the foundation deformability leads to an increase in the periods of free oscillations of the system and an increase in seismic resistance [118]. Therefore, it is advisable to lay high-modulus concrete in the contact zone of high dams, the height of the zone of which is determined by computational analyses, and the feasibility of protection cementation of the foundation especially in the zone under the upstream face of the dam requires appropriate justification.

Many examples are known of exploited gravity and buttress dams built on very deformable foundations. In Section 3.2.2, Rappbode and Bayna Basht dams built on soft foundations were mentioned. On the Bratsk and Ust-Ilim dams with a ratio of foundation and concrete deformation modules within 0.5–1.1, the openings of the contact seam were significantly higher than on the Krasnoyarsk dam with a modulus ratio of 0.3.

On the Kelnbrein arch dam of double curvature located on strong weakly deformed gneisses, intense crack formation in the contact zone began even during the filling of the reservoir (see Section 8.3.3 and Figure 8.34) and posed a direct threat to the dam. In this regard, when the reservoir was empty, repairs were carried out to strengthen it with the device on the side of the downstream face of the concrete block, which receives part of the hydrostatic load. These works cost $160 million, exceeding the initial cost of the dam.

Model studies conducted at the Massachusetts Institute of Technology (USA) showed that with a uniform distribution of the deformation modulus of the rock foundation, the arch dam SSS practically does not change when the displacements of the toes do not exceed 15% of the maximum displacement in the key [253].

The consequences of disturbance of the contact zone of the foundation can be divided into the following main groups [79]:

- the formation and opening of cracks (decompaction) and as a consequence an increase of the rock foundation permeability (multi-arch dam Girot in France, gravity dam Avon in Australia, arch dams Morrow Point in the USA, Les Toules and Pont Gal in Switzerland, Canales in Spain, and Schlegeis in Austria);
- erosion of the filler of fissures and interlayers leading to an increase in the permeability of the rock foundation (Agular gravity dam in Spain and Kariba arch dam between Zambia and Zimbabwe);
- reduction of strength and deformation characteristics of rocks during water saturation (gravity dams San Francis in the USA and Pieve di Cadore in Italy);
- increased uplift in the foundation and abutments due to the filtration properties of the rock (arch dam Malpasset in France and arch gravity dam Boulder in the USA) as well as due to the colmatage of the drainage system (gravity dam Hivassi in Japan and Douglas in the USA and arch dam Elsbert in South Africa).

According to the CME observations, in the first year after filling the reservoir of the Sayano-Shushensk dam, the decompression zone at the foundation spread on 27 m from the upstream face towards the downstream to a depth of 45 m, and the maximum opening of the contact joint was 13.6 mm [124,158]. During drawdown and filling the reservoir, the foundation under the upstream face was in a continuous process with the opening of existing fissures and the formation of new ones. The opening of the contact joint has reached drainage and therefore the filtration discharges have increased.

The constant increase in filtration discharges associated with the formation of new filtration paths during decompression of the foundation and fracture of the material filling the fissures led to the fact that the diagram of uplift on the bottom (Figure 4.1) remained much less than the calculated one [125].

The formation and opening of interblock joints on the upstream face in the lower part of the dam led to progressive filtration through the first pillar and influenced the working scheme of the arch-gravity dam: the load from the high consoles was redistributed to the archs, and the arch effect intensified. A number of authors believe that the state of the contact zone in the bed part of the dam and the opening of the contact joint are affected by the "hanging" of the first pillars of the bed part of the arch dam on the bank parts. There are limited vertical movements of the dam after

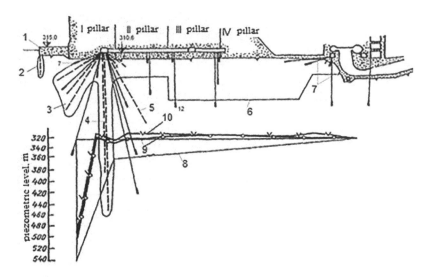

Figure 4.1 Uplift at the foundation of section 33 of the Sayano-Shushensk dam:
1 – apron, 2 – short cementation curtain, 3 – connecting cementation,
4 – deep cementation curtain, 5 – drainage curtain, 6 – protection cementation, 7 – piezometers; 8, 9, and 10 – diagrams of uplift according to SR
[11] and actual at LUS-540 m and at LUS-500 m, respectively.

monolithization, which cannot synchronously follow the foundation settlements under the effect of the weight of water reservoir [52].

The diagram of vertical movements of the bed section of the Sayano-Shushensk
dam (see Figure 3.39) shows that during the construction of the dam from 1977 to 1986,
settlements were observed under the influence of the weight of concrete and the initial
rise of the reservoir level. After filling the reservoir in 1986, the water saturation of the
rock mass began, and the rise of the foundation stabilized by 2001 [151].

According to field observations [186], the length of the opening of the contact joint
on the bed sections of the Bratsk and Ust-Ilimsk gravity dams was up to 15 m. The
change in the opening of the contact joint of section No. 30 of the Bratsk dam as
the reservoir is filled is shown in Figure 4.2 [96]. With increasing height of the dams,
the danger of opening the contact seam increases.

The analysis of the processes of connection of the dam with the foundation as a
single system "dam-foundation-reservoir" is associated with significant difficulties
due to multifactor, diversity, variability, and insufficient knowledge of the interaction
mechanism as well as the limited initial data. Despite the improvement of mathematical modeling methods in the calculation models of the "dam-foundation-reservoir"
system, it is still not possible to reproduce real engineering-geological conditions,
their change during filling of the reservoir, and the interaction of the structure with
the surrounding medium during construction and long-term operation.

Operating experience data from field observations of the SSS of operated dams
and their foundations show that their behavior may differ from the designed one. The
actual parameters of the decompaction zones at many dams exceeded the design
leading to disturbance of the conditions of normal operation.

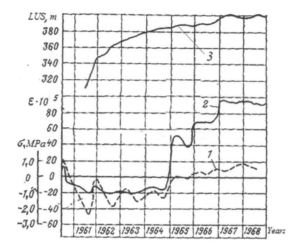

Figure 4.2 Opening of the contact joint of the Bratsk dam (section No. 30): 1 – stress in the concrete near foundation; 2 – opening of the contact joint; and 3 – upstream level.

Therefore, when designing after conducting design studies when choosing the final structural solutions and methods of construction, it is necessary to analyze and take into account the materials of operation similar in design, construction, and operation of dams in relation to the natural conditions of this site.

An analysis of accidents and damage to concrete dams shows that the most frequent cause of their destruction is precisely disturbances in the contact zone. As the English proverb says: "The chain is not stronger than the weakest link", the most important problem for concrete dams is to ensure reliable operation of the contact zone with the rock foundation. It should be noted that the possibility of a further increase in the height of concrete dams above 300 m directly depends on an effective solution to this problem.

4.2 Engineering measures to improve rock foundation

To neutralize the adverse features of the rock foundations and ensure reliable operation of the "dam-foundation" system, a set of engineering protection and anti-filtration measures are implemented which include [163] the following:

* protection cementation including cementation of weakened zones, faults, and large fissures;
* embedding by a mining way of faults lying at a depth, large fissures, caverns, weakened zones, or interlayers by filling with concrete or reinforced concrete (continuous embedding, gratings, and dowels);
* embedding by concrete of exits to the surface of faults and large fissures within dam bottom;
* arrangement of underground supporting structures;
* protection with prestressed anchors and anchors without stressing;
* creation of an anti-filtration circuit as a part of cementation and drainage curtains;

- device apron in front of the dam;
- local antifiltration measures including additional drainage, device of additional cementation curtains in the area of exits to the surface of faults, and large fissures outside the dam.

4.2.1 Preparing of rock foundations

The preparation of the rock foundation of the dam includes the following:

- working out of the pit to the marks of the dam bottom;
- treatment and cleaning the surface of the rock;
- protection against temperature and other atmospheric effects before the dam concreting;
- providing cohesion of the dam bottom with the rock foundation.

The depth of cutting of the dam into the rock foundation determined by the mark of the dam bottom on bed and on the bank abutments depends on the engineering-geological conditions of the rock mass weathering (preservation), fissuring (the nature of the discontinuity of the mass), changes of physico-mechanical properties in depth, dam height, and loads transferred by the dam to the foundation.

The classification of rock masses according to fissuring is given in Table 3.2 (see Section 3.2.1) and according to the degree of weathering in Table 4.1 [157].

The following zones are distinguished based on the data of engineering–geological surveys and studies in the rock mass:

- highly weathered destroyed rock to the stress state, which must be completely removed;
- highly fissured weathered rock to be fully or partially removed;
- moderately fissured, slightly weathered relative to the preserved rock satisfying (taking into account the possibility of improvement) the conditions of stability and strength of the "dam-foundation" system, and ensuring a reliable antifiltration contour and design filtration regime;
- slightly fissured.

Practice shows that the full removal of a strongly fissured rock is not always justified, since in addition to increasing the volume of rock excavation and concrete, it leads to a decompression of the rock mass. Increased permeability and deformability cannot be the cause of rock removal. An example of an extremely insignificant penetration

Table 4.1 Classification of rock masses by degree of weathering

The degree of weathering	Coefficient of weathering, k_w
Highly weathered	Less than 0.8
Weathered	0.8–0.9
Weakly weathered	0.9–1
Unweather	1

k_w – the ratio of the densities of weathered and unweather rock samples.

into rock abutments is the Toktogul dam (see Figure 2.23) and the Kowsar dam in the lower part (see Figure 2.22).

In the presence of weak interlayers in the foundation and abutments, horizontal or weakly inclined large fissures with lower strength, and deformation indicators, it is advisable in addition to the option of their removal to consider the possibility of protection of these areas with concrete dowels, plugs, gratings, etc. (see Section 4.3).

The bank abutments of the dam are characterized in most cases by high fissuring (including the presence of large fissures of on-board rebuff extending parallel to the bank slope) and the heterogeneity of the rocks in height and the presence of weak interlayers. In this regard, it is recommended to consider options with the carried out engineering measures to reduce excavation of the pit especially with high and steep slopes. The laying of the slopes of the pit within the bank abutments should ensure the stability of both the abutments and the dam during construction and operation, including the stability of the rock slope above the crest of the dam.

When performing drilling and blasting works during the development of pits, it is necessary to ensure the safety of the dam foundation through the use of special measures: the arrangement of a protective layer, the use of contour blasting, etc. [65,163].

In foundations composed of rocks the strength of which decreases significantly under atmospheric effects (siltstones, mudstones, shale, etc.), a protective layer is developed immediately before the concrete is laid.

4.2.2 Protection cementation

Protection cementation is used to equalize the deformation properties of the near-surface zone of the foundation and bank abutment dams to increase strength, water tightness, and to some extent slide resistance. Protection cementation is also used to "treat" rock foundations weakened by geological defects; in the presence of fissures with clay aggregate, in most cases, flushing of the fissures precedes, which allows increasing the slide characteristics along it.

Protection cementation of the rock foundation is contributed to the reduction of deformation heterogeneity of rock at dams such as Kambamba (Angola), Pieve di Cadore (Italy), Kirdzhali (Bulgaria), Dez (Iran), Castignon (France), and many others.

The heterogeneity of the deformation properties of the rock foundation was a decisive factor in choosing the type of dam Gomal Zam (Pakistan) located in a narrow canyon [196]. The rocks composing the site had high strength but contained many fissures of unloading, which greatly increased their deformability in height. If the deformation modulus of the rock of the slope base was 10,000 MPa, then at the upper marks, it decreased by 10 times. Works on the study of engineering-geological conditions and the possibility of improving the deformation characteristics of the rock foundation and the choice of the type of dam took 30 years.

Of the considered variants for the dam: rock-fill, gravity, with a massive buttress and two side arches, arch, and arch-gravity, after analyzing all aspects of the problem and the results of the research, a variant of arch gravity dam made of rolled concrete was adopted. At the same time, surface protection cementation had to be performed on the left and right banks to a depth of 50–60 m and was very symbolic in the bed. The cost of protection cementation was 40% of the total construction cost.

Protection cementation in the zone of the greatest compressive stresses transmitted by high gravity dams to the foundation and bank abutments of arch dams is of great importance especially under difficult engineering-geological conditions with the presence of tectonic disturbances, faults, severely disturbed rocks, strata, etc.

Protection cementation is effective in fissured rocks especially in layered rocks with developed tectonic fissuring or as a result of unloading.

In the presence of highly fissured decompressed rocks, protection cementation is envisaged over the entire area of the dam foundation.

The depth of cementing depends on the degree of fissuring, heterogeneity of the rock foundation, and compressive stresses transmitted to the foundation by a dam; for gravity dams, it is usually 6–10 m, and for arch dams, it is much larger.

Protection cementation in some cases allows increasing the strength and water tightness of the contact joint between the dam bottom and the foundation, which is especially important under the upstream face where tensile deformations take place under operating conditions. When it is performed, the strengthening of the connection zone of the cementation curtain with the dam body is achieved.

The Kambamba arch dam, 90 m high with 350 m crest length, was built on sandstones interspersed with clay shales [163,255]. At low marks, sandstones passed into breccias and conglomerates, which were located on bedrock granites. The layers of sedimentary rocks are almost horizontal and only occasionally broken by folds, but due to the large difference in the plasticity of sandstones and shales, their mutual displacement during the formation of folds led to the formation of sliding surfaces along the contacts between shales and sandstones. High deformation and strength heterogeneity of the mass at the elevation and abutments of the dam were revealed during the survey; however, when excavating the pit under the dam, it became clear that the negative influence of engineering-geological conditions on the dam safety was underestimated.

As a result, in order to ensure the reliability of the bank abutments of the dam, it was decided to consolidate the rock mass using protection cementation and make the following changes to the dam design:

- to improve the stability conditions of the dam, its shape in the plan was corrected in such a way as to direct the resulting arch force deep into the slope so the circular outline of the arches in the plan was replaced by three-center with increased radii in the bank parts;
- a significant heterogeneity of the dam abutments on the deformability of sandstones and shales caused tensile stresses in the dam body; for perception of it, on the dam bottom 2 and 3, rows of reinforcing meshes were installed with a cross section from $620\,cm^2$ at the crest to $1,000\,cm^2$ in the lower parts of the slope.

Protection cementation at a depth of 30 m was carried out from the cementation and drainage galleries of the dam (upper tier), from two adits traversed parallel to the slopes (lower tier) and from a horizontal adit under the riverbed.

The performed protection cementation was very effective and led to a significant increase of the deformation modulus while aligning their deformation properties (Table 4.2).

The inhomogeneous deformability of the foundation affects not only its value but mainly the size of the weakened zones.

Individual protection cementation of faults and large fissures in the foundation of dams are made in order to partially restore the monolithic of the rock foundation broken by a major discontinuity. As a rule, such cementation is made from wells crossing the surface of the cemented fault.

Table 4.2 Results of deformation modulus tests *in situ* (MPa)

	Sandstones		Clay shales
	In the vertical direction	In the horizontal direction	
Before cementation	1,200–10,500	7,100–18,200	700–2,900
After cementation	6,000–26,900	20,700–50,400	3,900–11,100

4.3 Protection of rock foundations with large fissures and faults

4.3.1 Embedding by mining way

Mining type of protection is performed if it is necessary to transmission ensure compressive or slide forces along a weakened surface or from one part of the rock foundation to another as well as to reduce the heterogeneity of the deformability of the foundation, if for some reason these goals cannot be achieved by cementation. Embedding by the mining method is carried out in various structural varieties: continuous embedding and arrangement of gratings and dowels [101,163].

The cementation of the contact between the concrete of the embedment and the rock as well as the cementation of the rock adjacent to the embedment is necessary for all types of embedding. In most cases, the embedment device in the rock foundation entails the formation of a backwater filtration flow. Therefore, when constructing embedding, drainage of the adjacent zone and diversion of filtration water are usually provided.

Continuous embedding is applied in the case of a significant amount of force to be transmitted through the disturbance, for example, when close to the dam bottom or in cases where the disturbance has a volumetric nature (cavity, node intersecting fissures, etc.). Combined solutions are possible when in the immediate vicinity of the dam, a continuous embedding is used, which in the less stressed zone passes into the grating. Continuous embedding was applied at the dam foundation: Nagawado (Japan, see Figure 4.3), Kurobe-4 (Japan, see Figure 4.4), and Montenard (France, see Figure 4.5); a continuous embedding in combination with a grating was made to strengthen the right-bank fault (see Figure 4.9) and four tectonic fissures at the foundation of the Inguri arch dam.

It should be noted that in many cases, local strengthening measures of rock masses were not provided by the project but were carried out already during the construction process and even after completion of construction. There are many such examples in engineering practice, the most characteristic of which are: Kariba dam, Kirdzhali, and Keban (Turkey).

Grates are a system of horizontal adits passable as a rule along the strike of a fissure or a weakened zone and inclined adits (shafts) passable along their fall; adits and

Table 4.3 Opening of major fissures in the foundation of Nagawado arch dam (Japan)

Fissure, designation	Left bank			River bed			Right bank	
	M	R	I	J	K	A	D	S
Opening, mm	0.1–2	0.15–0.2	0.35–4	0.3	0.5	0.3	0.3–1	0.3–0.4

shafts are filled with concrete. The thickness of the grates is buried in a healthy rock on both sides of the disturbance. Embedding in the form of gratings is used in those cases when the magnitudes of the forces that must be transmitted through the surface of the disturbance do not require a continuous embedding device. Embedding in the form of gratings was used in the foundations of arch dams: Montenard, Kurobe-4, and the Torrehon gravity dam (Spain); in the latter case, the grate was made in the plane of a dipping weakened zone in contrast to the first two dams where the grates was made to strengthen vertical or close-to-vertical faults.

The dowels are shafts or adits filled with concrete (reinforced concrete) drilled in the surface of the fault usually in the direction of the fall. The dowels are buried in a healthy rock on both sides of the disturbance. In order to prevent the dowels from turning, they are sometimes anchored into an adjacent rock. Embedding with dowels is easier compared to continuous embedding or embedding with gratings and is used with a relatively moderate amount of force to be transmitted through the disturbance surface. Dowelling was used at the foundation of the arch dams Aldeadavila and Santa Eulalia (Spain) and Kambamba. Unlike the first two dams, where the dowels were arranged in steeply dipping faults, horizontal and slightly inclined dowels were arranged at the foundation of the Kambamba dam along the surfaces of the most powerful layers of shale. In addition to static functions, the dowels that are part of the main cementation curtain also performed antifiltration functions.

The concrete embedding of fissures in the foundation of a high arch dam Nagawado is one of the first examples in engineering practice.

Nagawado arch dam (Japan) is 155 m high and 355 m long along the crest built on a foundation composed of biotite granites and hornfelses with many subvertical fissures on both banks parallel to the riverbed and having a significant opening (Table 4.3) [202].

Calculations and experimental studies have shown the need to embedding of fissures to ensure the required stability of abutments. A complete embedding of the fissures was performed using horizontal adits and inclined shafts specially passed along the fissures and then cementation of the fissures (Figure 4.3).

The natural aggregate was removed from the fissures by water jets under a pressure of 100 atm through a nozzle with a diameter of 15–17 mm and discharge $2.2 \text{ m}^3/\text{min}$. Fissures were washed from parallel galleries with a cross section of 2×2.5 m, passed in the plane of the fissure with an interval of 10 m. The void formed in the fissure was immediately filled with concrete. In total, $22,000 \text{ m}^3$ of aggregate was removed with in this way and replaced by concrete. The working out of fissures was carried out sequentially from site to site and ended with concreting of the passed sections of the fissures.

In addition, rock abutments from the surface were reinforced with prestressed anchors from 40 to 90 m long with a working load of each anchor of 332 tons. The total applied load was 40 thousand tons per massif volume of 250 thousand m^3.

The displacements of the dam abutments during the filling of the reservoir did not exceed 5 mm.

Kurobe-4 arch dam (Kurobegawa, Japan) was originally projected as an arch dam of conventional design with a height of 185 m and a crest length of 475 m. However, large fissures and caverns in the rock foundation found during the construction of the dam at the upper marks of both abutments prompted the designers to introduce significant changes in dam project [206].

In accordance with the specified geological situation, it was recognized that both abutments on the upper 60 m are unreliable for the perception of the forces transmitted

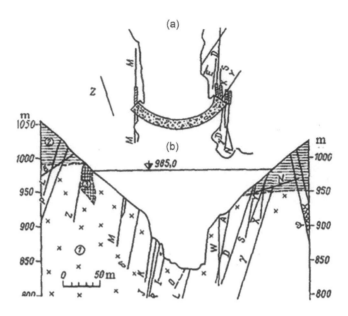

Figure 4.3 Nagawado arch dam (Japan): (a) section plan and (b) geological section along the riverbed.

from the arch dam. Therefore, it was decided to erect a dam with a "diving" crest and connecting it with the banks using gravity wings, which are deflected in the direction towards the upstream (see Figure 8.20).

Large fissures and caverns found in the immediate vicinity of the dam in biotite granites were cleared and filled with concrete. At the same time, four faults in the right-bank abutment were embedded using concrete gratings. For its device, horizontal adits along the strike of the fissures and inclined (close to vertical) shafts along the dip of fissures were passed (Figure 4.4). The largest gratings in the surfaces of two fissures required the installation of six tiers of adits and reached 60 m in height.

For a larger fissure, but also more distant from the dam heel, continuous grating was applied in a relatively small area. The total volume of fissures embedding amounted to 9,900 m^3. The caverns in the left-bank abutment were completely filled with concrete.

Arch dam Montenard (France) was 155 m high and was built in 1962 on the river Drak in a very deep and narrow canyon (the heights of the right and left banks were 450 and 350 m, respectively) formed by very strong limestones but cut along a number of fissures [101,141]. Large fissures extended parallel to the canyon and posed a danger to the stability of the dam in connection with which it was decided to embed them.

The "Beatrice", "Berenice", "Aglaya", "Clotilde", and "Claudette" fissures contained caverns, which were significant enough for them to be effectively strengthened by individual cementation. The boundaries of the protected part of the fissures were determined by the active zone of the foundation limited, according to the authors, by isolines of stresses normal to the fissures surface of 6 kg/cm^2 (the long-dashed line in Figure 4.5). Fissures were cemented at a low pressure of 6–8 kg/cm^2 to fill the caverns and at a higher pressure of 20–25 kg/cm^2 at the contact of the fissures with the rock mass.

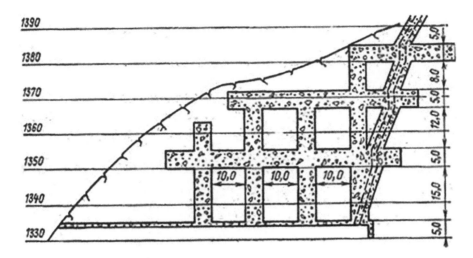

Figure 4.4 Embedding of Yellow fault in the abutment of Kurobe-4 arch dam.

The work was carried out in three stages: the first stage of cementation was carried out sequentially by horizons through wells drilled from several working chambers with a step of 5 m on a square grid during the concreting of the dam. The second stage was carried out with a fully constructed dam, when it served as a stop against the displacement of the slopes under the influence of efforts from cementation.

As a result of cementation, fissures were filled with mortar, which together with the rock fragments of the natural aggregate formed concrete connecting the wings of the faults. Special exploratory adits showed a good filling of the cavity of the fissures with a mortar. The third stage work was the final cementation of the contact between the formed concrete and the rock surfaces.

The largest crack "Red Julie" with a thickness of up to 0.7–1 m located in the right-bank abutment was filled with plastic red clay difficult to grout. Therefore, the protection of this fissure was decided to lead the mining method. At marks 385–415 m, a concrete grate was constructed consisting of four horizontal tunnels and five vertical shafts passed in the plane of the fissure respectively along its strike and dip (Figure 4.5).

The adits and shafts had a cross-section of 5 m^2 each; the distance between the adits was about 10 m and between the shafts was 15 m. Both adits and shafts were filled with concrete after sinking. To strengthen the contact of the concrete-rock, cementation was performed, which was carried out from the upper adit located at 415 m.

Inguri arch dam (Georgia) is 271.5 m high and built on the river Inguri in difficult geological conditions in a seismic area. The foundation is composed of limestones and dolomites with a fall of bedding at an angle of 50°–60° towards the downstream.

Limestones are quite strong (uniaxial compression strength of 80–90 MPa) but strongly fissured with deformation modules in the unloading zone of 4,000–8,000 MPa and 13,000 MPa and more in a healthy rock. On the right bank 110 m below the dam crest, there is a tectonic fault with a displacement of the order of 100–120 m but without signs of modern movements. In addition to the fault, the site crosses about two dozen large fissures with an opening of more than 10 cm, and there are six systems of fissures in the rock mass.

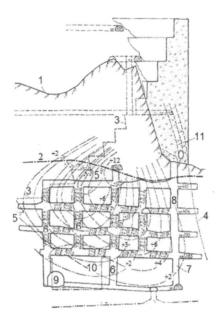

Figure 4.5 Embedding of the "Red Julie" fissure in the right-bank abutment of Montenard arch dam: 1 – exit of the fissure on the slope surface, 2 – boundary between the zone in which cementation is possible and the zone where it is not feasible, 3 – projection of the downstream face dam on the plane of the fissure, 4 – projection of the upstream face dam, 5 – galleries, 6 – vertical shaft, 7 – shaft passed from below, 8 – well, 9 – inlet tunnel HEP, 10 – projection of the contour of the HEP building on the fissure plane, and 11 – gallery running along the bottom of the dam.

Analysis and studies have shown the need to embed the fault and four tectonic fissures with an opening of more than 25–30 cm and filled with clay material. The purpose of embedding the fault was as follows:

- to ensure the transmission of shear stresses through a fault;
- to reduce the deformation heterogeneity of the rock mass in the area of action of forces from the dam;
- to neutralize possible differential displacement along the fault during the construction of the dam, filling the reservoir, and an earthquake.

Studies of the effectiveness of tectonic fault embedding were performed on two dimensional models of equivalent material at a scale of 1:500 [20]. In the second half of the 60s of the last century, when the dam was designed, the capabilities of computer analysis were modest, so the spatial problem was reduced to a system of two dimensional in six design sections (Figure 4.6). The designed sections were drawn through the planes of the lines of action of the forces from the dam R and the radial heels of the arches at the calculated mark [101].

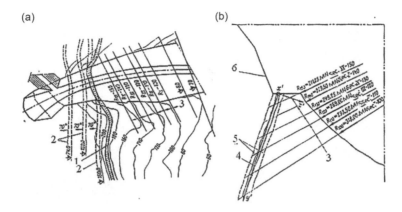

Figure 4.6 **Right-bank abutment of Inguri arch dam: (a) plan and (b) cross-section along the axis of the dam; I – fault exit to the slope surface, 2 – fault tracks at elevations, 3 – calculated sections, 4 – fault, 5 – zone decompression, and 6 – bottom of the dam.**

The foundation SSS in the fault zone was determined from the forces transmitted from the dam, the filtration forces at the foundation, and the dead weight of the rock mass.

Fault embedding sizes were determined by successive approximations. First, the SSS was calculated for nonembedded fault, while the fault and the surrounding decompression zone with the corresponding deformation characteristics were reproduced in the calculated sections. The possibility of transmitting shear forces along the fault plane was tested, and shear strength was analyzed using the slide resistance parameters of the fault filler material $tg\varphi_f = 0.65$ and $C_f = 0.1$ MPa. Figure 4.7a shows the isolines of the slide strength coefficients θ_{min} for nonembedding fault. Given the large size of the fault and the possibility of tectonic movements along it, the height of the arch dam, and in connection with this, the responsibility of protection measures, a high value of the allowable slide strength coefficient of 2 was adopted. The θ_{min} values in Figure 4.7a are much less than the permissible values, and therefore fault embedding was necessary.

A SSS analysis was performed for several embedment variants. The embedment was reproduced with concrete, and the strength was checked on the rock mass adjacent to the embedment, since the concrete is stronger than the rock mass. The parameters of slide resistance of the rock mass were taken as $tg\varphi_r = 1$ and $C_{sr} = 0.8$ MPa. Figure 4.7b shows the isolines θ_{min} for the selected variant of the fault embedding; the values of θ_{min} within the embedding are calculated based on the stresses and parameters $tg\varphi_r$ and C_r for the rock mass, and the values of θ_{min} outside the embedding are calculated based on the stresses and parameters $tg\varphi_f$ and C_f for the fault filler.

From a comparison of the isolines θ_{min} for the nonembedded (Figure 4.7a) and embedded (Figure 4.7b) fault, it can be seen that the values of θ_{min} outside the embedding almost did not change due to the insignificant effect of the forces from the dam on the SSS of the rock mass compared to the dead weight of the mass. It means that the forces from the arch dam are perceived by the rock mass in the embedment area. Taking this fact into account, it was accepted that it was possible to allow the values

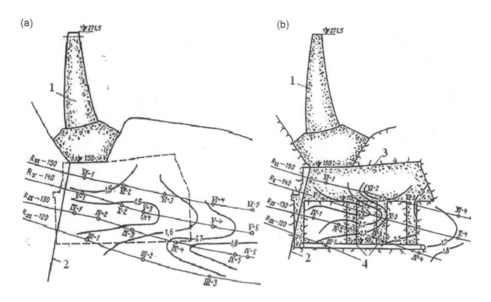

Figure 4.7 The isolines θ_{min} for nonembedded (a) and embedded (b) fault: 1 – Inguri arch dam, 2 – cementation curtain, 3 – continuous part of the embedding, and 4 – embedding in the form of pillars.

of θ_{min} outside the embedding to be less than 2. As a result of checking the integral slide strength at the embedding contact with the rock mass, a slide safety factor of 4.35 was obtained.

In addition to checking the strength of the stress state, an analysis was made of the displacements of the heel of the arch dam in the design sections for the case of homogeneous continuous foundation, for nonembedded fault, and for all embedded variants (Figure 4.8).

With the accepted variant of fault embedded, the displacements of the dam heel practically coincide with the displacements for homogeneous continuous foundation, i.e., the fault embedded provided the work of foundation as homogeneous continuous.

In addition to this, considerations of strength in determining the size and design of the embedding condition of work were taken into account. So, in order to connect the fault embedding with the cementation curtain, the embedding was shifted to the upstream, although this is not necessary on the strength conditions. Since in the lower part of the embedment θ_{min} exceeds the permissible strength factor, it was decided that between the 315 and 360 m marks, it was not necessary to complete the embedment but in the form of columns 10×8 m in size as shown in Figures 4.7b and 4.9. In the range of 360–390 m, the embedment was made with a continuous thickness of an average of 10 m. The volume of concrete of the continuous embedment was 33 thousand m^3 and grating was 18 thousand m^3.

This design of the embedding provided a smooth change in the hardness in continuous embedding and grating and a decrease in stress concentration in the rock mass.

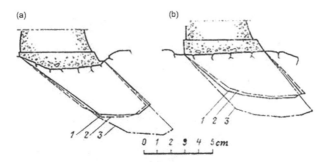

Figure 4.8 Diagrams of displacements of heel arch dam (a) and (b) design sections V-140 and VI-150; 1 – homogeneous continuous foundation, 2 – embedded fault, and 3 – nonembedded fault.

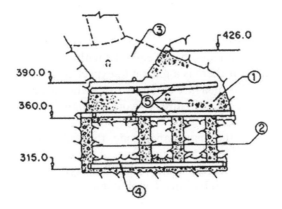

Figure 4.9 Embedding of the tectonic fault in the right-bank abutment of the arch dam Inguri: 1 – continuous embedding, 2 – vertical elements of the grating, 3 – special design of the dam saddle, 4 – horizontal elements of the grating, and 5 – galleries.

In order to strengthen the rock mass in the adit zone, protection cementation was performed. In order to reduce filtration forces from the upstream side of the embedding, borehole drainage was performed.

The fault was cleared from the surface to a depth of 30–40 m bringing the width of the trench at the saddle to 10 m. A special saddle construction was constructed in the recess (see Figure 8.37), which ensured the strength of the arch part of the dam with a possible displacement along the fault plane of more than 10 cm (for details see Section 8.3.2).

4.3.2 Embedding of faults and large fissures within dam support

The concrete embedding of faults and large fissures within the dam support is aimed at ensuring the joint work of the parts of the rock foundation divided by this disturbance under the action of forces transmitted from the dam. The measure consists in clearing

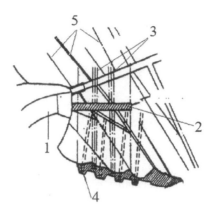

Figure 4.10 Kawamata arch dam. Strengthening the left-bank abutment: I – dam, 2 – transfer wall, 3 – prestressed anchors, 4 – retaining wall, and 5 – tectonic disturbances.

the disturbance to a certain depth and embed it with concrete followed by cementation on the contact surfaces if there is a reason to expect significant shrinkage of concrete. In terms of its scope, this measure can vary in a rather wide range – from shallow surface cutting of relatively small fissures to works of significant volume when embedding large disturbances.

With limited sizes of the disturbance zone, the embedment depth is assigned constructively or by analogy; for substantive zones of large extent, an analysis justification is carried out. As examples of large-scale works of this type, one can cite the embedding of fault in the foundation of arch dams: Kambamba, Kawamata (Japan, see Figure 4.10), and Rozhanel (France).

4.3.3 Device underground support structures

Underground supporting structures are usually used to transfer forces from arch dams through the surface of weakened or broken rocks to a healthy rock mass. Sometimes, along with geological factors, the use of underground support structures is dictated by the unfavorable direction of the forces transmitted from the arch dam as well as by the topographic conditions when, in a deep canyon with steep banks, the trench device is associated with large economically inexpedient excavations.

The most common constructive solution of underground support structures is the so-called "transfer walls" oriented in the plan approximately along the directions tangent to the upper arches of the dam. Another constructive solution is represented by underground buttresses arranged under the dam. Also possible are solutions such as horizontal "piles" (adits filled with concrete or reinforced concrete). Underground supporting structures were implemented on arch dams: Kawamata (Figure 4.10), Tviriviren (South Africa), Kariba (see Figure 4.11), and Kirdzhali (see Figure 4.12).

Kawamata arch dam (Japan) has a height of 120 m and a crest length of 137 m. The very strong rock liparites lie in the foundation of the dam severely tectonically disturbed [203].

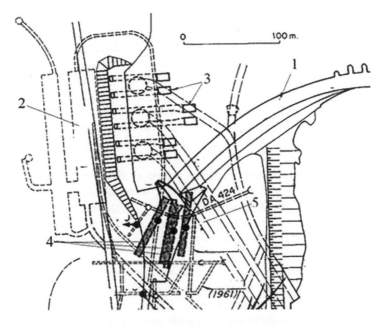

Figure 4.11 Kariba arch dam. Plan. Strengthening the right-bank abutment: 1 – dam, 2 – HEP building, 3 – water inlets, 4 – buttresses, and 5 – layer mica.

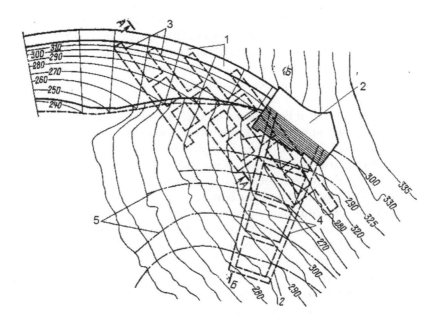

Figure 4.12 Kirdzhali arch dam (Bulgaria). Plan. Strengthening the left-bank abutment: 1 – arch dam; 2 – pier; 3 – frame No. 1, 4 – frame No. 2, and 5 – horizontals of the tectonic zone.

During the construction of the dam, great attention had to be paid to the left-bank abutment within which the F-30 fault and a series of minor disturbances were located. The situation was aggravated by the topographic conditions of the abutment since the dam rested on a protrusion of the left bank with the direction of forces from the dam almost parallel (in plan) to the horizontal of the day surface.

A number of protection measures were developed, including embedding the F-30 fault with a concrete plug 4–10 m thick and arranging the transfer wall from the arch dam heel through the weakened zone of the protrusion deep into the rock mass (Figure 4.10).

Work on the device of the transmitting wall was carried out from five adits of which later anchors were tensioned and installed to fix the wall and compress the left-bank rocky protrusion.

The concrete wall was buried in the rock mass at 20–25 m from the heel of the dam. The width and height of the wall, respectively, were 2.8–3.5 m and 70 m. The dimensions and direction of the wall were established on the basis of model studies. The depth of penetration was determined from the condition of ensuring the safety factor of stability of the left-bank abutment under the action of the force from the dam and its own weight of not less than 4.

Kariba arch dam (between Zambia and Zimbabwe) on the Zambesi river 128 m high and 617 m long was commissioned in June 1959 [248].

Already in the initial period of operation, it became obvious that the rock quality on the right bank is significantly worse than what was planned in the project based on field studies. At the end of 1959, during the construction of the road to the crest of the dam on the right bank, a layer of mica was found oriented parallel to the abutment dam.

Additional geological and geotechnical surveys showed that quartzites covering the main gneiss massif on the right bank have insufficient strength to perceive the load from the arch dam. In July 1961, it was decided to build buttresses in the right-bank abutment of the dam to transfer forces to a healthy rock (Figure 4.11).

The height of the erected four concrete buttresses reached 38–48 m, with thickness 6 m. Buttresses were made mainly by the mining method by sinking and filling concrete with adits. The buttresses together with the part of the dam resting on them formed a kind of underground gravity pier that transfers pressure to preserved gneisses. After carrying out these works in 1963, the water level in the reservoir was raised to the NHL.

Investigations and strengthening measures conducted since 1961 required costs amounting to 11% of the initial cost of construction [248].

Kirdzhali arch gravity dam (Bulgaria) was 103 m high and was built in 1961 in a site composed of strongly fissured gneisses with interbeds of biotite and schists. When conducting surveys of large tectonic disturbances in the foundation, it was not detected, and the foundation deformation module equal to 15,000 MPa was adopted in the project.

The complex geological structure of the foundation was revealed during the construction process almost at the end of the dam construction when the left-bank pier unexpectedly gave a lot of settlement and roll. In addition, talc lenses were discovered on the right bank at a depth of 7–12 m and on the left bank at a depth of 20–30 m under the dam bottom. Three weakening zones were also identified on the right bank capturing the base of the slope, which threatened the stability of the abutment dam, and also a tectonic fissure at a depth of 15–20 m with a dip angle of 70° and an opening from several cm to 1–2 m. According to the results of additional field, the deformation modulus of the rock mass on the right bank was reduced to 8,000–15,000 MPa, on the left bank to 1,000 MPa near the dam crest, and 3,000–4,000 MPa at the base of the slope.

The recommendations of JSC "Institute Hydroproject" were adopted to strengthen the rock foundation developed according to the results of analysis and model studies of the left-bank abutment and the entire dam together with its foundation. The stability analysis of the left-bank abutment of the arch dam showed that the reliability of the dam was not ensured either at the NHL mark of 328.5 m or at a lowered level of the reservoir of 315 m: the slide safety coefficient was, respectively, 1.10 and 1.20. The results of analysis stability of the left-bank slope showed that it is also in a state close to the limit [41,101].

To increase the reliability of the arch dam, it was recommended to erect two reinforced concrete frames from underground buttresses interconnected by vertical reinforced concrete beams at the dam foundation on the left bank (Figure 4.12).

Frame No. 1 consisting of four longitudinal buttresses 4 m thick oriented in the direction of forces from the dam serves as a support for the arch dam within the marks from 250 to 300 m and transfers forces from the dam to a healthy rock below the tectonic zone. Due to the deepening of the calculated slide surfaces and the involvement of a larger volume of rock mass, it was possible to increase the stability coefficient of the left-bank abutment to 1.80, which was recognized as sufficient under the conditions of the Kirdzhali arch dam. Frame No. 2 consists of two buttresses 3 m thick crossing the tectonic zone in the direction of dip. With the help of this frame, it was possible to significantly increase the stability of the slope above the tectonic zone and improve the conditions of support of the left-bank pier.

In addition to deep cementation of the foundation to a depth of 40–75 m, cementation of a talc lens from the research adit in the foundation, and protection of surface cementation of the entire dam foundation, drainage wells on the right bank were drilled from the same adit [41].

4.3.4 Strengthening by anchors

The strengthening of the rock foundation by steel anchors is aimed at ensuring the joint work of the rock foundation parts separated by geological disturbances and increasing the stability of the foundation and slopes or improving the stress state of the rock mass, thereby increasing its strength.

In modern practice, both active (prestressed) and passive (not stressed) anchors are used, but given the particular sensitivity of the arch dam to the ductility of the foundation, preference should be given to active anchors; the bearing capacity of passive anchors is realized only after a significant displacement of the protected part of the foundation occurs.

Passive anchors are used mainly as a temporary measure during the construction period as well as for protecting rock slopes outside the zone of dam support.

Rock foundations possess as a rule relatively high resistance to compressive forces, poorly resist tensile and slide forces, and the worse, the more they are broken by fissures. In this regard, the main idea of strengthening the rock foundation with active anchors is to increase its resistance to tensile and slide forces due to compression. The magnitude and direction of the preliminary compression force by the anchors are determined on the one hand by the stress state of the rock mass and on the other by its geological structure and orientation of existing discontinuities. In determining the direction of anchor forces, the constructive and production aspects are also subject to consideration.

The preload force is determined taking into account both elastic and inelastic deformations in the rock mass leading to unloading of anchors. When determining the pretensioning forces and the depth of the anchor, the adverse effect exerted by them on the stress state of the rock mass in the fixing zone should take into account.

Practical tasks that are solved by managing the stress state of the rock foundation with the help of active anchors are quite diverse and often intertwined at one object. However, these tasks can be divided into three groups:

a. compression of the rock foundations of dams in order to increase stability;
b. compression of the rock foundations of dams in order to improve the stress state and increase strength;
c. compression of rock masses in order to increase the stability of the slopes of the pit, the banks of the reservoir, and the slopes in the downstream of the dam.

When compressing rock foundations with active anchors, an increase of friction force along the potential slide surface is achieved. Additional holding force can also be created in a direction parallel to the slide surface and opposite to the direction of the possible slide. Examples of preliminary compression of the rock foundation in which these two methods are used is the arch dam of Santa Eulalia (Spain).

In some areas, inelastic (plastic) deformations in the rock foundations of arch dams may arise. Most often, this situation occurs at the foundations with an unfavorable direction of the forces transmitted from the arch dam and with insufficient cutting of the dam into the rock foundations in the area of the downstream face. Improving the stress state of the rock foundations at the downstream face of the arch dam is achieved through engineering measures such as the following:

• compression of the rock mass at the downstream face of the arch dam using prestressed anchors (for example, the Vaiont arch dam, see Figure 1.31);
• device of a concrete slab at the downstream face at the lower elevations, the purpose of which is to increase the dam cutting into the foundation (Figure 4.13);
• combined protection with a concrete slab and anchors passive or active (for example, Inguri arch dam, see Figure 8.23).

As already mentioned, during the design of the Inguri arch dam, it was not possible to study the SSS of the rock foundation at the downstream face in spatial conditions; therefore, the calculations were performed in plane sections (see Figure 4.6). Analysis of the strength of the rock mass in solving the elastic task showed that the local strength coefficients θ_{min} are less than the permissible 1.3 at the downstream face and even less than unity (Figure 4.13a). After considering several variants for strengthening, in the form of a concrete slab, passive and active anchors, and a combination of a slab with anchors, a variant was selected, for which the values and distribution of θ_{min} are presented in Figure 4.13c. In this case, the heel displacements were almost equal to the displacements in the elastic task (the difference is within 1%). This indicates that the real rock foundation corresponded to the elastic foundation hypothesis [101] adopted in the analysis of an arch dam SSS (at that time, in the analysis of the SSS of

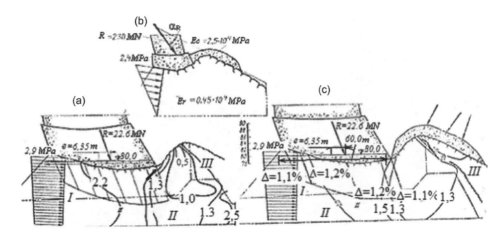

Figure 4.13 The isolines θ_{min} and the diagrams of displacements of the Inguri heel arch dam before (a) and after (c) protection measures in downstream; (b) design scheme.

arch dams, the ductility of the foundation was taken into account, but the stresses in it were not analyzed).

Active anchors allow you to create three-dimensional compression in the pre-stressed area, which leads to an increase in the bearing capacity of the rock foundation; at the same time, the aim of increasing the stability of the abutment of the dam is achieved. Examples of precompression of the rock foundation with active anchors are arch dams: Kawamata (see Figure 4.10), Tvirivieren, Nagawado (see Figure 4.3), Jose Antonio Paez (Venezuela), and Inguri (see Figure 2.22).

Compression of rock slopes in the downstream weakened by major disturbances with the help of active anchors was performed on arch dams: Chirkey (see Figure 2.30), Al Atasar (Spain, see Figure 4.14), Grancharevo (Yugoslavia, see Figure 4.15), Aldead-avila (Spain), Des (Iran), Chaudanne (France), and Tashien (island Taiwan).

Anchoring of rock masses is considered economically feasible and effective when the total anchor force does not exceed 15%–20% of the weight of the mass being strengthened [73].

El Atazar arch dam (Spain), 134 m high, was built in 1972. Fractured Silurian schists and quartzites lie at the foundation of the dam. When digging a pit under the dam on the left bank, a series of collapses occurred due to the unfavorable orientation of the existing fissure systems and the presence of a large tectonic disturbance. To remove the uplift in the abutment, four drainage galleries with a drainage well system were passed towards the upstream. However, the presence of clay material in the fissures prevented drainage of the massif, which threatened the stability of the massifs in the abutment [79].

To ensure the stability of the left-bank abutment, it was decided to use reinforced concrete anchor belts with prestressed anchors. However, taking into account that the unstable zone is located in the upstream in the zone of the variable level of the reservoir, it was decided to strengthen it with a reinforced concrete lattice "sewn" to the bank with prestressed anchors with a working force of 230 tons (Figure 4.14).

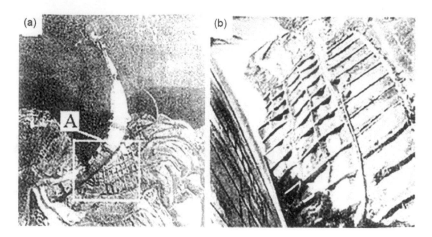

Figure 4.14 Strengthening the left-bank abutment of the El Atazar arch dam: (a) plan and (b) detail A.

Fissures parallel to the bed on the right bank threatened the stability of the pit under the water well, and therefore, before the excavation of the pit, the right-bank adjoining was reinforced with nine reinforced concrete anchor belts with 50 m long anchors and a working force of 250 tons.

A total of 1,000 anchors with forces of 250 tons and 1,000 anchors with forces of 20 tons were used. To stabilize the load in the anchors with a force of 250 tons, it took from one to three years and only after that, they were finally cemented.

Grancharevo arch dam (Yugoslavia) was 123 m high and 450 m long along the crest. It was built in 1964. The foundation of the dam is composed of fractured limestones, and karst was noted on the left bank. Between the layers of limestone falling at an angle of 7°–20° towards the right bank, there were interlayers containing clay.

During the development of the pit, a landslide occurred in the lower part of the left-bank abutment near the heel of the dam with a volume of 8,000 m³. This landslide, as well as the possibility of further creeping of the left-bank slope, posed a serious threat to the reliability of the left-bank rock mass as a dam abutment [256]. In this regard, it was decided to arrange a retaining wall in the lower part of the left-bank slope directly affected by the landslide and to protect the rest of the slope between the dam and the spillway chute with prestressed anchors. Prestressed anchors were placed at the intersection of reinforced concrete beams located on a rectangular grid on the surface of the slope (Figure 4.15).

In total, 90 anchors were installed with a bearing capacity of 200 tons each. Anchors were installed at an angle of 20° to the horizon to a depth of 40–60 m. Each anchor consisted of 55 cores with a diameter of 7 mm.

Chirkey arch dam (Russia) built in 1977 is 231 m high and is located in a narrow canyon with almost vertical banks composed of limestones with interlayers of clay and marl. The seismicity of the area is estimated at 9 points. Layered limestones with clay interlayers with a thickness of about 10 cm and a fall of 10°–12° towards the riverbed lie at the foundation of the left bank. Several systems of fissures were found with a

Figure 4.15 Strengthening the rock mass in the left-bank abutment of Grancharevo arch dam.

fall towards the slope at angles from 40° to 80° filled with calcite and clay. The main threat to the stability of the abutments was represented by fissures of the side rebuff parallel to the bed especially on the left bank where their opening on the crest of the slope reached 0.5 m.

The presence of a dangerous situation in the left-bank abutment was confirmed by a dump of 5,500 m^3 of rock mass, which occurred when the depth of the pit under the dam reached 100 m. Detailed analysis made after this showed that to ensure the normative safety of the slopes and stability of the left side of the pit under a seismic effect of 9 points, it is necessary to apply a holding load of 330 MN (33 thousand tons) to the rock mass.

Given the steepness of the slopes and the deep laying of fissures, it turned out to be impossible to use the classical methods of strengthening the mass with single anchors or anchor belts. A method was developed and applied for strengthening packages of anchors embedded in galleries. In the depths of the mass, galleries parallel to the slope were passed connected to the slope surface by transverse adits. On each of the horizons on the slope surface and in longitudinal galleries, reinforced concrete beams interconnected by packages of prestressed anchors were placed (Figure 4.16).

Each package consisted of 32 anchors (4 × 8) with a diameter of 56 mm made of steel with a yield strength of 650 MPa. Each anchor was stretched to a load of 520 kN (52 t), and thus, the total load applied by the package of anchors was 166 MN (1,660 t).

To ensure the required safety factor $k = 1.5$ in addition to anchor packages in galleries, inclined passive anchors 25 m long were installed at the base of the slope, intersecting clay layers at an angle of 60°. In each well, three rods with a diameter of 40 mm were cemented from steel of a periodic profile. Forces in these rods should have appeared in the presence of slope displacements along clay interlayers.

The measurements and observations showed that prior to anchoring the rock mass, constant displacements towards the bed at a speed of 0.1–0.6 mm/day were observed; after completion of the strengthening work in October 1971, these displacements almost completely stopped.

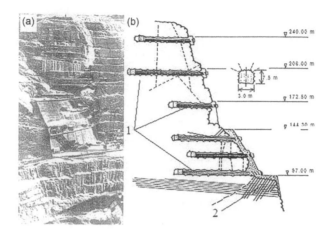

Figure 4.16 Strengthening the left-bank abutment of the Chirkey arch dam (a) and (b) vertical section; 1 – packages of prestressed anchors installed in galleries and 2 – passive anchors.

Hongrin arch dam (Switzerland) built in 1970 consists of two (North and South) arch dams of double curvature (see Figure 8.22). In the North dam with a height of 125 m, the right-bank abutment consisting of thin layers of limestone with layers of black slate, unlike other zones, is characterized by a large number of tectonic disturbances, higher water permeability, and a reduced modulus of deformation (on average 9 GPa). In such geological conditions, in the presence of local deepening in the downstream to ensure the stability of the right-bank abutment, 57 prestressed anchors were installed with a stress of 150–1,600 kN and embedded into a depth of 38 m; the local depression in the rock abutment was filled with concrete with the installation of 42 prestressed anchors in addition which drainage from deep wells was performed. However, the operation of most of the drainage network turned out to be less efficient than envisaged in the project which led to increased water leakages in the right-bank abutment.

After 40 years of operation, it is obvious that the prestressing forces of the anchors decreased as a result of relaxation.

In the course of the studies conducted in 2006, the parameters of the rock mass of the right-bank abutment were clarified and then used in the numerical model, with the help of which the analysis of abutment stability was performed. The analysis results showed the decisive role of existing prestressed anchors in ensuring the stability of the right-bank abutment [220].

To protect the strengthened areas of the rock foundation from the effects of filtration water, in most cases, additional drainage was carried out in the area of protected measures (embedding fissures with concrete and compression using active anchors).

A similar drainage was performed in the foundation of the arch dams Tviriviren, Santa Eulalia, Aldeadavila, Torrejon, and Des. An additional grouting curtain is often arranged to reduce the impact of the filtration flow on the rock mass being strengthened (Tvirivien arch dam).

Fraile arch dam (Peru) is 72 m high located in a narrow gorge with almost vertical slopes. The dome dam in the upper part rested on the piers on both banks.

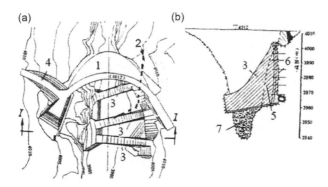

Figure 4.17 Fraile arch dam: (a) plan and (b) section I-I; 1 – dam 2 – construction tunnel, 3 – buttress, 4 – wall, 5 – collapsed part of the bank, 6 – anchors, and 7 – alluvium.

The left-bank pier directly perceived the pressure of the water, and the right-bank pier was protected by a special wall.

The emergency condition was created in April 1958 3–4 months after the beginning of filling the reservoir. In the downstream, in the area of the left-bank dam abutment, rock mass collapse occurred with a volume of about 15 thousand m^3. As a result of the collapse, the outlet tunnel was cut, and a powerful stream of water with a flow rate of 60 m^3/s began to break out of the collapsed rock. A real threat to the stability of the dam was created, which required the emptying of the reservoir (Figure 4.17).

It was assumed that the collapse occurred after fissures appeared in the unreinforced concrete lining of the tunnel due to seismic shocks preceding the collapse. As a result, water from the tunnel penetrated into fissures in the downstream of the left-bank abutment, the pressure of which combined with the forces of gravity and the unfavorable stress state from the forces transmitted by the dam brought the slope to a limit state.

After emptying the reservoir, a complex of repair work was carried out which included a concrete slab anchored into the remaining part of the left-bank slope, buttresses with which the left-bank slope was rested against the right bank, and tunnel lining and slope protection cementation restored.

4.4 Antifiltration measures and connection of concrete dams with the foundation

4.4.1 Ensuring water resistance foundation. Antifiltration measures

Filtration of water of the dam foundation leads to force on the dam and in the fissures, loss of water from the reservoir, suffusion of the material of the aggregate fissures, and reduction of the strength and deformability of the rocks of the foundation.

Compressive stresses in the rock mass can significantly affect filtration: when compressing the fissures, the flow rate can sharply decrease (which in turn can lead to an increase in uplift as happened in the bank abutment of the Malpasset arch dam in

France), and the presence of tensile stresses almost always leads to a sharp increase in filtration due to the opening of existing and the appearance of new fissures.

The increased permeability of rock foundations and bank abutments is one of the main factors leading to accidents at concrete dams; therefore, it is important to develop the necessary antifiltration measures during design. During operation, monitoring of filtration is necessary, and the prompt implementation of additional antifiltration measures is carried out if necessary.

Existing methods to prevent increased permeability of rock mass at the foundation and in the abutments of concrete dams are as follows:

• the creation of antifiltration curtains, their strengthening, or the arrangement of additional curtains in operating conditions;
• construction of aprons in the upstream of dams;
• embedding of concentrated filtration ways.

The arrangement of antifiltration curtains in the form of cementation and/or drainage curtains is the main protective measure created with the aim of

• reduction of uplift and weighing pressure in the foundation and bank abutments of the dam, thereby increasing the stability of the dams;
• reduction of filtration discharges through the HP to prevent excessive losses of water from the reservoir;
• protection of rocks from mechanical and chemical suffusion.

The antifiltration curtain is usually placed at the foundation as close as possible to the upstream face of the concrete dam and, if there is an apron, also under the apron's tooth.

With a small thickness of water-permeable rocks, the cementation curtain is brought to the roof of the water-resistant rocks, and when they are deep, it is usually brought to rocks characterized by specific water absorption of 0.01–0.05 L/min. The thickness of the cementation curtain depends on the properties of the rocks in the foundation (the presence of fissures, karst, readily soluble rocks, dam height, and head) and is determined by the critical head gradient in the curtain. The critical head gradients in the curtain depending on the specific water absorption within the curtain are given in Table 4.4 [11,157].

Curtains are formed by one, two or three rows of wells. Usually in the upper part of the curtain near the contact with the dam, where the head drop of the filtration flow is greatest in the curtain, several rows of shallow wells are arranged. The distance between the wells in the row is on average 2–4 m.

Table 4.4 Critical head gradients

Specific water absorption of rock in the curtain q_c, $L/(min \cdot m^2)$	Critical head gradient in the curtain I_{cr}
Less than 0.01	35
0.01–0.05	25
0.05–0.10	15

The fissuring usually decreases with depth and head gradients fall so the number of rows of wells decreases with depth and one row of wells is brought to the design depth.

The depth of the curtain can reach 100–150 m (Inguri dam), 220 m (Kasseb dam in Tunisia), and more. In bank abutment of high dams, well depths are reduced due to their completion from adits located 40–60 m in height (Inguri and Chirkey dams). At the bottom of high dams, galleries are almost always arranged for cementing and drilling wells regardless of general construction work and ensuring accessibility to the cement curtain during the operation of the dam. The cementation pressure at domestic dams is usually 3–4 MPa, and higher pressures are applied at foreign dams.

Well cementing wells are usually drilled in successive descending zones. Less commonly in relatively disturbed rocks, full-depth drilling (Chirkey dam) is used, followed by cementation by ascending zones.

Deterioration in the quality of the antifiltration curtains at the foundation of dams can occur due to the following reasons:

- leaching of curtains and walls depending on their composition, hydraulic gradients, and aggressiveness of filtering water;
- migration of particles (suffusion) creating passages in the curtains and removal of particles from the foundations;
- aging of curtain materials and increase in the resistance of drainage systems;
- fissures arising in the foundation crossing the antifiltration elements and reaching the drainage device.

The drainage curtain is an effective measure to change the filtration regime and reduce the filtration forces in the dam foundation and abutments.

The drainage curtain is vertical or inclined in the form of a series of drainage wells, which are drilled from galleries in the dam and adits at the foundation. The depth of the drainage curtain is determined depending on the depth of the cementation curtain, the degree of fissuring and water permeability of the foundation rocks, and the presence of highly permeable interlayers and it is usually 0.5–0.7 of the depth of the cementation curtain. With a layered foundation alternating between waterproof and aquifers, drainage wells must cross the aquifers.

Under favorable engineering-geological conditions, strong and low permeability (at $K_f < 0.1$ m/day), and nonsuffusion foundations, the cementation curtain becomes ineffective and can be eliminated. So, a number of dams were erected only with drainage curtains, for example, in Spain, there are about a dozen gravity dams and three arch ones. Among them are the Bembesar gravity dam, 102 m high (1969), on the paleozoic schists, and the arch dam Matalavilla, 113 m high (1967), on columnar quartzites.

The length of the antifiltration curtain deep into the bank abutments, its direction in plan, and its depth depend on the engineering-geological and filtration conditions, the effect of bypass filtration on the loss of water from the reservoir, and the stability of rocks of the bank slopes. As a rule, the direction of the bank curtain is perpendicular to the hydro-isohypses of the filtration flow.

For high dams located in canyon-shaped sites, bank curtains are usually made of two or more tiers arranged from bank tunnels – galleries. So, on the Vaiont arch dam (Italy), the bank curtains are made from eight adits (see Figure 1.31) at various levels in the limestone mass [163].

At the foundation of the Chirkey (see Figure 2.30) and Inguri (see Figure 1.29) arch dams, several tiers of drainage tunnels were made of which drainage wells were drilled; on the Inguri dam the lower section of the construction tunnel and research adits are used as drainage. It is advisable to place drainage wells if possible closer to the upstream outside the zone of high compressive stresses.

Under favorable engineering-geological and filtration conditions, the burial of the curtain into the bank mass may be insignificant.

An example of a decrease of uplift at the foundation and bank abutment of a dam under operating conditions is the Hoover gravity dam in the USA built in 1935 on slightly fissured andesitic tuff breccias. Initially, cementing and drainage curtains with a depth of 30–45 m were made at the foundation. When filling the reservoir to NHL, there was a sharp increase in uplift, which exceeded the design one and became dangerous for the stability of the dam. In 1939, the filtration discharge was 150 L/s. Repair work was carried out until 1946 and consisted of deepening the existing drainage curtains to 120 m and arranging a new cementation curtain. Figure 4.18 shows uplift before and after additional antifiltration work, a comparison of which shows its effectiveness. Filtration discharges after completion of repair work decreased to 30 L/s.

Below are some nonstandard examples of strengthening of antifiltration curtains in the foundation and bank abutment concrete dams during their operation.

Canelles dome dam (Spain). Complex and lengthy antifiltration works were carried out on the arch dam Canelles (Spain), 151 m high, built in 1958 [71,187]. The foundation of the dam is composed of karst limestones, which at a depth of 130 m are underlain by a layer of "Kapa Negra" marls with a thickness of 20–30 m, which the designers accepted as waterproof (Figure 4.19).

In April 1958, at a water level in the reservoir 100 m below the NHL, filtration was recorded in the floor of the HEP engine room. Despite the cementation work performed, the filtration discharge increased reaching 8 m³/s when the water level was reached 40 m below the NHL. Moreover, many sources were found on the left bank below the

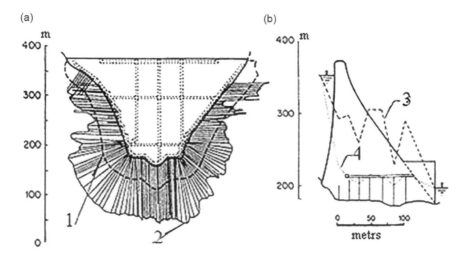

Figure 4.18 Hoover gravity dam: (a) 1 and 2 – antifiltration curtain before and after increasing depth; (b) 3 and 4 – uplift before and after increasing curtain depth.

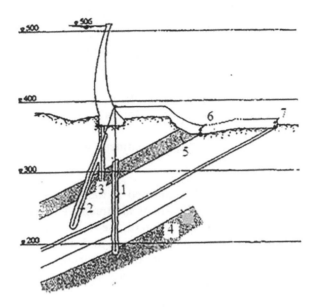

Figure 4.19 Canelles arch dam. Antifiltration measures in the left-bank abutment of the dam: I – cement curtain according to the initial design, 2 – cementation curtain with a slope of 35° towards the upstream, 3 – additional cementation curtain, 4 – layer marl "Capa Negro", 5 – layer marl "Margas Ondas", 6 – exits of filtration water, and 7 – concrete mount stilling pool.

"Capa Negro" interlayer. Due to the fear of erosion at the foundation of the dam, the reservoir level was lowered to the level of 426 m, which was kept during repair work. Initially, to study the situation in the left-bank abutment, galleries were made that revealed a large number of karst caves. A three-row cementation curtain was installed on the "Margas Ondas" interlayer area. For cementation, pure cement, sand, and clay cement mortars were used. Detected karst cavities were filled with concrete. During the construction of the left-bank curtain, 116 km of wells were drilled with an average consumption of 303 kg of cement per 1 m of well.

When refilling the reservoir, filtration discharges decreased but still amounted to 1.5 m³/s. Additional studies have established that residual filtration occurs in the riverbed of the dam.

Cementation in this area at flow velocity (up to 60 km/h) with an unloaded reservoir was of the greatest difficulty. It was decided to make an inclined curtain (pos. 2 in Figure 4.19). Fissure embedding (6% from cementation volumes) was carried out under a pressure of 0.6 MPa using specially developed materials. Sixty percent of the injection areas were treated with pure cement mortars. For cementing the riverbed part, about 23 km of wells were drilled with an average absorption of 856 kg of cement per 1 m of well. After the completion of the full volume of additional cementing works, as a result of which the area of the cementation curtain reached 210 thousand m² against 130 thousand m² according to the initial design, the filtration discharge was reduced to 60 L/s.

Abutment filtration problems were also recorded on the right bank [187]. Field studies of foundation fissured and slide strength in combination with analysis showed that the forces caused by the filtration flow in combination with the forces transmitted by the dam to the rock mass reduce the reliability of the right-bank abutment of the dam. The analysis considered rock blocks formed on the right bank by steeply falling systems of latitudinal strike fissures in combination with horizontal planes.

To strengthen the right-bank abutment dam, it was decided (Figure 4.20)

- to reduce filtration forces in the abutment;
- to change the direction of the resultant forces transmitted by an arch dam to the rock abutment;
- to strengthen and load the right-bank abutment, which has a concave outline from the side of the downstream.

To achieve these goals, a side wall connected to the upstream face of the dam with an elastic joint was erected in the right-bank abutment. The purpose of this wall was to deviate the resultant force from the dam on the right bank into the interior of the mass as well as to create additional cementation and drainage curtains (BED) with their removal from the tension zone where the AB curtain was located (Figure 4.21). In addition, the abutment was loaded with concrete and reinforced with prestressed anchors with a total force of 45 thousand tons.

Repair work was carried out at the HP for 16 years during which the HEP worked with a load of 70% of the design.

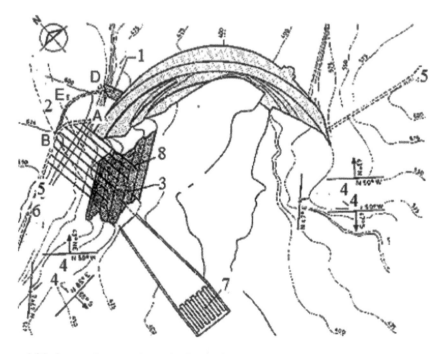

Figure 4.20 Strengthening the right-bank abutment of the Canelles arch dam: 1 – side wall, 2 – spillway gallery, 3 – concrete, 4 – fissure systems, 5 – cement curtain, 6 – drain curtain, 7 – stilling pool, and 8 – prestressed anchors.

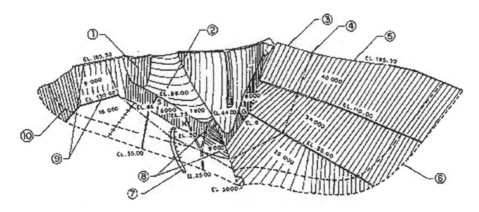

Figure 4.21 Antifiltration curtain at the foundation of the Oymapinar arch dam (axonometric): 1 – auxiliary arch dam, 2 – concrete lining, 3 – zone full embedding, 4 – border between full and fast embedding, 5 – zone quick-embedding, 6 – border cementation curtains, 7 – outlet tunnel, 8 – cementing wells, 9 – cementation galleries, and 10 – contact cementation curtains with lime shale.

Oymapinar arch dam (Turkey) is 185 m high and built on heavily karst limestones. Already during geological surveys, large karst caves were discovered at the foundation, which required the development of engineering measures in the project.

It was decided to abandon the classic vertical antifilter curtain based on its inefficiency due to the inability to achieve a waterproof horizon. Therefore, a special cementation curtain was created at the foundation of the dam and in the upstream in the form of an overturned dome or "trough" which closed under the bottom of the reservoir and rose in banks almost to the crest of the dam. On the downstream side, the "trough" was connected to the dam with a shallow vertical cementation curtain, and on the upstream side, it reached waterproof clay shales.

Thus, closed cementation curtain was created at the foundation, which "hugged" the reservoir (Figure 4.21). The total area of the curtain was 180 thousand m^2. After filling the reservoir in 1984, the maximum filtration discharge did not exceed 200 L/s [79].

El Cajon arch dam. A similar solution was used in the construction of the arch dam El Cajon in Honduras (Figure 4.22), where wells were drilled and cemented from 11.5 km of cementing adits that were drilled in abutments and under the bottom of the reservoir.

The curtain area amounted to 530 thousand m^2 and required the drilling of 485 km of cementing wells in which 100 thousand tons of cement was injected under pressure up to 5 MPa with a distance between the wells of 5 m; the cementation criterion was the maximum absorption of 50 kg of cement/m.

4.4.2 Closing the focused filtration ways

Often in rocks (especially carbonate), there are concentrated filtration ways along large fissures, which can lead to leaching of the filler of these fissures with dangerous consequences.

Lake Lanyo arch dam (USA), 19 m high, (see Figure 8.7) built in 1925 on sedimentary rocks, was collapsed for this reason. Stratums of rocks in the left-bank abutment of the dam extended along the canyon with a fall towards the river at an angle of 30°.

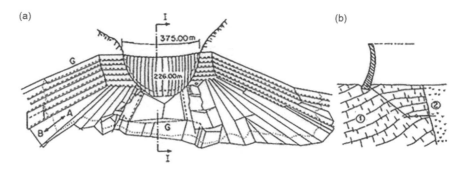

Figure 4.22 View from the upstream to the antifiltration curtain at the foundation of the arch dam El Cajon Dam: (a) view from the upstream and (b) section along I-I.

Such mass structure caused water filtration in uncemented rocks. At full head, the filtration rate exceeded critical gradients, and erosion of the aggregate of fissures in the left-bank abutment began. As a result, the left-bank abutment subsided and then its collapse.

Aguilar gravity dam, 48 m high (Spain), built in 1963 had a similar problem. A zone of tectonic disturbance filled with clay material was discovered in the left-bank abutment dam composed of limestones and marl. During the first filling of the reservoir, erosion of clay aggregate was recorded, and the discharge of flow filtration was 50 L/s. In 1965, cementation work was performed to eliminate filtration. Filtration was eliminated, and injection wells were used as drainage wells [79].

Often, polymeric materials (epoxy, vinyl, and other resins) or hot bitumen (asphalt) is used to eliminate concentrated filtration, which allows plugging a fissure or cavity, after which it is cemented with a solution.

Of the 30 dams built in the USA in the Tennessee River Valley, 21 dams are located on karst limestones. To close the concentrated filtration ways at the foundation, hot bitumen was used; in most cases, this had to be done after completion of the dam construction and filling of the reservoir. So, for example, on the **Tims Ford dam** built in 1970, during the first filling of the reservoir, six concentrated filtration ways appeared on the left bank (30 m^3/min) and two on the right bank (5 m^3/min). All of them were covered with hot bitumen and cement.

At the **Great Falls dam**, 4,432 m^3 of bitumen and 7,064 m^3 of concrete were pumped into the foundation to block 96 concentrated filtration ways [65].

Stewartville gravity dam, 63 m high, was built in 1948 in Canada on massive crystalline rocks with the inclusion of zones of disturbed limestone; during the operation, unsuccessful attempts were repeatedly made to reduce the filtration in the foundation discharge, which had consumed 4.5 m^3/min by 1974 [230]. In 1982, attempts were again made to inject cement mortar but also to no avail. An examination of the wells with a television camera revealed two concentrated filtration zones: one cavity at the concrete-rock contact 5 m wide and 100 mm opening and the second in the 3 m wide bed with 200 mm opening. The high flow rate in the cavities did not allow their cementation.

To stop the filtration, hot ($t = 110°$) bitumen was pumped through one of the wells into the cavity under a pressure of 0.3 MPa. After a few minutes, a cement mortar

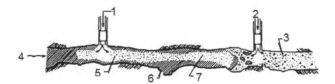

Figure 4.23 Scheme of sequential cavity closing by hot bitumen and cement: 1 – well for hot injection bitumen, 2 – well for cement grout, 3 – cement mortar, 4 – stream of water, 5 – cold bitumen, 6 – hot bitumen in cavities and fissures, and 7 – hot bitumen substitute cooled concrete.

(Figure 4.23) was pumped into the cavity blocked by bitumen through another well to finally close the cavity and fill in small fissures.

The injection of cement was necessary since bitumen is subject to creep under high hydrostatic pressure of water. Within one day both cavities were closed using calling $11 \, m^3$ of bitumen. Then the cavities were cemented, while only 15% of the amount of cement that was injected to no avail during the previous two months was used. This technology has significantly reduced filtering (Figure 4.24).

Sayano-Shushensk arch-gravity dam of the HEP is built on strong ($R_c = 119$–$135 \, MPa$) deeply metamorphosed crystalline schists with a deformation modulus of 10,000 to $28,000 \, MPa$. The rigidity of the foundation, the large width of the site, the erection and loading of the dam with an incomplete profile, and other factors led to the appearance of zones of concrete tension on the upstream face of the dam and to the decompaction of the foundation under the upstream face.

After the reservoir began to fill when the head on the dam was 60% of the design value, the first signs of opening a contact joint between foundation and number of sections of the dam appeared. When the head on the dam reached 82% of the design value in 1985, the concrete-rock contact opened under the entire bed part of the dam to a depth of more than $27 \, m$ towards the downstream. Tensile strains in concrete led to cracks (opening of horizontal joints) at marks 344–359 m and concentrated filtration up to 460 L/s through the upstream face concrete.

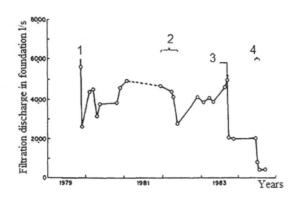

Figure 4.24 Decreased filtration at the foundation dam Stewartville due to the injection of bitumen: 1 – apron in the upstream, 2 – apron in the upstream and grouting, 3 – injection with bitumen, and 4 – injection with bitumen.

The decompaction of the foundation occurred to a depth exceeding 60 m. Filtration through the dam concrete and through the foundation reached 520 and 549 L/s, respectively, against 100–150 L/s according to the design.

The absence of bottom outlets did not allow emptying the reservoir, and therefore, the restoration of water resistance was carried out at a head of more than 200 m. The high level of filtration did not allow the use of traditional solutions for injection. Experimental work on repairing the dam body using epoxy mortars of the French company "Soletanche" was started in 1995. After completion of the filling of cracks in the body of the concrete dam in 1998, work was continued on filling the cracks at the contact of the dam with the foundation.

As a result of the work performed, the filtration discharge through the dam body and rock foundation was reduced by 99.5% and 78%, respectively.

4.4.3 Constructions of dam connection with foundation

As noted in Section 4.1, when a cement curtain is installed in the foundation under the upstream face of the dam, the deformation modulus of the foundation increases, due to which the decompaction zone increases, the contact zone SSS deteriorates, and the concentration of compressive stresses under the downstream face increases.

An analysis of design studies and natural observations of the SSS of the contact area with the traditional solution of the underground circuit prompted a change in attitude towards the cementation curtain and drainage. At the XII International Congress on Large Dams [127], attention was paid to the strengthening of the role of drainage as the most effective and reliable solution to reduce uplift.

At the same time, the cementation curtain plays the role of leveling the filtration properties of the foundation, filling the caverns and fissures in the foundation, and reducing the filtration pressure and discharges. In favorable engineering-geological conditions of strong, low permeability, and nonsuffusion foundations (at $K_f < 0.1$ m/day), it is advisable to consider the possibility of excluding it from the underground circuit.

Bembesar gravity dam, 102 m high, was erected on the Paleozoic shales (1969) with drainage without cementing curtains; the **arch dam Matalavilla** with a height of 113 m was erected on columnar quartz (1967) in Spain also with drainage without a cement curtain [65,187].

Poor filtration unloading of the dam foundation due to the water tightness of the foundation outside the decompaction zone increases the filtration forces, while at the same time, the effective operation of the drainage avoids in many cases the consequences of opening fissures in the contact zone. The filtration regime of the base of the **Ust-Ilim dam** is characterized by very low uplift due to the influence of nonproject drainage, which was spontaneously formed by wells for the CME installation.

An almost rectangular uplift plot was observed in the decompacted zone of the foundation of the station sections of the **Bratsk dam**, which caused additional decompaction of the foundation. In the sections where drainage wells were drilled, the uplift (Figure 4.25) and the opening of the contact joint sharply decreased [68,96].

In the global practice of designing concrete dams, there are two trends:

- prevention of tensile stresses in the contact zone;
- assumption of a limited decompaction zone with a special construction of dam connection with the rock foundation.

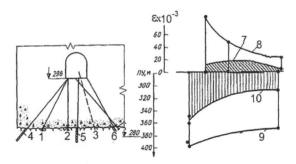

Figure 4.25 The effect of drainage on the opening of the contact joint in section No. 36 of the Bratsk dam: (a) vertical section on gallery near upstream face, (b) plots of uplift and opening of contact joint: 1–3 – strain gauges; 4, 5, and 6 – piezometers; 7 – opening with an uplift plot 10 (drainage open); and 8 – the same with plot 9 (drainage is closed).

For gravity dams, the influence of filtration pressure on the SSS of the contact zone and the stability of the dam is great. If we imagine that it was possible to completely remove the uplift on the dam foundation, then the coefficient safety on slide will increase by 25%–30%, compression zone will form at the contact of the upstream fact of the dam with the foundation, and the maximum compressive stresses in the foundation under the downstream face will significantly decrease.

For high gravity dams, an improvement of the contact zone SSS and reduction or elimination of tensile deformations at the contact of the upstream face with the foundation can be achieved by increasing the volume of concrete and reducing the inclination of the downstream face ($m=0.8–0.9$) or the inclination of the upstream face to the side upstream. Such decisions were made for the high dams made of RCC dam built in the last decade: Longtan, $h=216.5$m (see Figure 2.21), Guangzhao, $h=196$m, and Dzhenge, $h=131$m, in China; Miyegase (see Figure 1.24) and Urayama, $h=155$m, Hasan, $h=123$m, and Satsunaigawa, $h=114$m, in Japan; Miel-I, $h=188$m (see Figure 2.20), and Porce-2, $h=123$m, in Colombia (see Figure 1.25); and Ralko, $h=155$m, and Pangue, $h=113$m, in Chile.

The same effect can be achieved in dams of a flattened profile due to a significant increase in concrete volume and the use of a water load acting on the upstream face of the dam.

However, as calculated studies of the SSS and field observations show, it is practically impossible to avoid tensile stresses in the foundation under the upstream face of the dam. With increasing dam height, tensile stresses increase. At the same time, such constructive measures as perimeter joints, joint notches do not provide a significant improvement of the foundation of SSS under the upstream face.

According to the observation results [33], a decompaction zone with tensile stresses up to +3.4 MPa, fissure openings, and increased filtration during normal operation of the cementation and drainage curtains was formed at the foundation under the upstream face of the arch dam Inguri. Serious violations took place at the foundation under the upstream face of the arch dam Kelnbrein and the arch gravity dam Sayano-Shushensk.

Elimination of tensile stresses at the contact of gravity dams with the foundation and reduction of fissuring formation under the upstream face and in the zone of the cement curtain with a decrease in filtration pressure in the dam foundation can be achieved by installing a concrete apron with a main antifiltration curtain in front of the dam. In addition, the apron device will eliminate the hydrostatic pressure of the water reservoir on the upstream of the dam in the apron area, thereby improving the working conditions of the dam and the contact zone.

Concrete aprons of various designs cut off by deformation joint from the dam and removing part of the hydrostatic pressure were used on a number of dams including the arch dams Zillergrundl, 186 m high, and Kelnbrein (see Figure 8.34), 200 m high, in Austria, the arch-gravity dam Bor, 120 m high (Figure 4.26c), the arch-buttress dam Roseland, 150 m high, in France (see Figure 1.38), the gravity dams Krokstremen in Sweden (Figure 4.26a) and Rauschenbach in Germany (Figure 4.26b), and the gravity RCC dam Mujib in Jordan (see Figure 7.18) [79].

A variant of a lightweight gravity dam 124 m high with a screen under seismicity (9 points) is shown in Figure 4.27 [27].

Structural solutions in which the concrete apron and the dam separated by a deformation seam are not interconnected and perceive force effects independently can affect the reliability of their joint work as a single structure; therefore, they require careful study. The inclusion in the dam design of an additional critical element – the deformation joint with seals, the operation of which determines the stability and strength of the structure as a whole – requires special calculation justification. The reliability of the dam should be analyzed taking into account the possibility of an emergency violation of the seals of the deformation joint.

To improve the SSS of the contact zone, a solution is used with a device at the foundation under the upstream face of the dam of a vertical joint-slot, the purpose of which is to separate the compressed part of the rock mass under the dam in the downstream from the extended one in the upstream [101].

The dam was mated with the rock foundation using a vertical joint slot during repair work on the arch dam Schlegeis, 131 m high, in Austria [101,233] erected in a relatively wide site with L/h greater than 4 (see Section 8.3.2 and Figure 8.49) [233].

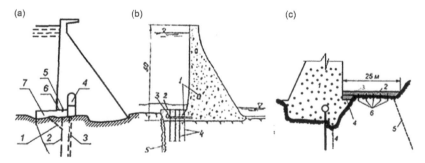

Figure 4.26 Dams with concrete apron separated by deformation joint from the dam: (a) Krokstremen gravity dam: 1 – fissure, 2 – anchor, 3 – drainage curtain, 4 – gallery, 5 – bitumen key, 6 – plate from stainless steel, and 7 – concrete apron; (b) Rauschenbach gravity dam: 1– gallery, 2 – concrete apron, 3 and 4 – stressed cords, 5 – cementation curtain; and (c) Bor arch gravity dam: 1 – dam, 2 – concrete apron, 3 – rib key, 4 – key, 5 – cement curtain, and 6 – anchor.

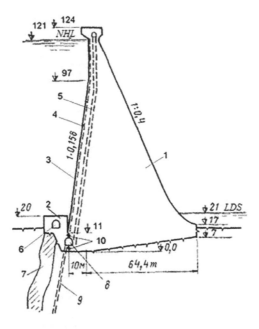

Figure 4.27 Lightweight dam with screen and apron: 1 – dam; 2 – apron; 3 – screen; 4 and 5 – the first and second rows of the drainage in dam; 6 – cement gallery; 7 – cement curtain; 8 – drainage gallery; 9 – drainage curtain; and 10 – bitumen keys.

Kapanda gravity dam, 110 m high, was erected in Angola according to the project of JSC "Institute Hydroproject". A canyon with a height of 60 m with a ratio of width to height of $L/h = 2.5$ is composed of slightly fissured strong sandstones weakened by fissures' side rebuff, bedding, and interbeds of mudstones (Figure 4.28). The dam within the canyon part is designed continuous so as to use the spatial effect.

As a result of solving the elastic spatial problem of the dam-foundation system, it was obtained that the contact stresses under the upstream face in the steep section are tensile and exceed the tensile contact strength (Figure 4.29a). The main tensile strains in the foundation reaching +1.95 MPa significantly exceeded the tensile strength of sandstones. At zero strength, the tensile contact of the upstream face with the foundation opened almost the entire length, and the opening ranged from 0.06 mm in the bed part to 0.36 mm in a steep bank abutment. A small opening of the contact is associated with a large rigidity of the "dam-foundation" system and therefore did not lead to any noticeable increase in compressive stresses in the dam at the downstream face.

Investigations of the filtration regime showed that opening of the contact led to a slight increase in uplift on the dam bottom, while filtration discharges doubled compared to continuous contact, under conditions of low permeability sandstones its small and did not exceed 0.2 L/s per 1 run. m.

Opening of contact led to a decrease in the main tensile stresses in the base under the upstream face up to +1.2 MPa but exceeded the tensile strength of sandstone. When the subvertical joint slot was installed under the upstream face to a depth of 6–8 m, not only tensile stresses on the contact (Figure 4.29b) but also the main tensile stresses in the

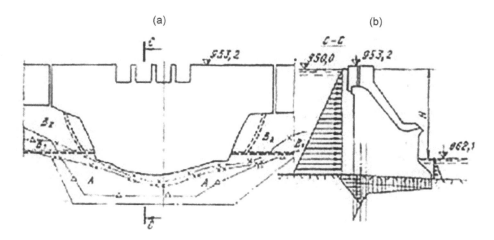

Figure 4.28 Kapanda Gravity dam (Angola): (a) view from the upstream and (b) section along C–C (design scheme).

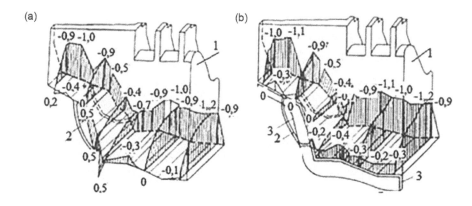

Figure 4.29 Kapanda gravity dam (project). Contact stresses (MPa): (a) with continuous foundation and (b) with a subvertical joint slot; I – dam, 2 – tectonic fissure, and 3 – joint slot, 8 m deep.

foundation under the upstream face of the dam disappeared, which were displaced under the mouth of the joint slot where the danger of uncontrolled decompaction is less relevant.

The design of the joint slot should not have caused complications during the construction work: it could have been formed by explosions in wells with specially selected charges or by continuous drilling as was done on the arch dam Schlegeis in Austria (see Section 8.3.2)

A training ground was set up at the Kapanda HEP construction site to test the technology for creating a joint slot in a rock foundation, but the contractor, construction company Odebreht, did not make this decision "as violating the rock foundation".

A joint slot in the form of a trench, which is filled with a material that significantly differs in its dynamic properties from the rock foundation, can also perform the functions of seismic isolation [91,118].

4.4.4 Proposals for dam construction of connection with rock foundation

To increase the reliability of the joint work of concrete apron with a dam, structural solutions [89,223,233] with flexible connection of apron with a dam have been proposed (Figure 4.30).

In the connection variant in Figure 4.30a, the apron in the form of a concrete mass is combined into a single structure with a dam by an arch ceiling. Under the ceiling, a drainage cavity is formed, which, if necessary, can be connected to the downstream. To obtain additional load on the ceiling, earth load may be laid. An arch ceiling acting as a spacer structure transfers horizontal and vertical forces from its own weight and water load to the apron and dam. The stability of the concrete apron, which accepts horizontal hydrostatic pressure and significant filtration pressure, is ensured by water loading, its own weight, and the holding forces transmitted by the arch ceiling.

Without significantly increasing the volume of concrete apron, it is possible to obtain a fairly uniform diagram of compressive stresses along its bottom, which allows for a favorable SSS of the contact zone. The stability of the dam despite the reduction of its own volume is achieved by reducing the hydrostatic pressure as well as by completely removing the filtration pressure on the bottom of the dam. A concrete apron can be made in the form of a concrete slab anchored to the foundation if necessary, and cementation and drainage curtains are made from the cavity formed by the arch ceiling.

In the dam design (Figure 4.30b and c), a gallery is made in the dam body, which by a lower inclined deformation joint separates the apron from the dam body. In this embodiment, the possibility of mutual movements of the apron and the dam is provided. By increasing the width of the apron, it is possible to increase the effect of the weight of the water load and increase the holding moment relative to the center of gravity of the dam bottom. In this case, a fairly uniform distribution of normal compressive stresses along the bottom of the apron and dam can be obtained.

To ensure a favorable stress state in the arch ceiling, the central angle of the arch should be about 150° with a ratio of thickness to axial radius of 0.2–0.3. To increase water resistance, the arch ceiling can be covered with a geomembrane made of PVC film. The height of the concrete apron can be 0.07–0.15 of the dam height and the width of 0.2–0.3 of the width of the bottom of the dam.

Figure 4.30c shows the dam design with the screen device on upstream face.

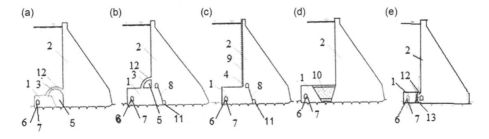

Figure 4.30 Structural solutions with flexible connection of apron with a dam: 1 – concrete apron, 2 – dam, 3 – arch ceiling, 4 – flat ceiling, 5 – drain cavity, 6 and 7 – cementation and drainage curtains, 8 – deformation joint, 9 – screen, 10 – soil apron, 11 – drainage, 12 – geomembrane, and 13 – asphalt concrete.

To neutralize the mutual movements of the apron and dam, a design is proposed (Figure 4.30d) with a device between the apron and the dam cavity in the form of a trapezoid with a smaller lower base filled with poorly permeable soil (clay and loam) with drainage at its base. Under the influence of hydrostatic pressure on the soil apron, which has the shape of a wedge spacer, compressive forces arise in it ensuring the joint work of all elements. Take into account the permissible pressure gradient $I=8$–10, the thickness of the soil apron can be 0.1–0.12 of the height of the dam. The laying of the faces of the apron and dam within the soil apron can be 0.2–0.5 and the total width of the concrete and soil apron can average 0.3–0.4 of the width of the bottom of the dam. In the junction of the apron to the upstream face, it is advisable to lay more plastic material. The drainage layer at the base of the soil part of the apron together with the cementation and drainage curtains of the concrete apron provide an almost complete perception of the filtration pressure by the apron and an almost head free regime at the foundation of the concrete dam.

The calculation results for a dam with a height of 80 m under the action of the main operating loads and a concrete and rock deformation modulus ratio of 1 showed (Figure 4.31) the absence of a decompaction zone at the dam foundation and apron, a fairly uniform diagram of compressive stresses at the apron base, and a decrease in compressive stresses from the downstream face of the dam. In sections of the arch ceiling from the side of the upstream face of the dam, only compressive stresses were obtained.

Computational studies [91] of the SSS of apron from loam (deformation modulus $E=50$ MPa, Poisson's ratio $v=0.35$, $\varphi=18°$, and $C=2$ t/m²) taking into account plastic deformations showed that under the influence of hydrostatic compression and plastic deformations, compaction and hardening of the apron earth occurred; due to

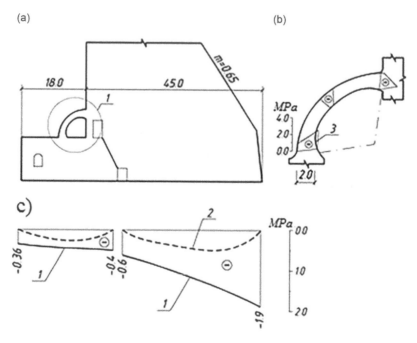

Figure 4.31 Stresses (c) at the contact of the apron and dam (a) with the foundation and at the arc ceiling (b): 1 and 2 – plots of normal σ_y and tangent τ stresses on the contact, 3 – plots of normal stresses in an arch ceiling; "+" – compression.

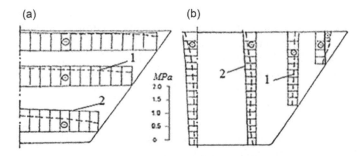

Figure 4.32 Plots of stresses (MPa): (a) vertical σ_y and (b) horizontal σ_x in the earth apron of the dam: 1 – calculation according to the theory of elasticity, 2 – calculation according to the theory of an elastoplastic body; "+" - compression.

compressive stresses in the earth, its reliable operation is achieved in conjunction with concrete apron and dam (Figure 4.32).

The flexible part of the apron can be performed by filling the trapezoidal cavity between the concrete apron and the dam with asphalt concrete (Figure 4.30e) characterized by high water resistance, durability, workability, and ductility. Asphalt concrete can be rolled or cast. At the same time, the height of the trapezoidal cavity will be significantly less given that the permissible head gradients for asphalt concrete are many times higher than for earth materials (loam and clay). To improve the connection of asphalt concrete with concrete surfaces, it can be coated with asphalt mastic before laying asphalt concrete. Under the influence of water pressure on the expanded surface of the asphalt concrete, it is compressed, which ensures the water tightness of the contact of the asphalt concrete with the concrete surfaces of the dam and apron.

The uniform distribution of normal compressive stresses at the contacts of the asphalt concrete with the apron and dam and the reduction of uplift allow optimizing the underground circuit, improving the high dam SSS, and improving the reliability of both individual elements (cement curtain, apron) and the overall bearing capacity of the "apron-dam-foundation" system compared to traditional construction.

In such dams, more favorable conditions for the mutual adaptation of the structure and the rock foundation are provided, especially in difficult engineering-geological conditions with a heterogeneous foundation, with significant deformations of the reservoir bed, which is typical for large reservoirs with high dams.

The dam with apron is less sensitive to load changes caused by significant seasonal fluctuations in the reservoir level during operation.

To assess the effectiveness of such a dam design, the stress state and stability of the dam with an apron and classical profile at a height of 100–200 m and various characteristics of the rock foundation were calculated. In addition to this, static seismic loads of various intensities were taken into account.

The analysis of the calculation results shows a significant improvement in the working conditions of the contact zone of the dam with the apron:

- at the foundation of the concrete apron and dam, only compressive normal stresses act, which make up (0.2–0.3) σy from the upstream face of apron and (0.1–0.4) σy

from the upstream face of dam, where σy is the maximum compressive stress at the foundation;

• seismic resistance of the dam increases, when seismic effects of the SSS of the apron contact zone change slightly;

• on the whole, a less thick profile and a decrease of the stiffness of the dam working together with the apron and the section between them of viscous and plastic material allow improving its dynamic characteristics, while reducing the total volume of concrete to 10%–15%.

In the variant (Figure 4.30e), a seam-notch on the upstream face of the dam can be made on the section of the cavity with asphalt concrete (Figure 4.33).

In this design, under the influence of hydrostatic pressure, the asphalt concrete is compressed in the vertical and horizontal directions with the provision of waterproof contact between the asphalt concrete and the upstream face of the dam over the entire height. In the case of the opening of the contact joint between the dam and the foundation, the bitumen under pressure is extruded from the asphalt concrete and the open section is sealed. When installing a seam-notch, working conditions improve. The effectiveness of using such solutions in each case should be justified by special design studies, including filtration and SSS system "apron-dam-foundation".

The possibility of applying new design solutions should be justified during the design by carrying out appropriate calculations and analysis of the features of the construction and reliability in specific environmental conditions of the dam site.

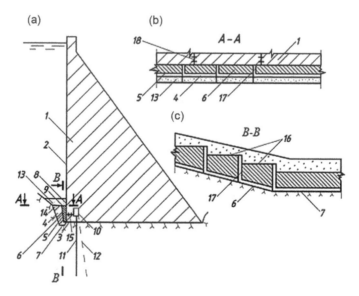

Figure 4.33 A dam with a cavity near upstream face expanding from the bottom to the top and filled with asphalt concrete: (a) cross section; (b) the plan for AA; (c) section along BB: 1 – dam, 2 – upstream face, 3 – bottom of the dam, 4 – concrete facing of the slope with face 5, 6 – expanded trapezoidal joint filled with asphalt concrete, 7 – foundation, 8 – lower part of upstream face of the dam, 9 – top of extended joint, 10 – gallery, 11 and 12 – cementation and drainage curtains, 13 – protective layer of earth, 14 – top of the facing in the form of concrete wall, 15 – joint notch with seals, 16 – stage extended joint on inclined part, 17 – transverse walls, and 18 – key.

Chapter 5

Concrete and reinforcement

5.1 Concrete

Building materials for concrete and reinforced concrete dams must satisfy the requirements of SR [15] and the requirements of this chapter.

In dams and their elements, depending on the working conditions of the concrete in the individual parts, four zones must be distinguished, the concrete of which must be presented with the requirements (for details, see Section 7.3.2) for the following:

- water resistance;
- frost resistance;
- strength.

Waterproof is the highest water pressure, at which it does not leak through test samples of 180 days of age.

The concrete mark for waterproof is assigned depending on the head gradient and the temperature of the water in contact with the structure, °C, according to Table 5.1 taking into account aggressive water medium according to SR [15]. Head gradient is the ratio of the maximum head, m, to the thickness of the structure or the distance from the upstream face to the drainage, m.

The age (hardening time) of concrete, corresponding to its design class in terms of compression and tension strength and waterproof mark, should be assigned taking into account the construction and filling of the reservoir according to SR [15].

As a rule, the age of monolithic concrete of dams corresponding to its class in terms of strength and water resistance should be taken equal to 180 days and age in frost resistance – 28 days.

For concrete dams with a height of more than 60 m and a concrete volume of more than 500 thousand m³, the indicated age in strength and water resistance should be taken, as a rule, equal to one year.

Concrete waterproof marks (W2; W4; W6; W8; W10; W12; and W14) should be assigned depending on the pressure gradients in accordance with the requirements of SR [15] according to Table 5.1.

When protecting the upstream face with waterproofing (screen), the waterproof of concrete may be taken a mark lower than that of a mark with an unprotected upstream face.

For parts and elements of dams periodically washed by water, when a stream of water with entrained sediments acts on concrete, and also when the requirements for

Table 5.1 Waterproof concrete marks

Temperature water, °C	Concrete marks for water resistance at head gradients			
	Up to 5 inclusive Over 5	Up to 10 Over 10	Up to 20 Over 20	Up to 30
Up to 10°C inclusive	W2	W4	W6	W8
Over 10°C up to 30°C	W4	W6	W8	W10
Over 30°C	W6	W8	W10	W12

Note: For structures with a head gradient above 30, a concrete mark of waterproofing W14 and higher should be assigned.

cavitation resistance are presented to concrete, the concrete mark for waterproof is accepted no less than W10, for frost resistance – no less than F200, and for compressive strength – class no less than B25.

In aggressive water media, the concrete waterproof mark should be taken one step higher than that required by SR [15].

In structures located under water in the northern building and climatic zone, the concrete marks for waterproof should not be lower: for concrete structures W4 and for reinforced concrete structures W6. At the same time, the hardening time (age) of concrete corresponding to its design mark for waterproof should be taken equal: for reinforced concrete structures – 28 days and for massive structures erected in warm formwork – 60 days.

Frost resistance of concrete is characterized by the largest number of alternating freezing and thawing cycles maintained by 28-year-old samples when tested without reducing concrete strength by more than 15%.

Concrete marks for frost resistance (F50; F75; F100; F150; F200; F300; F400; F500; F600; F700; F800; and F1000) should be assigned depending on the climatic conditions of the dam construction area and the estimated number of alternate freezing and thawing cycles in a year in accordance with the requirements of SR [15] according to Table 5.2.

Class of concrete in compressive strength should be taken according to Table 5.3 corresponding to the value of guaranteed concrete strength with a security of 0.95. In massive structures, it is allowed to use concrete B27.5; B30; B35; and B40 with guaranteed strength values with a security of 0.90.

For the inner zone of concrete gravity dams, it is allowed to use concrete with guaranteed strength values with a security of 0.85. The projects envisage the following classes of concrete in compressive strength: B5; B7.5; B10; B12.5; B15; B17.5; B20; B22.5; and B25. The class of concrete in axial tension strength should be taken according to Table 5.4 depending on the values of the design resistance of concrete $B_t0.8$; $B_t1.2$; $B_t1.6$; B_t2; $B_t2.4$; $B_t2.8$; and $B_t3.2$. This characteristic is established in cases where it determines the strength of structures and is controlled in production.

The number and zonal placement of different classes of concrete in the structure should be taken so that at each stage of the dam construction, no more than four classes of concrete are simultaneously laid; an increase in their number is allowed only with proper justification.

Table 5.2 Appointment of marks on frost resistance

Climatic conditions	For a zone of variable water level and spillway face with the number of cycles of alternate freezing and thawing						
	Up to 25	26–50	51–100	101–150	151–200	201–250	51–300
Moderate	F50	F100	F150	F200	F300	F400	F600
Severe	F100	F150	F200	F300	F400	F600	F800
Especially severe	F200	F300	F400	F500	F600	F800	F1000

Notes:
1. Climatic conditions are characterized by the average monthly temperature of the coldest month:
   moderate – above minus 10°C;
   severe – from −10°C to −20°C inclusive;
   especially severe – below −20°C.
2. The average monthly temperatures of the coldest month for the construction area are determined according to SR 131.13330 as well as according to the hydrometeorological service.
3. With the simultaneous effect of freezing and thawing in aggressive media, concrete of higher marks/frost resistance is used: when exposed to a weakly and moderately aggressive media – by one step and when exposed to a highly aggressive media – by two steps.

Table 5.3 Standard and design resistance of concrete on compression

Concrete class for compressive strength	Standard and design concrete resistance, MPa (kg/cm^2)					
	Standard: design resistance for the limiting state of the second group			Design resistance for the limiting state of the first group		
	Axial compression $R_{bn}, R_{b,ser}$	Axial tension $R_{btn}, R_{bt,ser}$		Axial compression R_b	Axial tension R_{bt}	
		vibrated	rolled		vibrated	rolled
1	2	3	4	5	6	7
B5	3.5 (35.7)	0.55 (5.61)	0.39 (3.98)	2.8 (28.6)	0.37 (3.77)	0.26 (2.65)
B7.5	5.5 (56.1)	0.70 (7.14)	0.58 (5.92)	4.5 (45.9)	0.48 (4.89)	0.39 (3.98)
B10	7.5 (76.5)	0.85 (8.67)	0.78 (7.96)	6 (61.2)	0.57 (5.81)	0.52 (5.35)
B12.5	9.5 (96.5)	1 (10.2)	0.95 (9.70)	7.5 (76.5)	0.66 (6.73)	0.63 (6.42)
B15	11.3 (115)	1.15 (11.7)	1.10 (11.2)	8.9 (91)	0.75 (7.65)	0.73 (7.45)
B17.5	13 (133)	1.27 (13)	1.23 (12.6)	10.3 (105)	0.83 (8.41)	0.80 (8.20)
B20	14.9 (152)	1.40 (14.3)	1.38 (14.1)	11.7 (120)	0.90 (9.18)	0.90 (9.15)
B22.5	16.7 (170)	1.50 (15.3)	-	13.1 (134)	0.97 (10)	-
B25	18.5 (189)	1.60 (16.3)	-	14.5 (148)	1.05 (10.7)	-
B27.5	20.2 (206)	1.70 (17.3)	-	15.8 (161)	1.12 (11.4)	-
B30	22 (224)	1.80 (18.4)	-	17 (173)	1.20 (12.2)	-
B35	25.5 (260)	1.95 (19.9)	-	19.5 (199)	1.30 (13.3)	-
B40	29 (296)	2.10 (21.4)	-	22 (224)	1.40 (14.3)	-

For concrete dams with a concrete volume of more than 1 million m^3 along with the concrete classes specified in SP [15] for compression strength, intermediate class values should be accepted. The characteristics of these concretes (design and standard resistance, elastic modulus, etc.) should be taken by interpolation.

Table 5.4 Standard and design resistance of concrete axial tension

Concrete tension strength class axial tensile	Standard and design resistance of concrete at axial tension MPa (kg/cm²)	
	Standard: design resistance for the limiting state of the second group $R_{btn,bt}$, $R_{bt,ser}$	Design resistance for the limiting state of the first group R_{bt}
$B_t0.8$	0.8 (8.1)	0.62 (6.32)
$B_t1.2$	1.2 (12.2)	0.93 (9.49)
$B_t1.6$	1.6 (16.3)	1.25 (12.7)
B_t2	2 (20.4)	1.55 (15.8)
$B_t2.4$	2.4 (24.5)	1.85 (18.9)
$B_t2.8$	2.8 (28.6)	2.15 (21.9)
$B_t3.2$	3.2 (32.4)	2.45 (25)

Strength classes of concrete are determined by the stress state of the material of structures in specific sections, and the composition of concrete must meet the requirements for frost resistance, strength, and waterproofness by the time of dismantling of the concrete mass, if such a requirement is indicated in the design.

The design resistances of concrete dams at the age of 180 days (or 1 year) should be determined based on the design resistances of concrete established during design, required by the time the building is loaded with operational loads, taking into account the actual age that the concrete will have by the specified time, and the conditions for erecting the dam, according to the following formulas:

on compression

$$R_b = T_{b\tau} / (\gamma_{\tau c} \gamma_\eta) \qquad (5.1)$$

$$R_{b,ser} = R_{b\tau,ser} / (\gamma_{\tau c} \gamma_\eta) \qquad (5.2)$$

on tension

$$R_{bt} = R_{bt\tau} / (\gamma_{\tau t} \gamma_\eta) \qquad (5.3)$$

$$R_{bt,ser} = R_{bt\tau,ser} / (\gamma_{\tau t} \gamma_\eta) \qquad (5.4)$$

where

R_b, R_{bt} and $R_{b,ser}, R_{bt,ser}$ are the design concrete compressive and tension strengths of concrete, respectively, for the limiting states of the first and second groups at the age of 180 days (or 1 year);

$R_{b\tau}, R_{bt\tau}$ and $R_{b\tau,ser}, R_{bt\tau,ser}$ are the concrete compressive and tension strengths, respectively, for the limiting states of the first and second groups, required by the dam calculations for strength by the time the structures are loaded with operational loads;

$\gamma_{\tau c}$ and $\gamma_{\tau t}$ are the coefficients that take into account the effect of concrete age on its compressive and tensile strength, respectively, determined according to Table 5.5;

γ_η is the coefficient taking into account the difference in concrete strength of control samples and structures and equal to

Table 5.5 Coefficients $\gamma_{\tau c}$ and $\gamma_{\tau t}$ influences of concrete age on its strength

Concrete age by the time the structure is loaded, year	Coefficients		
	$\gamma_{\tau c}$ at compression for area		$\gamma_{\tau t}$ at tension
	With an average annual outdoor temperature of 0°C and higher	With a negative average outdoor temperature	
0.5	1/0.9	1/0.9	1/0.9
1	1.1/1	1.05/1	1.05/1
2	1.15/1.1	1.10/1.05	1.10/1.05
3 and above	1.20/1.15	1.15/1.10	1.15/1.10

Notes:
1. The numerator shows the values of the coefficients $\gamma_{\tau c}$ and $\gamma_{\tau t}$ for concrete age 180 days and the denominator – for concrete age 360 days.
2. For sectional cutting, the coefficient $\gamma_{\tau c}$ should be taken as for areas with an average annual temperature of detectable air of 0°C and higher.
3. For class I dams, the coefficients $\gamma_{\tau c}$ and $\gamma_{\tau t}$ are recommended to be specified through experimental studies of concrete of the accepted compositions.

1 – for mechanized manufacture, transportation, and supply with the distribution and compaction of concrete mix by hand vibrators;

1.1 – with automated preparation of concrete mix and fully mechanized transportation, laying, and compaction of concrete mix.

The design concrete resistances for the limiting states of the second group $R_{b,\mathrm{ser}}$ and $R_{bt,\mathrm{ser}}$ are introduced into the calculation with the coefficient of concrete working conditions $\gamma_{bi} = 1$.

The density of heavy concrete ρ in the absence of experimental data can be taken according to Table 5.6.

Initial modulus of elasticity of concrete E_b of massive structures under compression and tension should be taken according to Table 5.7.

Ultimate elongation (ultimate relative deformation) is adopted according to Table 5.8.

Shear modulus of concrete is taken equal to $G = 0.4\,E_b$.

Initial transverse strain coefficient (Poisson's ratio) is taken to be for massive structures – 0.15 and for bar and plate structures – 0.20.

Table 5.6 Heavy concrete density ρ

Aggregate density (g/cm³)	Average concrete density ρ (g/cm³), with maximum aggregate size (mm)				
	10	20	40	80	120
2.60 ÷ 2.65	2.26	2.32	2.37	2.41	2.43
2.65 ÷ 2.70	2.30	2.36	2.40	2.45	2.47
2.70 ÷ 2.75	2.33	2.39	2.44	2.49	2.50

Table 5.7 Initial modules of elasticity under compression and tension E_b

Method of compaction	Concrete cone settlement (cm)	Maximum size of coarse aggregate (mm)	Initial modules of elasticity under compression and tension E_b 10^{-3}, MPa (kg/sm²) at concrete class in compressive strength												
			B5	B7.5	B10	B12.5	B15	B17.5	B20	B22.5	B25	B27.5	B30	B32.5	B35
1	2	3	4	5	6	7	8	9	10	11	12	13	14	15	16
Vibrated	Up to 4	40	23 (235)	28 (285)	31 (315)	33.5 (340)	35.5 (360)	37 (380)	38.5 (395)	39.5 (405)	41 (420)	42 (430)	43 (440)	44.5 (455)	46 (470)
		80	26 (265)	30 (305)	34 (345)	36.5 (375)	38.5 (395)	40 (410)	41.5 (425)	42.5 (435)	43.5 (445)	44.5 (455)	45 (460)	46.5 (475)	47.5 (485)
		120	28.5 (290)	33 (340)	36.5 (365)	38.5 (390)	40.5 (415)	42 (430)	43.5 (445)	44.5 (455)	45.5 (465)	46.5 (475)	47 (480)	48.5 (495)	49.5 (505)
	4-8	40	19.5 (200)	24 (245)	27 (275)	29.5 (300)	31.5 (320)	33 (335)	34.5 (350)	36 (365)	37 (380)	38 (385)	39.5 (405)	41 (420)	42.5 (435)
		80	22.5 (230)	28 (285)	30 (305)	32.5 (330)	34.5 (350)	36 (370)	37.5 (380)	39 (400)	40 (410)	41 (420)	42 (430)	44 (450)	45.5 (465)
		120	24.5 (250)	29 (295)	32.5 (330)	35 (355)	37 (380)	38.5 (395)	40 (410)	41 (420)	42 (430)	43 (440)	44 (450)	45.5 (465)	46.5 (475)
	8-16	40	13 (135)	16 (165)	18 (185)	21 (215)	23 (235)	25.5 (260)	27 (275)	28.5 (290)	30 (305)	31.5 (320)	32.5 (330)	34.5 (350)	36 (365)
		80	15.5 (160)	19 (195)	22 (225)	24.5 (250)	26.5 (270)	28.5 (290)	30 (305)	31.5 (320)	33 (335)	34 (345)	35 (360)	36.5 (370)	37.5 (385)
		120	17.5 (180)	21.5 (220)	24.5 (250)	27 (270)	29.5 (295)	31 (315)	32.5 (330)	34 (345)	35 (350)	36 (365)	37 (375)	38 (390)	39 (400)
	Over 16	40	–	13 (135)	16 (165)	18 (185)	21 (215)	23 (235)	25.5 (260)	27 (275)	28.5 (290)	30 (305)	31.5 (320)	32.5 (330)	34.5 (350)
		80	–	15.5 (160)	19 (195)	22 (225)	24.5 (250)	26.5 (270)	28.5 (290)	30 (305)	31.5 (320)	33 (335)	34 (345)	35 (360)	36.5 (370)

Method of compaction	Concrete cone settlement (cm)	Maximum size of coarse aggregate (mm)	Initial modules of elasticity under compression and tension E_b 10^{-3}, MPa (kg/sm²) at concrete class in compressive strength											
			B5	B7.5	B10	B12.5	B15	B17.5	B20	B22.5	B25	B27.5	B30	B32.5
1	2	3	4	5	6	7	8	9	10	11	12	13	14	15
Rolled	-	Along concreting layers												
		40	20.5 (210)	25 (255)	28 (285)	30 (310)	32 (325)	33 (340)	35 (355)	36 (365)	37 (375)	38 (385)	39 (400)	40.5 (415)
		80	23 (235)	27 (275)	30.5 (310)	33 (335)	35 (350)	36.5 (375)	38 (390)	39 (400)	40 (410)	41 (420)	42 (430)	44 (450)
		Cross concreting layers												
		40	16 (165)	18.5 (190)	20.5 (210)	22 (225)	23.5 (240)	25 (255)	26 (265)	27 (275)	28 (285)	29 (295)	30 (305)	31.5 (320)
		80	18 (185)	20.5 (210)	22.5 (230)	24 (245)	25.5 (260)	27 (275)	28 (285)	29.5 (300)	30.5 (310)	31.5 (320)	32.5 (330)	34 (345)

Table 5.8 Ultimate tensile strength of concrete

Concrete cone settlement (cm)	Maximum size of coarse aggregate (mm)	Ultimate tension strength of concrete ε_{lim} 10^5 at concrete class in compressive strength												
		B5	B7.5	B10	B12.5	B15	B17.5	B20	B22.5	B25	B27.5	B30	B35	B40
Up to 4	40	3.5	3.7	4	4.2	4.5	4.8	5	5.3	5.5	5.8	6	6.5	7
	80	3	3.2	3.5	3.7	4	4.3	4.5	4.8	5	5.3	5.5	6	6.5
	120	2.7	3	3.2	3.5	3.7	4	4.2	4.5	4.7	5	5.2	5.7	6.2
4–8	40	4	4.2	4.5	4.7	5	5.3	5.5	5.8	6	6.3	6.5	7	7.5
	80	3.5	3.7	4	4.2	4.5	4.8	5	5.3	5.5	5.8	6	6.5	7
	120	3.2	3.5	3.7	4	4.2	4.5	4.7	5	5.2	5.5	5.7	6.2	6.7
Over 8	40	6	6.2	6.4	6.5	6.7	6.9	7	7.2	7.4	7.6	7.7	8	8.5
	80	5°	5.2	5.4	5.6	5.8	6	6.2	6.4	6.6	6.8	7	7.5	7.8
	120	4.5	4.7	4.9	5.1	5.3	5.6	5.8	6	6.2	6.5	6.7	7	7.5

The strength of concrete structures designed for use in especially severe climatic conditions (with an average monthly temperature of the coldest month – minus 20°C and below) by the time of concrete freezing should be as follows:

- for nonmassive elements of dams in the zone of variable water level and in the zones of the structure in contact with frozen soil – not less than 100% of the design strength; for other zones and parts of the dam – not less than 70% of the design strength;
- for massive dam elements in zones of variable water level and concrete contact with frozen soil – not less than 70% of design strength and in surface and underwater zones – not less than 50% of design strength.

The class of concrete and monolithic mortar must not be lower than the class of concrete of monolithic structures if the latter is not lower than B25. In other cases, the class of concrete and monolithic mortar should be one step above the concrete class of the monolithic structure.

For the construction of dams, portland cement, sulfa-resistant portland cement with mineral additives, and pozzolanic portland cement should be used, and for underwater and internal zones, in addition, slag portland cement should be used.

It is recommended to limit the content in clinker C_3A to 8%. The number of cement marks should be, as a rule, no more than two or three and should be limited to one or two cement suppliers.

For dams of I and II classes, it is recommended to develop special specifications for cement, coordinating and approving them in the prescribed manner.

The calculated values of the slide characteristics of the concrete masonry along the construction joints are given in Table 5.9.

To ensure the required frost resistance of concrete of marks F200 and higher, as well as to increase the density and waterproof of concrete and its technological properties, the use of surface-active and complex additives should be provided in accordance with SS 26633.

For the internal zones of gravity and arch gravity dams, it is necessary to consider the possibility of using hard concrete mixture compacted by rolling (rolled concrete).

Table 5.9 The design values of the slide characteristics concrete masonry at construction joints

Shear characteristic	Structures with sectional cutting into blocks				Structures with pillar cutting into blocks	
	Class concrete vibrated		Class rolled concrete		Class vibrated concrete	
	B5 ÷ B17.5	B20 ÷ B40	B5 ÷ B17.5	B20 ÷ B30	B5 ÷ B17.5	B20 ÷ B40
Coefficient of friction (gearing) $tg\varphi$	1.1	1.2	1	1.1	1	1.1
Cohesion C, MPa	0.3	0.4	0.2	0.3	0.1	0.2

Table 5.10 provides recommendations on the use of concrete additives for various parts of hydraulic structures in various climatic conditions.

5.2 Cement

The following types of cements are used for hydraulic concrete: portland cement and its varieties – with moderate exothermy, plasticized, hydrophobic, sulfate-resistant, slag portland cement and pozzolanic portland cement. Technical requirements for cements are established in SS 10178, SS 22266, and SS 33174.

By mechanical strength, cements are divided into marks: 300, 400, 500, 550, and 600.

The type of cement is assigned taking into account the zonal breakdown of the structure into concrete classes and the degree of massiveness of the structure. For underwater and underground concrete, concrete of the inner zone, any of the listed cements can be used but mainly slag portland cement and pozzolanic portland cement, as well as portland cement and slag portland cement with the addition of fly ash. Moderate exothermy, sulfate-resistant, plasticized, and hydrophobic portland cement is used for concrete of a zone with a variable water level. When substantiating slag portland cement, bulk concrete may be found above the water level.

Special technical requirements that establish the optimal mineralogical composition of clinker, grinding fineness, type of mineral additives, and their optimal content should be developed using cement for concrete structures with a concrete work volume of more than 250 thousand m^3, dams of I–III classes, as well as other critical structures in

Table 5.10 Areas of rational use of additives for concrete HS

Parts of structures	Name of additives			
	Of the plasticizing effect	Of the air-involving action	Of the plasticizing and air involving action	Of the retarding action
	LST S-Z	LHD, SDO, START (SVEK)	PFLH	SP
1. Concrete and reinforced concrete of HS – parts located in the zone of a variable water horizon:				
a) in especially severe climatic conditions;	- -	+	±	-
b) in severe climatic conditions;	⊕ -	+	+	(+)
c) in moderate climatic conditions.	⊕ -	+	+	(+)
2. Parts constantly under water	+ -±	+	+	(+)
3. Parts constantly under water	+ +	+	+	(+)
4. Parts of internal zones	+ +	+	+	(+)
5. Concrete of water conduits and other structures experiencing tensile stresses	+ ±	+	±	(+)
6. Cavitation-resistant and wear-resistant concrete	+ -	-	±	(+)

Parts of structures	Name of additives				
	Integrated action		Accelerating action	Microfiller	
	LST+ LHD or LST+ SDO or LST+SHB (SVEC)	S-Z + LHD or S-Z+ SDO or S-W+SHB (SVEC) S+PVLH	HK	Fly ash	Micro
1. Concrete and reinforced concrete of HS – parts located in the zone of a variable horizon of water:					
a) in especially severe climatic conditions;	+	±	(+)	-	-
b) in severe climatic conditions;	+	±	(+)	-	-
c) in moderate climatic conditions.	+	±	(+)	-	-
2. Parts constantly under water	+	±	+	+	(+)
3. Parts constantly under water	+	+	(+)	+	-
4. Parts of internal zones	+		(+)	+	-
5. Concrete of water conduits and other structures experiencing tensile stresses	+	±	-	-	-
6. Cavitation-resistant and wear-resistant concrete	+	±	-	-	+

Notes:
1. The + sign indicates the advisability of introducing the additive; ± – the additive can only be used after an appropriate feasibility study;
2. (+) – the additive can only be used as a timing regulator setting in combination with another additive providing a set of requirements for concrete in each particular case;
3. – indicates that the additive is used in normal or in high dosage.

particular, working in harsh and especially harsh climatic conditions, where the specifics of work require more stringent regulation of the composition of the supplied cement.

5.3 Sand

For hydraulic concrete, natural sands are used according to SS 8736 or enriched mixtures of grains of hard and dense rocks or artificial mixtures obtained by crushing these rocks according to SS 31424, as well as sand from blast-furnace and ferroalloy slag of ferrous metallurgy according to SS 5578.

By grain size, sand is divided into three fractions:

- large – size modulus $M_{kr} \geq 3.5$–2.5;
- medium – $M_{kr} \geq 2.5$–2;
- small – $M_{kr} \geq 2$–1.5.

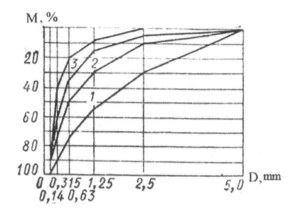

Figure 5.1 Grain composition of sands allowed for use for hydrotechnical concrete: 1 – large sands, 2 – medium sands, 3 – fine; M – complete residues on control sieves, % by weight; D – diameter of the holes of the control sieves, mm.

Table 5.11 Permissible content of harmful impurities in sand

Admixtures	For concrete		
	Zones of variable water level	*Of the underwater and internal zones*	*Of the surface*
Clay, silt, and fine dust fractions, % by mass, not more than:	2	5	3
including clay, % by weight, not more than:	0.5	2	2
Organic impurities determined by the staining method	Liquid coloration is not darker than the color of the standard according to SS 8735 75		
Sulfate and sulfur compounds in terms of SO₃, % by mass, not more than	l	l	l
Opal and other amorphous changes in silica	At the stage of prospecting for quarries, it is necessary to examine the sand for the content of these impurities in it, potentially capable of reacting with the alkali of cement		
Mica, % by weight, not more than	l	3	2

Note: For concrete l m thick located in the zone of variable water level, the content of clay, silt, and fine dusty fractions determined by elutriation is allowed not l% by weight.

The grain composition of sand for hydraulic concrete according to GOST 10268-80 should be within the limits shown in Figure 5.1. The permissible content of harmful impurities in the sand is given in Table 5.11.

When using fine sands, it is necessary to add surface-active substances (air-entraining or plasticizing).

5.4 Large filler

For hydraulic concrete, gravel, crushed stone, and crushed stone from gravel are used as a large aggregate. Coarse aggregate is divided into the following fractions:

- at $D_{max}=20$ mm into two fractions: 5–10 and 10–20 mm;
- at $D_{max}=40$ mm into three fractions: 5–10, 10–20, and 20–40 mm;
- at $D_{max}=70$ mm into four fractions: 5–10, 10–20, 20–40, and 40–70 mm;
- at $D_{max}=120$ mm into five fractions: 5–10, 10–20, 20–40, 40–70, and 70–120 mm.

It is allowed to use gravel and crushed stone of fractions: 5–15, 15–30, 30–60, and 60–120 mm at individual objects of power construction in coordination with the Ministry of Energy.

Aggregates with a grain size of more than 120 mm can be used for a feasibility study.

Requirements regarding the permissible content of harmful impurities in coarse aggregate are given in Table 5.12; requirements for the physical and mechanical properties of coarse aggregate are given in Table 5.13.

Frost resistance of crushed stone and gravel for concrete of HS should be as follows:

- from 0°C to −10°C Mp3 100;
- from −10°C to −20°C Mp3 200;
- below −20°C Mp3 300.

For concrete of HS with frost resistance mark Mp3 300 and higher, operated in a zone of variable water level, the use of gravel as a large aggregate is not allowed.

Table 5.12 Permissible content of harmful impurities in large aggregate

Admixtures	For concrete	
	Zones of variable water level and surface	Underwater and internal zones
Clay, silt, and fine dust fractions, determined by coloring, % by weight, not more than:	1	2
Organic impurities	Color not darker than the color of the standard according to SS 8269-76	
Sulfate and sulfur compounds in terms of SO_3, % by weight, not more than:	0.5	0.5
Opal, opal-like rocks, and other minerals	At the stage of quarrying, it is necessary to study aggregates for the content of impurities in them that can react with the alkali of cement	

Table 5.13 Requirements for physico-mechanical properties of coarse aggregate

Admixtures	For concrete	
	Zones of variable water level	Underwater, internal zone and surface
Gravel of crushed stone mark by strength, MPa (kg/cm2), not lower:		
from igneous rocks	100 (1,000)	80 (800)
from sedimentary rocks	80 (800)	60 (600)
from gravel	Dr -8	Dr -12
The ratio of crushed stone mark in strength to concrete class, %:		
from igneous and metamorphic rocks	300	250
from sedimentary rocks	250	200
The density of the grains of the rock, including pores, g/cm^3, not less than:	2.5	2.3
Water absorption for crushed stone, % no more:		
from igneous and metamorphic rocks	0.5	0.8
from sedimentary rocks	1	2
The content of grains of weak rocks in gravel and gravel	5	10

5.5 Water

Water for washing aggregates and preparing concrete should not contain impurities that impede the normal setting and hardening of concrete and contribute to corrosion of the reinforcement.

It is not allowed to use swamp, wastewater contaminated with impurities (salts, acids, oils, etc.), and water having a pH of less than 4.

5.6 Features of rolled hydrotechnical concrete

Unlike ordinary concrete, rolled concrete is laid out from a rigid concrete mixture with a low cement content and a high pozzolana content, layers with a height of mainly 0.3–0.4 m, and compaction with vibratory rollers. Its physical and mechanical properties depend not only on its composition, but also on the technology of erection and vibro-rolling.

Based on the long-term practice of using rolled concrete in concrete dams, detailed recommendations have been developed on the rational composition of concrete for various zones of dams and its physico-mechanical characteristics, on the use of cement together with pozzolan, on aggregates, and on the technology described in the Bulletin of the ICOLD: RCC Dams. Bulletin ICOLD, No. 126, 2017 [245].

5.7 Reinforcement

For reinforcing reinforced concrete structures of HS, reinforcing steel should be used, which belongs to one of the following types:

- bar reinforcing steel;
- hot-rolled – smooth, class A-I, periodic profile of classes A-II, A-III, A-IV, A-V; thermally and thermomechanically hardened;
- of a periodic profile of classes At-IIIC, At-IVC, At-VSC; hardened by a hood of class A-IIIv;
- wire reinforcing steel;
- ordinary cold drawn wire – of a periodic profile of class BP-I.

For embedded parts and connecting plates, rolled carbon steel should be used, as a rule. Reinforcing steel classes A-IIIB, A-IV, and A-V are recommended for prestressed structures.

Regulatory and calculated resistances of the main types of reinforcement used in reinforced concrete structures of HS depending on the class of reinforcement should be taken according to Table 5.14.

Table 5.14 Standard and design resistance of reinforcement

Reinforcement class diameter (mm)	Standard resistance and design resistance to tension reinforcement for limit states of the second group, MPa (kg/cm²), R_{sn}, $R_{s,ser}$	The design resistance of the reinforcement for limit states of the first group, MPa (kg/cm²)		Compression
		Tension		
		Longitudinal R_s	Transverse (clamps, bent rods), R_{sw}	R_{sc}
Bar reinforcement				
A-I	235 (2,400)	225 (2,300)	175 (1,800)	225 (2,300)
A-II	295 (3,000)	280 (2850)	225 (2,300)	280 (2,850)
A-III				
6-8	390 (4000)	355 (3,600)	285 (2,900)	355 (3,600)
10-40	390 (4000)	365 (3,750)	290 (3,000)	365 (3,750)
A-IV	590 (6000)	520 (5,200)	405 (4,150)	400 (40,00)
A-V	785 (8000)	680 (6,950)	545 (5,550)	400 (4,000)
Hardened hood class A-III with control:				
Stresses and extensions	540 (5,500)	490 (5,000)	390 (4,000)	200 (2,000)
Only extensions	540 (5,500)	450 (4,600)	360 (3,700)	200 (2,000)
Wire reinforcement				
Bp-I, 3	410 (4,200)	375 (3,850)	270 (2,750)	375 (3,850)
Bp-I, 4	405 (4,150)	365 (3,750)	265 (2,700)	365 (3,750)
Bp-I, 5	395 (4,050)	360 (3,700)	260 (2,650)	360 (3,700)

Notes:
1. In welded frames for clamps made of class A-III fittings, the diameter of which is less than 1/3 of the diameter of the longitudinal rods, R_{sw} equal to 255 MPa (2,600 kg/cm²).
2. In the absence of adhesion of reinforcement with concrete, R_{sc} is zero.

Table 5.15 **Coefficients of operating conditions**

Constructions	Coefficients of reinforcement operating conditions	
	Symbol	Meaning
Reinforced concrete element	γ_{s2}	1.1
Steel-reinforced concrete structures (open and underground)	γ_{s3}	0.9

Note: If there are several factors acting simultaneously, the product of coefficients of the working conditions is introduced into the calculation.

When calculating the reinforcement according to the main tensile stresses, the design resistance of the reinforcement should be taken as for longitudinal reinforcement on the action of the bending moment.

Coefficients of working conditions of nonstressed valves should be taken in accordance with Table 5.15 and for tensile reinforced valves – according to current regulatory documents.

The coefficient of reinforcing working conditions when calculating the limit states of the second group is taken to be equal to unity.

Chapter 6

Analysis of concrete dams

6.1 Mechanic of continuous media

The justification of the reliability of decisions when designing concrete dams is made using engineering calculations, experimental and numerical studies of the SSS of the "reservoir-dam-foundation" system, hydraulic, filtration, and temperature conditions in construction and operation.

In modern conditions, when performing calculations and numerical studies, methods of continuum media mechanics are used, which allow one to take into account the real properties of materials of structures and foundation rock (Table 6.1). This approach is governed by the design standards of SR [8–16].

The sections of continuum media mechanics, which are used when performing computational studies of concrete dams and other HSs, include hydromechanics, theory of filtration, theory of heat conduction, and mechanics of a solid deformable body [152]. The following is a general description of these sections of continuum media mechanics.

Mechanics of a solid deformable body [169] allows one to obtain a mathematical description of the occurrence and change in time of stresses, strains, and displacements in a solid deformable body under the action of loads applied to it. The solutions to the problems of mechanics of a solid deformable body are used to obtain the SSS of the "construction–foundation" system, on the basis of which the strength of the dam and its elements as well as the strength and stability of the foundation is studied.

Distinguish between static and dynamic problems of mechanics of a solid deformable body. Static tasks are considered in cases when the body under the influence of external loads that do not change in time is at rest, or in cases where the changes in external loads in time occur slowly enough so that at any moment in time it would be possible to consider the static state of the body.

If dynamic loads are applied to the body (seismic, explosive loads), it is necessary to consider dynamic problems associated with the propagation of elastic or elastic–plastic waves.

Depending on the material properties taken into account, it is customary to consider the following branches of the mechanics of a solid deformable body.

Theory of elasticity is the simplest and at the same time the main branch of mechanics of a solid deformable body [25,169]. In the theory of elasticity, it is believed that bodies have perfect elasticity, which is understood as the ability of a body, which has undergone deformations under the action of the loads applied to it, to completely restore its original shape after removing these loads. Moreover, in accordance with

Table 6.1 Mechanics of continuous media

Section	Mechanics of a solid deformable body				Structures mechanic	Material resistance
Object of study	Body				Rods	Sections
Hypo-thesis	Body continuity, natural unstressed state of the body				1. The neutral axis bends and does not change its size 2. The sections are flat and perpendicular to the neutral axis 3. The absence of stress between the layers 4. Linear connection between stresses and strains	Linear connection between stresses and strains
Branch	Classical theory of elasticity	Applied theory of elasticity	Theory of plasticity	Theory of creep	—	—
Hypotheses	1. Spherical isotropy 2. Uniformity 3. Linear relationship between stresses and strains 4. Displacements are small compared to body size 5. Relative elongations and slides are small compared to unity	1. Anisotropy 2. Nonuniformity 3. Nonlinear relationship between stresses and strains	1. Anisotropy 2. Nonuniformity 3. Stresses and deformations do not depend on time	1. Anisotropy 2. Nonuniformity 3. Stresses and deformations are time dependent	—	—

Hooke's law, a linear relationship between the stresses and strains arising in the elastic body due to the application of loads and impacts to it is assumed.

It should be noted that almost all structural materials, including concrete as well as rock foundation of dams, possess elastic properties at relatively low values of effective stresses.

Elastic characteristics of the foundation, such as the deformation modulus and Poisson's ratio, are determined as a result of laboratory and field studies. The elastic modulus and Poisson's ratio of concrete, determined on the basis of special studies, depend not only on the composition, but also on the age of the concrete since there is an increase in the elastic modulus and a decrease in the Poisson's ratio in time. When performing calculations of concrete dams for the construction period, it is important to take into account the time-dependent changes in the elastic characteristics of concrete, considered as elastic hardening material.

The well-known analytical solutions of the classical theory of elasticity allow us to study SSS in spatial conditions, but their use is limited due to the difficulty of approximating real structures. In addition, the hypothesis of a linear relationship between stresses and strains does not meet the needs of the designer.

The theory of plasticity is an important branch of the mechanics of a solid deformable body. It considers elastic–plastic bodies in which not only elastic strains proportional to stresses can arise, but also plastic strains associated with stresses by a nonlinear dependence [81,116]. Note that for the most elastic–plastic materials, the change in strain during unloading occurs in accordance with the linear relationship between stress and strain. After unloading, elastic–plastic bodies do not restore their original shape because the so-called residual or plastic deformations remain in them. Taking into account the development of plastic deformations in high stress zones is important for the correct assessment of the SSS of dams and their foundations.

The concrete bodies of the dam and the rock foundation considered in the framework of the theory of plasticity are considered to be elastic–plastic bodies. To describe the elastic–plastic properties of such bodies, data from special laboratory and field studies are used, as a result of which nonlinear relationships between stresses and strains are obtained.

Creep theory makes it possible additionally take into account deformations developing in time in loaded bodies at constant stresses. *Stress relaxation* is directly related to creep, which is a decrease in stresses in a loaded body over time with constant deformations. Almost all building materials in particular concrete and rock have creep. Therefore, creep accounting is important in determining the SSS of concrete dams especially during the construction period, when creep deformations in young concrete are manifested to a large extent.

A linear creep theory is usually considered, in accordance with which creep strains linearly depend on stresses and are damped functions of time. To describe the properties of creep, a creep measure is used, which represents a nonlinear damped dependence of the relative deformation of a material sample caused by a unit stress on time. The parameters of the creep measures of the rock base are determined in laboratory and field studies.

On the formation of SSS, filtration, and temperature regimes of massive HS, significant and sometimes decisive influence is exerted by the construction sequence. This is due to the following circumstances.

A massive HS cannot be erected instantly. In the process of erecting such a structure, a gradual change in its shape and size occurs, and the static indeterminacy of the "dam-foundation" system increases. In addition, during the construction process, a reservoir may be filled. Simultaneously with the change in the shape and size of the structure, a gradual application of loads from its own weight, hydrostatic pressure, temperature effects, and other SSS occurs. The HS being erected changes until it completely formed together with the end of the construction. The final SSS of the structure differs from that, which would have occurred with the instantly erected structure.

When reproducing in the calculations the sequence of construction of structures, one should take into account the NSS of the foundation, which is determined on the basis of data from engineering and geological surveys and studies or by calculation.

Consideration of the impact of the sequence of construction of concrete dams on their SSS is widely covered in the technical literature, for example [64,69,183].

In solving problems of continuum mechanics, numerical methods are usually used: finite differences (FDM), finite elements (FEM), and boundary elements (BEM). Of these methods, the most effective is the FEM, which allows us to take into account the structural features of concrete dams, heterogeneity of the foundation, and other factors. Computer programs ANSYS, ABAQUS, MIDAS, etc. are widely used to solve practical problems. Using these programs, it is possible to carry out SSS studies with reproduction of the sequence of dam construction and reservoir filling, filtration, and temperature regimes under conditions substantially close to real ones: taking into account inelastic effects, creep, and ductility of materials.

The dimensions of the calculated area of the "dam-foundation" system should be assigned on basis of the Saint-Venant principle so that the boundary conditions on the circuit, which limit the calculated area, do not affect the determined parameters (displacement, stress, temperature, filtration head) of the dam and its adjacent grounds.

In developing the design scheme for the "dam-foundation" system, the following recommendations [60,62,64] on determining the size of the foundation can be used. The planned dimensions of the calculation area of the foundation should be taken at least $5B_d$, in depth from the bottom of the structure – at least $2B_d$, where B_d is the characteristic size of the structure. When making calculations of concrete gravity dams, usually in the framework of solving two-dimensional problems, the value of B_d is taken equal to the width of the dam along the sole.

Using the **method of resistance of materials**, simple calculations are made to check the strength of the cross sections of the elements of structures according to the forces and moments determined by the methods of **structural mechanics**; constructions and structures are approximated by the rod elements. Many computer programs have been developed for calculating rod systems; in design practice, MicroFe, SCAD, etc. are the most popular. As a rule, elastic systems are considered in these calculations, and the linear relationships between stresses and strains are used.

Hydromechanic allows us to get a mathematical description of the behavior of resting and moving water. Approaches and methods of hydromechanics are used to determine the hydraulic regime of hydraulic structures [26].

In the framework of *hydrostatics*, which is a section of hydromechanics that describes the behavior of a liquid at rest, the loads on the surface of the dam and its foundation from hydrostatic pressure can be determined.

In the framework of *hydrodynamic*, the behavior of a moving fluid is described; with its help, for example, the flow conditions can be determined when water flows through the dam and hydrodynamic loads on the structure can be found. An important problem that can be solved in the framework of hydrodynamics is the propagation of wind waves in the reservoir and the interaction of these waves with the dam. Finally of particular interest is the problem of the propagation of seismic waves in the water of the reservoir, their interaction with the dam and the foundation, which is important for assessing the seismic resistance of structures [112].

When solving the problems of hydrodynamics in relation to concrete dams, water is often considered as an ideal incompressible fluid. With significant head, it becomes necessary to take into account the viscosity and compressibility of water.

The basic differential equations and information on approaches and methods for solving stationary and nonstationary problems of hydromechanics are given in many works, for example [87,185]. As applied to HS, various solutions to the problems of hydrodynamics are considered in a number of works, for example [268].

It should be noted that in the design of HS and in particular concrete spillway dams in addition to hydromechanical calculation methods, the simplified hydraulic methods are widely used. The main points of such methods as applied to HS are described in a large number of works, for example [65,67].

Filtration theory allows us to get a mathematical description of the movement of water in the pores and cracks of the rock foundation and concrete dam. The solution of problems of the theory of filtration is necessary to determine the filtration regime of the dam in various operating conditions. As a result of solving filtration problems within the computational domain "construction–foundation", filtration heads, gradients of filtration heads, and filtration discharges, and if necessary the position of the depression surface of the filtration flow are determined. These data are used to assess the filtration strength of the dam body and its foundation. In addition, these data make it possible to determine the filtration hydrodynamic loads within the computational area, which is necessary for performing SSS calculations of the dam and assessing its strength.

The main filtration characteristic used in solving problems of the theory of filtration is the filtration coefficient. The values of the filtration coefficient for concrete dam and rock foundation are established on the basis of special laboratory and field studies [84]. In necessary cases, the filtration heterogeneity of the dam foundation is taken into account.

In these works [138,185], the basic differential equations of the theory of filtration, and information on approaches and methods for solving problems are given. As applied to the HS, solutions to problems in the theory of filtration are given in Ref. [21]. When designing small dams as well as at the preliminary stages of designing high dams, the simplified methods for solving filtration problems are used.

Theory of thermal conductivity allows us to get a mathematical description of the distribution of heat in bodies or the transfer of heat from one body to another. As applied to concrete dams, the solution of problems of the theory of thermal conductivity is given for the "construction–foundation" system. The results of solving these problems are necessary to determine the thermal regime of the dam in the construction and operational periods. These results are the initial data for performing SSS calculations of the dam, assessing its strength, and also for assessing the crack resistance of the concrete masonry of a structure under temperature effects.

When solving the problems of the theory of thermal conductivity, thermophysical characteristics such as specific heat capacity, thermal conductivity, and thermal diffusivity are used. The values of these characteristics for the dam concrete and the rock mass of the foundation are established on the basis of special laboratory and field studies. In necessary cases, the thermophysical heterogeneity of the dam and foundation is taken into account. Data on the time-decaying rate of heat dissipation of concrete during cement hydration is determined using laboratory tests.

Solutions to various problems of the theory of thermal conductivity applied to concrete HS considered in [39,177].

6.2 Loads and effects

6.2.1 Combinations of loads and effects. Design cases

When calculating the SSS strength and stability of concrete dams [9,12,14], the following main loads and effects are taken into account (Figure 6.1):

- Dead weight of the structure G_d;
- Hydrostatic pressure W_1 from the side of the upstream and W_2 from the side of the downstream and the bed of the upstream and downstream;
- Weighing P_v and filtration P_f water pressure on the dam bottom;
- Temperature effects T;
- Hydrodynamic loads on the spillway dam when water flows W_w pass through it;
- Ice pressure P_i;
- Sediment pressure P_{ws};
- Pressure of wind waves P_w;
- Seismic hydrodynamic pressure of water and sediment from the upstream side W_{s1};
- Seismic hydrodynamic pressure of water from the downstream side P_w;
- Loads from technological equipment.

The loads acting on concrete dams differ in the type of action, origin, duration, and repeatability.

By the nature of the action, *static* and *dynamic* loads are considered.

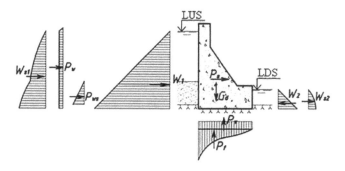

Figure 6.1 Main loads on a concrete gravity dam.

The duration of the action of the load is divided into *permanent* and *temporary*. Permanent loads operate throughout the entire period of operation of the structure, while temporary loads may be absent in certain periods. Temporary loads, in turn, are divided into *long-term*, *short-term*, and *special*.

Constant loads include the following:

1. Own weight of the structure, including the weight of permanent technological equipment (locks, lifting mechanisms, etc.), the location of which does not change during operation;
2. Hydrostatic pressure on the upstream face of the dam and the upstream bed at a normal retaining level (NHL) in the reservoir;
3. Hydrostatic pressure on the downstream face of the dam and the downstream bed at a water level in the downstream, corresponding to the passage through the HS of the minimum discharge for technological and environmental requirements;
4. Weighing and filtration pressure of water on the dam bottom and the foundation at the NHL in upstream and the minimum water level in the downstream and the normal operation of the antifiltration and drainage devices;
5. Weight of the soil sliding with the dam and the lateral pressure of the soil from the upstream and downstream.

Temporary long-term loads and effects include the following:

6. Pressure of sediment deposited in front of the dam;
7. Temperature effects determined for a year with an average amplitude of average monthly temperatures.

Short-term loads include the following:

8. Hydrostatic pressure on the dam downstream face and on the bed downstream at a water level in the downstream, corresponding to the passage through the HS of the design water discharge at the NHL in the reservoir (instead of load 3);
9. Weighing and filtration pressure of water on the dam bottom and the foundation at NHL in the upstream and the water level in the downstream in accordance with load 8 (instead of load 4);
10. Hydrodynamic loads on a spillway dam when design water discharge passes through it at NHL in a reservoir;
11. Ice pressure determined at its average perennial thickness;
12. Pressure of wind waves determined at the average long-term wind speed;
13. Loads from hoisting, reloading, transport devices, and other structures and mechanisms (bridge and overhead cranes, etc.);
14. Loads from floating bodies.

Special loads and effects include the following:

15. Hydrostatic pressure on the dam upstream face and the bed upstream at a forced retaining level (SRL) in the reservoir (instead of load 2).

16. Hydrostatic pressure on the dam downstream face and the bed downstream at the maximum estimated level of the downstream, corresponding to the passage of the maximum design water discharge through the HS at the SRL in the reservoir (instead of loads 3 and 8).

17. Weighing and filtration pressure of water on the dam bottom and the foundation at the SRL in the upstream and the maximum design water discharge in the downstream (instead of loads 4 and 9).

18. Hydrodynamic loads on the spillway dam when design water discharge passes through it at the SRL in the reservoir (instead of load 10).

19. The weighing and filtration pressure of water on the dam bottom and the foundation, resulting from a malfunction of one of the antifiltration or drainage devices at NHL in the upstream and the minimum water level in the downstream (instead of loads 4, 9, and 17).

20. Weighing and filtration pressure of water on the dam bottom and the foundation, resulting from a malfunction of one of the antifiltration or drainage devices when the design water discharge is missed at NHL in the upstream and the water level in the downstream (instead of loads 4, 9, 17, and 18).

21. Temperature effects determined for a year with a maximum amplitude of monthly average temperature fluctuations, as well as for a year with a maximum monthly average temperature (instead of load 7).

22. Ice pressure determined at a maximum multiyear ice thickness with a security of 1% (instead of a load of 11).

23. Wind wave pressure determined at a maximum multiyear wind speed with a security of 2% for structures of I and II classes and 4% for structures of III and IV classes (instead of load 12).

24. Seismic effects (inertial loads, seismic hydrodynamic pressure on the dam faces), determined for the design earthquake (DE).

25. Seismic effects (inertial loads, seismic hydrodynamic pressure on the dam faces), determined for the maximum credible earthquake (MCE).

Design combinations of loads and effects. Design cases.

In the calculations according to the limiting states, the **usual** and **unusual** combinations of loads and effects are considered.

Usual combinations include permanent, temporary, and the long-term and short-term loads and effects.

Unusual combinations include loads and effects of the corresponding main combination and one of the special loads. Loads and impacts should be taken in the most unfavorable but possible combinations separately for the construction and operational periods.

Combinations of loads and effects correspond to **design cases** that should be considered when performing calculations.

For the construction period, one **design construction case** is usually considered corresponding to an unusual combination of loads and effects. In this case, the loads from the dead weight of the dam are taken into account, including the weight of the constant technological equipment, the location of which does not change during operation, temperature effects, seismic effects directed from the side of the downstream to upstream (reverse seismic).

For the **operational period**, it is necessary to consider **design cases** for the usual and unusual combinations of loads and effects.

With the **usual combinations of loads and effects** in the general case, it is necessary to consider the following operational design cases.

1st operational design case with the usual combination of loads and effects is the case of maximum head on the dam when wind pressure is taken into account, which is determined at the average long-term wind speed. The NHL in the upstream, the minimum estimated water level in the downstream, and the normal operation of the antifiltration and drainage devices are considered. The loads and *i* effects 1–7, 12–14 are taken into account.

2nd operational design case with the usual combination of loads and effects is the case of the maximum head on the dam when the ice pressure is taken into account, determined at its average long-term thickness. It is necessary to consider the NHL in upstream and the calculated water level in the downstream, and the normal operation of the antifiltration and drainage devices are considered. The loads and effects 1–7, 11, 13, and 14 are taken into account.

3rd operational design case with the usual combination of loads and effects is a flood pass when taking into account the pressure of wind waves, determined at an average long-term wind speed. It is necessary to consider the NHL in upstream and the calculated water level in the downstream, which corresponds to passage of the design water discharge through the HS at the NHL in the reservoir, and the normal operation of antifiltration and drainage devices. The loads and effects 1, 2, 5–10, and 12–14 are taken into account.

With **an unusual combinations of loads and effects** in the general case, it is necessary to consider the following operational design cases.

4th operational design case with an unusual combination of loads and effects is a flood pass when taking into account the pressure of wind waves, determined at an average long-term wind speed. The SRL in the upstream and the calculated water level in the downstream, which correspond to the maximum design water discharge through the HS at the SRL in the reservoir, and the normal operation of the antifiltration and drainage devices are considered. The loads and *i* effects 1, 5–7, and 12–18 are taken into account.

5th operational design case with an unusual combination of loads and effects is the case of maximum head on the dam when taking into account the pressure of wind waves, determined at the average long-term wind speed, and malfunction of one of the antifiltration or drainage devices. It is necessary to consider NHL in the upstream and the minimum estimated water level in the downstream. The loads and effects 1–3, 5–7, 12–14, and 19 are taken into account.

6th operational design case with an unusual combination of loads and effects is the case of maximum head on the dam when taking into account the ice pressure, determined at its average long-term thickness, and malfunction of one of the antifiltration or drainage devices. It is necessary to consider NHL in the upstream and the minimum estimated water level in the downstream. The loads and effects 1–3, 5–7, 11, 13, 14, and 19 are taken into account.

7th operational design case with an unusual combination of loads and effects is a flood pass when taking into account the pressure of wind waves, determined at the average long-term wind speed, and malfunction of one of the antifilter or drainage devices. It is necessary to consider the NHL in upstream and the estimated water level in the downstream area, which corresponds to the passage through the HS of the design water discharge at the NHL in the reservoir. The loads and effects 1, 2, 5–8, 10, 12–14, and 20 are taken into account.

8th operational design case with an unusual combination of loads and effects is the case of the maximum head on the dam, taking into account the pressure of the wind waves, determined at the average long-term wind speed, and the maximum temperature effects. It is necessary to consider NHL in the upstream and the minimum estimated water level in the downstream. The loads and effects 1–6, 12–14, and 21 are taken into account.

9th operational design case with an unusual combination of loads and effects is the case of the maximum head on the dam when taking into account the ice pressure determined at its average long-term thickness and maximum temperature effects. It is necessary to consider NHL in the upstream and the minimum estimated water level in the downstream. The loads and effects 1–6, 11, 13, 14, and 21 are taken into account.

10th operational design case with an unusual combination of loads and effects is a flood leak when taking into account the pressure of wind waves, determined at the average long-term wind speed, and maximum temperature effects. It is necessary to consider the NHL in upstream and the estimated water level in the downstream area, which corresponds to the passage through the HS of the design water discharge at the NHL in the reservoir. The loads and effects 1, 2, 5–10, 12–14, and 21 are taken into account.

11th operational design case with an unusual combination of loads and effects is the case of the maximum head on the dam when the maximum pressure of the wind waves is taken into account. It is necessary to consider NHL in the upstream and the minimum estimated water level in the downstream. The loads and effects 1–7, 13, 14, and 23 are taken into account.

12th operational design case with an unusual combination of loads and effects is the case of the maximum head on the dam when the maximum ice pressure is taken into account. It is necessary to consider NHL in the upstream and the minimum estimated water level in the downstream. The loads and effects 1–7, 13, 14, and 22 are taken into account.

13th operational design case with an unusual combination of loads and effects is a flood pass taking into account the maximum pressure of the wind waves, determined at the average long-term wind speed. It is necessary to consider the NHL in upstream and the estimated water level in the downstream area, which corresponds to the passage through the HS of the design water discharge at the NHL in the reservoir. The loads and effects 1, 2, 5–10, 13, 14, and 23 are taken into account.

14th operational design case with an unusual combination of loads and effects is the case of the maximum head on the dam taking into account the pressure of the wind waves, determined at the average long-term wind speed, and seismic effects at the level of the DE. The NHL in the upstream and the minimum design water level in the downstream and the normal operation of the antifiltration and drainage devices are considered. The loads and effects 1–7, 12–14, and 24 are taken into account.

15th operational design case with an unusual combination of loads and effects is the case of the maximum head on the dam when taking into account the ice pressure determined at its average long-term thickness and seismic effects at the level of the DE. The NHL in the upstream and the minimum design water level in the downstream and the normal operation of the antifiltration and drainage devices are considered. The loads and effects 1–7, 11, 13, 14, and 24 are taken into account.

16th operational design case with an unusual combination of loads and effects is a flood pass taking into account the pressure of the wind waves, determined at the average long-term wind speed, and seismic effects at the level of the DE. It is necessary to consider the NHL in the upstream and the calculated water level in the downstream, corresponding to the passage through the HS of the design water discharge at the NHL in the reservoir, and the normal operation of the antifiltration and drainage devices. The loads and effects 1, 2, 5–10, 12–14, and 24 are taken into account.

17th operational design case with an unusual combination of loads and effects is the case of the maximum head on the dam when taking into account the pressure of the wind waves, determined at the average long-term wind speed, and seismic effects at the level of MCE. The NHL in the upstream and the minimum calculated water level in the downstream and the normal operation of the antifiltration and drainage devices are considered. The loads and effects 1–7, 12–14, and 25 are taken into account.

18th operational design case with an unusual combination of loads and effects is the case of the maximum head on the dam taking into account the ice pressure determined at its average long-term thickness and seismic effects at the level of MCE. The NHL in the upstream and the minimum design water level in the downstream and the normal operation of the antifiltration and drainage devices are considered. The loads and effects 1–7, 11, 13, 14, and 25 are taken into account.

The **19th operational design case** with an unusual combination of loads and effects is a flood pass when taking into account the pressure of wind waves, determined at the average long-term wind speed, and seismic effects at the level of MCE. The NHL in the upstream and the design water level in the downstream, corresponding to the design water discharge through the HS at NPU in the reservoir, and the normal operation of antifiltration and drainage devices are considered. The loads and effects 1, 2, 5–10, 12–14, and 25 are taken into account.

Note that when performing calculations of concrete dams in a number of specific conditions possible to restrict ourselves to considering only some of the listed design cases. For example if special measures provide for the exclusion of ice pressure on the dam, it is possible to refuse to consider design cases in which ice loads are taken into account. For high dams, it is possible to refuse to consider operational design cases with special combinations of loads and impacts, including wave and ice loads, because usually the contribution of these loads is negligible.

6.2.2 Determination of loads and effects

The load due to the dead weight of the dam and its elements is defined as the product of the volume V of the specific gravity of concrete $\rho_b \times g$:

$$G = \rho_b \times g \times V \qquad (6.1)$$

where
  ρ_b – density of concrete, determined on the basis of testing samples of concrete;
  g – acceleration due to gravity.

At the preliminary design stage, the average density of concrete can be taken according to Table 5.6, depending on the density and maximum aggregate size [15].

In the absence of data on aggregate density, the average density of concrete is taken at an average aggregate density of 2,650–2,700 kg/m³. Typically, when performing approximate calculations, the density of concrete is assumed to be 2,400 kg/m³ and the density of reinforced concrete is 2,500 kg/m³.

Hydrostatic pressure loads on the dam surface are determined by the known hydrostatic formulas. Typically when performing dam calculations, these forces are decomposed into horizontal and vertical components from the upstream and downstream sides, respectively (Figure 6.2).

Horizontal W_1 and W_2 and vertical W_3 and W_4 components of hydrostatic pressure forces can be found as the corresponding areas of the horizontal and vertical pressure diagrams. The ordinates of the diagrams p_w at a point at a depth y from the water level are determined by the formula:

$$p_w = \rho_w \times g \times y \tag{6.2}$$

where

ρw – the density of water taken equal to $\rho w = 1{,}000\,\text{kg/m}^3$.

When the suspended sediment is contained in water, the density of water can be increased by 5%–10% or more.

The vertical components of the hydrostatic pressure forces of water can be directed both down and up.

The filtering water through cracks and pores of the rock foundation exerts a force on the bottom of the structure, called *uplift*, which is the sum of the weighing and filtration hydrodynamic pressure. The force of the uplift of the water at the bottom of the P_u dam is equal to

$$P_u = P_v + P_f \tag{6.3}$$

where

P_v – the force of the weighing pressure;
P_f – the force of the filtration hydrodynamic pressure.

Uplift is transmitted to the bottom of the dam not over the entire area, but only over the area of the cracks and pores of the rock foundation. This circumstance is taken

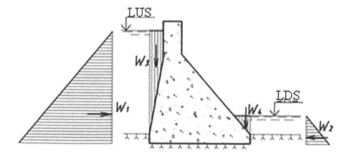

Figure 6.2 Hydrostatic pressure on the faces of a concrete dam.

into account by the coefficient of effective uplift area of the foundation $\alpha_{2,f}$. The value of the coefficient $\alpha_{2,f}$ is taken according to the results of special studies. Prior to performing these studies as well as in preliminary calculations, it is recommended to take $\alpha_{2,f} = 0.5$ in the compressed zone of the foundation and $\alpha_{2,f} = 1$ in the extended zone of the foundation [14].

The force value P_v is defined as the area of the plot of the weighing pressure, the ordinates of which p_v are equal to:

$$p_v = \rho_w \times g \times h_v \times \alpha_{2,f} \qquad (6.4)$$

where
   h_v – the difference between the downstream-level marks and the bottom point of the dam, in which the ordinate p_v is determined.

The value of the force P_f can be found as the area of the diagram of the filtration hydrodynamic pressure, the ordinates of which p_f are equal to:

$$p_f = \rho_w \times g \times h_f \times \alpha_{2,f} \qquad (6.5)$$

where
   h_f – the filtration head at the bottom of the dam in which the ordinate p_f is determined.

The values of filtration heads within the dam bottom are set based on the results of filtration calculations and studies.

For low (up to 60 m) concrete dams on rock foundations, it is allowed to determine the filtration heads h_f according to the simplified method using the diagrams shown in Figure 6.3 and Table 6.2 [14].

The values of filtration heads on the axis of the cement grout H_{as} and drainage H_{dr} are determined by the following formulas:

$$H_{as} = \alpha_{as} \times H_d, \, H_{dr} = \alpha_{dr} \times H_d \qquad (6.6)$$

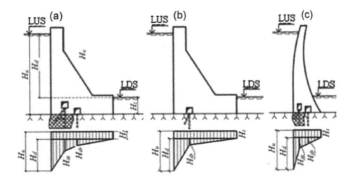

Figure 6.3 Diagrams of piezometric pressure on the bottom of a concrete dam: (a) and (c) with a cementation curtain and drainage, (b) with drainage.

Table 6.2 Coefficients α_{as} and α_{dr}

Dam type	Values $\alpha_{as} = H_{as}/H_d$ and $\alpha_{dr} = H_{dr}/H_d$ for load combinations					
	Usual and unusual during normal operation of the antifiltration and drainage devices			Unusual in the event of a mal-function of the antifiltration and drainage devices		
	Dams with grouting curtains		Dams without grouting curtains	Dams with grouting curtains		Dams without grouting curtains
	α_{as}	α_{dr}	α_{dr}	α_{as}	α_{dr}	α_{dr}
Massive gravity (Figure 6.3a and b) classes: I II III and IV	0.40 0.40 0.30	0.20 0.15 0.05	0.20 0.15 0.05	0.50 0.50 0.35	0.30 0.20 0.10	0.40 0.30 0.10
Arch (Figure 6.3c) classes I and IV	0.40	0.20	0.20	0.60	0.35	0.40

where

> H_d – the design head at the dam equal to the difference between the levels of the upstream and downstream;
>
> α_{as} and α_{dr} – coefficients whose values are taken depending on the type of dam, its class, and the calculated combination of loads according to Table 6.2.

The described approach to determining the hydrostatic pressure on the dam surface [14] can be used when calculating the stability of concrete dams of all classes as well as when calculating the strength of concrete dams of II, III, and IV classes on rock foundations.

For class I dams on a rock foundation, an updated methodology for determining the force effect of water is recommended.

In accordance with this technique, the intensity of water pressure on the external faces of the dam is $p_w \times (1-\alpha_{2d})$, where p_w is the hydrostatic pressure and α_{2d} is the co-efficient of effective uplift area for concrete. The value of $\alpha_{2,d}$ should be taken on the basis of special studies. When performing preliminary calculations in the compressed zone of concrete, the value $\alpha_{2,d}$ is taken to be $\alpha_{2d} = 0.5$ and in the extended, $\alpha_{2d} = 1$. In calculations of dams of II–IV classes, it is allowed to take $\alpha_{2d} = 0.5$ [14].

The intensity of water pressure on the free surfaces of the foundation in the upstream and downstream (loading of the foundation with water) is taken to be $p_w \times (1 - \alpha_{2f})$, where p_w is the hydrostatic pressure on the foundation and α_{2f} is the coefficient of effective area of the uplift of the foundation. This load may not be taken into account when calculating dams of all classes with a height of up to 60 m as well as dams of I and II classes with a height of more than 60 m – at the preliminary stages of design.

To assess the power effect of filtered water, the characteristics of the filtration flow (levels, heads, pressures, head gradients, flow) are determined. Moreover, for dams of I and II classes, numerical methods are used to solve problems of the theory of filtration.

The FEM is most often used which is implemented in a large number of computer programs, for example, ANSYS, ABACUS, and MIDAS. As a result of solving the filtration problem, the hydrodynamic pressure field p_{vf} can be obtained:

$$p_{vf} = \rho_w \times g \times \left(h_v + h_f \right) \tag{6.7}$$

where

h_v and h_f – see formulas (6.4) and (6.5).

Uplift in the form of surface forces with intensity $p_{vf} \times (\alpha_{2,f} - \alpha_{2,d})$ acts normally to the bottom of the dam.

Within the design area of the dam foundation, it is necessary to take into account the volume forces arising from the filtration of water in the foundation mass. The horizontal q_{fx} and vertical q_{fy} components of these volume forces are determined by the following formulas:

$$q_{fx} = -\frac{\partial}{\partial x}\left(p_{hd} \times \alpha_{2,f} \right), q_{fy} = -\frac{\partial}{\partial y}\left(p_{hd} \times \alpha_{2,f} \right) \tag{6.8}$$

Between the upper face and the drainage of the dam body, as well as between the dam bottom and the downstream level, the concrete is water-saturated. Filtration volumetric forces act in these zones. The horizontal q_{dx} and vertical q_{dy} components of these volume forces are equal

$$q_{dx} = -\frac{\partial}{\partial x}\left(p_{hd} \times \alpha_{2,d} \right), q_{dy} = -\frac{\partial}{\partial y}\left(p_{hd} \times \alpha_{2,d} \right) \tag{6.9}$$

When performing SSS calculations of dams taking into account the indicated filtration volumetric loads, the values of the specific gravities of the foundation and concrete should be taken in a water-saturated state.

When conducting computational studies of the gravity dam of the Boguchan HEP, the volumetric filtration forces at the foundation were taken into account in the form of surface forces on the cementation curtain (for more details, see Section 6.3.2). As a rule, the filtration coefficient of the cementation curtain is less than that of the surrounding rock mass; therefore, the gradients of volumetric filtration forces in it will be higher than in the massif; this is also facilitated by the decompression zone in the foundation under the upper face of the dam and in front of the cementation curtain. In this regard, it is justifiable to reduce the volumetric filtration forces to the surface applied horizontally to the cementation curtain.

Hydrodynamic pressure of a stream of water. When water flows skipped, the averaged and pulsating hydrodynamic pressure acts on the spillway face of the dam. To determine the hydrodynamic pressure, it is necessary to perform hydraulic calculations to determine the flow velocities and flow depths within the spillway face of the dam. The methods of such calculations are described in Ref. [87].

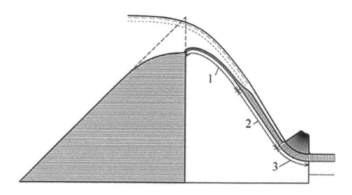

Figure 6.4 Scheme for determining hydrodynamic loads on a spillway dam: I – vertex; 2 – straight section; 3 – terminal section.

In different parts of the spillway of the dam, there is a different nature of the flow of water. It is customary to distinguish between a section within a spillway vertex, a rectilinear section of a spillway face, and an end section curved in a vertical plane (Figure 6.4).

On the spillway vertex which is usually outlined by the Krieger–Ofitserov coordinates, the flow of water differs little from the free fall of the stream. Therefore, in this section, the averaged hydrodynamic pressure of the flow on the dam is insignificant and can be neglected.

In the straight sections of the dam spillway, there is a changing movement of water. In this case, the pressure change in depth can be taken in accordance with the linear law, and the hydrodynamic pressure on the spillway face can be found by the following formula:

$$p = \rho_w \times g \times h \times \cos\theta \tag{6.10}$$

where
h – depth of flow in the considered section;
θ – angle of inclination of the spillway face of the dam to the horizon.

At the end section of the interface of the spillway face with a flat bed or in the area of the toe-springboard, a curved movement of water takes place. The value of the averaged hydrodynamic pressure on the concave curved sections of the dam spillway in this case can be found by the following formula:

$$p = \rho_w \times g \times h \times \frac{V^2}{g \times R} \tag{6.11}$$

where
h – depth of flow;
V – average flow velocity;
R – radius of curvature of the curved section.

Formula (6.11) is applicable for $R > (6 \ldots 8) \times h$.

On vertical upper face of the dam spillway, near vertex is affected by a slightly lower averaged hydrodynamic pressure of the water as compared with the hydrostatic (Figure 6.4).

The pulsating hydrodynamic pressure of the water flow on the spillway face of the dam can be determined on the basis of solving the problems of the mechanics of the turbulent boundary layer. In Ref. [87], formulas are given for determining the pulsating hydrodynamic pressure of a water stream.

Pressure of the wind waves. Under the influence of pressure, wind waves form on the surface of the reservoir and overtaking phenomena occur which should be taken into account in calculating the SSS, strength, and stability of concrete dams [12,14].

In the case of a high wave height and a relatively low dam, it may turn out that such pressure determines the main dimensions of the structure. *The overtaking phenomena are the drops and rises of the water level off the coast of the reservoir caused by currents generated by the wind.* In this case, the surface of the reservoir deviates from the horizontal and acquires a bias in the direction opposite to the direction of the wind. The height of the water-level rise in front of the dam due to surge phenomena should be taken into account when determining the mark of the crest of a deal dam by the condition that there is no overflow of water.

The following main wave-forming factors have a decisive influence on the parameters of the waves in the reservoir and the height of the wind surge of the water:

- Wind speed;
- Duration of the wind;
- Wave acceleration length depending on the size and shape of the reservoir;
- Depth of the reservoir;
- Relief and roughness of the reservoir bed.

Usually when determining the parameters of waves and the height of the wind surge of water, the most dangerous case of steady waves during prolonged exposure to wind is considered. As a rule, the calculations do not take into account the influence of the topography and roughness of the reservoir bed.

Wind waves that form on the surface of the water of the reservoir are considered as gravitational waves. These waves are divided into force, free, and mixed.

Forced waves are the waves that are under the direct influence of the wind. They are three-dimensional (spatial) waves.

Free waves or swell waves propagate after the cessation of wind due to inertial forces. They are two-dimensional or cylindrical waves.

Mixed waves arise as a result of superposition of forced and free waves. They relate to three-dimensional waves.

When the waves interact with the structure, their partial or complete reflection occurs and the so-called reflected waves arise. When the reflected waves are superimposed onto incident waves, *interference* waves are formed.

An important special case of interfering waves is *standing* waves, which are formed when several waves of constant height approach a structure with a vertical or steeply inclined surface. The height of the standing waves is approximately twice the height of the free waves at the same length.

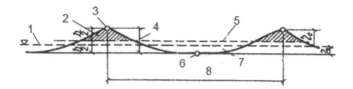

Figure 6.5 **Profile and wave elements.**

At a certain critical depth of water in the reservoir $H_{r,cr}$, the forced or free waves pass into the surf waves carrying a *breaker*.

With a sharp change in depth in front of structures or within it (in the case of a sloping structure), the waves roll over on a steep slope forming *breaking* waves.

Typically, the following basic elements and parameters of regular, two-dimensional free waves are considered (Figure 6.5):

1. *The wave crest* is the part of the wave located above the static level.
2. *The top of the wave* is the highest point of the crest of the wave.
3. *The depression (hollow)* of the wave is the part of the wave between two crests located below the static level.
4. *The sole of the wave* is the lowest point of the depression of the wave.
5. *Wave height h* is the vertical distance between the top and sole of the wave.
6. *The wave length λ* is the horizontal distance between two adjacent tops or soles of the wave.
7. *The middle wave line* is a horizontal line dividing the wave height into half.
8. *The design water level* in the reservoir.

     In addition, the following are considered:

- Steepness of the wave h/λ is the ratio of the height of the wave to its length;
- Wave period τ is the time after which the entire process of oscillation is repeated;
- Wave propagation velocity s is the velocity of the wave crest in the horizontal direction;
- *Wave front* is the line of the top of the wave crest in the plane.

The main design parameters of the waves in the reservoir include the height h, length λ, and wave period τ. It is believed that these parameters depend on the calculated wind speed V_w, the wave acceleration length L, and the average water depth in the reservoir H_r, m, at the considered level.

The most widely used method of determining the parameters of waves in reservoirs is proposed by S.S. Strekalov [12].

Most often, the calculated wave of 1% security in a system with a height of $h_{1\%}$ is considered, i.e., wave whose height is the largest of 100 waves.

As a result of the interaction of waves with a vertical or steeply inclined face of a concrete dam, the so-called standing waves are formed. The pressure of these waves on the structure can be determined by the method of V.K. Shtentsel [14]. The initial data for calculating the effect of standing waves on a concrete dam are wave height h of the security P_w in the wave system, average wavelength λ_m, average wave period τ, and water depth in the reservoir in front of the dam H_r. The calculation itself is reduced to determining the elevation or lowering of the wave surface at the structure, as well as the values of the wave pressure in depth at the calculated time points. In this case, the section of the reservoir in front of the dam is usually considered as a deep-water zone.

Figure 6.6 shows the plots of pressure of a standing wave on a vertical wall for three design cases.

The wave pressure force P_w can be defined as the area of the wave pressure plot. The point of application of this force can be found as the ratio of the static moment of the area of the wave pressure plot S_w to the force P_w.

According to the current SR [12,14], the maximum wind speeds of various annual probabilities of exceeding (or supply) p_w are taken as the **design wind speeds** depending on the type of calculations performed and the class of construction. The values of these speeds are determined by the results of statistical processing of observational data at weather stations.

The design wind speed should be determined at a height of 10m above the design water level in the reservoir. Moreover, if the observations are carried out at a different height, the design wind speed should be brought to a height of 10m using the coefficients given in the SR [12].

To determine the wave pressure on concrete dams, the parameters of the waves in the reservoir should be determined at the design wind speeds corresponding to the annual probabilities of exceeding p_w, which are taken depending on the considered combination of loads and the class of the dam:

* If the wave pressure is considered as a short-term load, then $p_w = 50\%$;
* If the wave pressure is considered as unusual load, then for dams of I and II classes $p_w = 2\%$, and for dams of III and IV classes $p_w = 4\%$.

When determining the level of the crest of a concrete dam, the wave parameters in the reservoir and the height of the wind surge of the water should be determined at the design wind speeds corresponding to the annual probabilities of exceeding p_w. These parameters are taken depending on the class of the dam, the combination of loads and effects considered, which corresponds to the calculated level of water in the reservoir:

* For the usual combination of loads and effects at NHL water for dams of I and II classes $p_w = 2\%$, and for dams of III and IV classes $p_w = 4\%$.

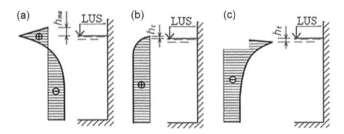

Figure 6.6 Plots of standing wave pressure on a vertical wall for three design cases [12].

- For an unusual combination of loads and effects at a forced SRL water in the reservoir for dams of I and II classes $p_w = 20\%$, for dams of III class $p_w = 20\%$, and for dams of IV classes $p_w = 50\%$.

Note that the parameters of the waves in the reservoir and the height of the wind surge of the water in front of the dam should be determined for the most unfavorable direction of wind speed and wavelength.

Ice pressure. When designing dams in areas with severe climatic conditions, it is necessary to reckon with the possibility of influencing the construction of significant ice loads. The estimated thickness of the ice formed on the surface of the reservoir in winter is determined by special calculations depending on the values of negative air temperatures and the duration of the frosty period, the flow of the reservoir, and other factors.

Distinguish between static and dynamic ice pressure. Static pressure of ice on a structure can arise due to the temperature expansion of a continuous ice cover with an increase in air temperature (and consequently of ice). In addition, static ice pressure can occur as a result of a collapse of an ice field or a blocking mass of ice under the influence of a current or wind. To prevent static ice pressure during operation, measures such as ice chipping and maintaining ice-free ice cover before construction are provided. Dynamic pressure on the structure is exerted by moving ice floes.

The intensity of ice pressure on the structure does not exceed the value

$$q_{ice} = 0.5 \times R_c \times h_{ice} \tag{6.12}$$

where

q_{ice} – the pressure of ice referred to the unit length of the dam;
h_{ice} – the estimated thickness of the ice, the value of which is determined depending on the climatic conditions of the area of construction of the HP;
R_c – the compressive strength of ice taken depending on the average daily air temperature T in the range from 0.45 MPa at $T = 0°C$ to 1.5 MPa at $T = -30°C$.

The resultant ice pressure force is considered applied below the calculated water level by $0.3 \times h_{ice}$.

In SR, Ref. [12] sets out a method for determining the loads and effects of ice for various design cases and various conditions for the interaction between ice and the structure.

Sediment pressure. The nature of the sediment depends on the depth of the reservoir and the flow rate of solid runoff. On mountain rivers near dams of small height, sand and sand–gravel deposits are deposited. At high dams that form deep reservoirs, silt particles, clays, and colloidal particles are deposited.

The estimated height of the sediment layer h_{ws} and the corresponding elevation of their surface z_{ws} are determined by water management calculations (Figure 6.7).

In the case of a vertical upper face of the dam, the horizontal pressure plot of sand and sand–gravel deposits is assumed to be triangular (Figure 6.7a); by excluding sediment friction forces along the upper face of the dam, the sediment pressure intensity p_{ws} at a depth of h_{ws} is determined by the following formula:

$$p_{ws} = (\rho_{ws} - \rho_w) \times g \times h_{ws} \times \xi_0 \tag{6.13}$$

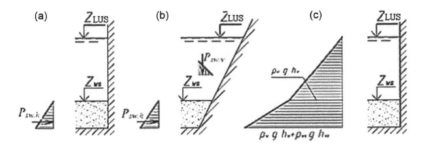

Figure 6.7 Plots of sediment pressure on the upper face of the dam: sand and sand–gravel deposits on the vertical (a) and inclined; (b) faces of the dam; (c) water pressure and sediment as heavy fluid.

where
ρ_{ws} and ρ_w – respectively, the density of sediment soil saturated with water and the density of water;
ξ_0 – the coefficient of lateral pressure of sediment soil equal to

$$\xi_0 = \frac{v_{ws}}{1 - v_{ws}} \tag{6.14}$$

v_{ws} – Poisson's ratio of sediment soil.

The values of the sediment soil characteristics ρ_{ws} and v_{ws} should be determined on the basis of special studies. When performing calculations at the preliminary design stages, the following values of ρ_{ws} and v_{ws} can be taken:

- For loose sands with a porosity coefficient, $e = 0.9$ $\rho_{ws} = 1,900\,\text{kg/m}^3$, $v_{ws} = 0.39$;
- For sands of medium density with a porosity coefficient, $e = 0.7$ $\rho_{ws} = 2,000\,\text{kg/m}^3$, $v_{ws} = 0.34$;
- For dense sands with a porosity coefficient, $e = 0.5$ $\rho_{ws} = 2,100\,\text{kg/m}^3$, $v_{ws} = 0.31$.

The horizontal and vertical sediment pressure will act on the inclined upper face of the dam. Plots of these loads as in the previous case are accepted in the form of triangles (Figure 6.7b); by excluding friction forces, the value of sediment pressure intensity p_{ws} at a depth h_{ws} is determined by the formula (6.13), in which the value of the lateral pressure coefficient of sediment soil ξ_0 is calculated by any of the following formulas:

$$\xi_0 = \frac{\cos^2\left[0.5\times\left(\alpha + \phi_{ws,r}\right)\right]}{\cos^2\left[0.5\times\left(\alpha - \phi_{ws,r}\right)\right]} \text{ or } \xi_0 = \frac{\sin^2\left(\alpha + \phi_{ws,r}\right)}{\left[\sin\left(\alpha\right) + \sin\left(\phi_{ws,r}\right)\right]^2} \tag{6.15}$$

where
α – the angle of inclination of the upper face of the dam to the horizon;
ρ_{ws} and ρ_w – as before the density of sediment soil saturated with water and the density of water;

$\varphi_{ws,\,r}$ – conditional value of the angle of internal friction of sediment soil, determined by the following formula:

$$\varphi_{ws,r} = \arcsin\left(1 - 2v_{ws}\right)$$ (6.16)

The intensity of the vertical pressure of sand and sand–gravel sediments $p_{ws,v}$ at a depth of h_{ws}, determined without taking into account the friction forces of the soil along the upper face of the dam, can be found by the following formula:

$$p_{ws,v} = p_{ws,h} \times \mathrm{ctg}(\alpha)$$ (6.17)

The horizontal P_{wsh} and vertical P_{wsv} components of the pressure force of sand and sand–gravel deposits are defined as the corresponding areas of pressure intensity plots p_{wsh} and p_{wsv}.

The pressure of sand and sand–gravel deposits on small dams located on pliable nonrocky foundations can be considered as active pressure. In the case of a vertical upper face of the dam, the sediment pressure force P_{ws} can be found by the Coulomb formula:

$$P_{ws} = \frac{1}{2} \times \left(\rho_{ws} - \rho_w\right) \times g \times h_{ws}^2 \times \mathrm{tg}^2\left(45° - \frac{\phi_{ws}}{2}\right)$$ (6.18)

where
φ_{ws} – the angle of internal friction of the sediment soil.

In deep low-flow reservoirs, small sediments in the form of clay, silt, and colloidal particles are deposited in front of the dam with practically no connections between them. In this case, the sediment layer is considered to be a layer of heavy fluid with a density ρ_{ws}, which does not have internal friction (Figure 6.7c). The pressure intensity of such a liquid p_{ws} at a depth of h_{ws} can be determined by the following formula:

$$p_{ws} = \rho_w \times g \cdot x_w + \rho_{ws} \times g \times h_{ws}$$ (6.19)

where
ρ_w – the density of water;
h_w – the depth of water above the surface of sediment.

The density of the p_{ws} containing heavy fluid in the formula (6.19) is determined on the basis of special studies. At the preliminary design stages, the p_{ws} value can be taken in the range from $p_{ws} = 1{,}200\,\mathrm{kg/m^3}$ to $p_{ws} = 1{,}500\,\mathrm{kg/m^3}$.

6.2.3 Temperature effects

The temperature regime of a concrete dam changes due to seasonal fluctuations in the ambient temperature (air and water in the reservoir), insolation of the downstream face of the dam, the influence of heat flow from the foundation, heating due to exothermic during hardening of concrete, artificial cooling, and other factors.

A time-varying temperature field causes thermal stresses to appear in the dam, which can cause cracks. Cracks that violate the continuity of the structure reduce its reliability and durability. In some cases, cracks lead to a change in the static scheme of structure and sometimes to a loss in the bearing capacity of the structure.

The formation of temperature fields and temperature stress fields in a concrete dam begins from the moment of its erection and continues throughout the construction and operation of the structure.

The concrete mix placed in the block usually has a temperature different from the temperature of the air and the base of the block. In summer, the temperature of the concrete mix is usually lower than the ambient temperature, and in winter, it is higher. Concrete laid in the block is subsequently heated as a result of exothermic during cement hydration. At the same time, there is a process of cooling the block from the surfaces due to the temperature difference between the surface of the block and the air. Concrete laying in adjacent and upper blocks and concrete care measures (watering the surface, artificial cooling) significantly affect the temperature conditions of the block and the block system. As a result, an unsteady highly nonuniform temperature field arises in the block under consideration and the block system. Under the influence of the temperature field, an unsteady stress field arises in the block and the block system. The formation and subsequent change in the temperature stress field is complicated by factors such as:

- Growth with age of the modulus of elasticity of concrete;
- Very significant at an early-age creep of concrete;
- Changing the static scheme of the structure at the time of laying concrete in adjacent and upper blocks.

The change in the temperature field and the field of temperature stresses does not end at the moment of completion of the dam construction and filling of the reservoir and continues until it cools completely after which it is determined only by seasonal fluctuations in the ambient temperature (air and water).

To evaluate the influence of temperature effects on concrete dams, two calculations must be made: the calculation of the time-varying temperature field in the dam and the calculation of temperature stresses in the structure.

Main factors affecting the formation of the temperature regime of concrete dams can be divided into two groups: natural (climatic) and production-technological.

Among the natural factors associated with the climatic conditions of the area where the concrete dam is located are:

- Varying time air temperatures;
- Varying time water temperatures in the reservoir;
- Varying time temperature of the water in the river during the construction period as well as the water temperature in the downstream of the dam during the operational period;
- Varying time groundwater temperatures at the dam foundation;
- Intensity of solar radiation;
- Geothermal flow from the bowels of the Earth.

A number of production-technological factors include the following:

- Heat in hardening concrete;
- Accepted sequence of erection and cutting of the dam into concrete blocks;
- Use of pipe cooling of concrete masonry;
- Watering the surface of concrete blocks with chilled water;
- Use of insulated formwork and tents when concreting dams in winter conditions.

Characteristics of changes in air temperature are determined on the basis of processing data from meteorological observations for a sufficiently long period of time (30–50 years). In the general case, the air temperature in each particular place can be considered as a random function of time.

One of the important characteristics is the *average annual air temperature $T_{a,m}$*, which depending on the location of the construction object can fluctuate over a fairly wide range. Below are the mean annual air temperatures for some HEP with high concrete dams.

"Great Energy Dam Ethiopian Renaissance" (Ethiopia)	28°C
Nam Chien (Vietnam)	19.2°C
Inguri (Georgia)	14°C
Toktogul (Kyrgyzstan)	8.3°C
Sayano-Shushensk (Russia)	0.5°C
Ust-Ilim (Russia)	−3.9°C
Zeya (Russia)	−4.5°C
Kankun (Russia)	−10.4°C

The average annual air temperature for each particular year is slightly different from the average annual temperature by 1°C–1.5°C. Therefore, *the average annual air temperature* can be considered as a fairly stable characteristic.

Monthly average temperatures are used to estimate air temperature. Figure 6.8 shows the examples of changes in long-term monthly average air temperatures for three HPs located in different climatic regions.

To describe the change in average monthly temperatures over the course of the year T_a, the law of harmonic oscillations in the form of a cosine wave is often applied, which allows the use of analytical solutions:

$$T_a = T_{a,m} + A_{T,a} \times \cos\left(\frac{2\pi}{t_y} \times t\right) \tag{6.20}$$

where

$A_{T,a}$ – amplitude of fluctuations of average monthly temperatures;
t_y – period of annual fluctuations ($t_y = 8{,}760$ hours);
t – time.

Monthly average air temperature values are used when calculating the temperature regime and thermal stress state of concrete dams during the operational period. In the

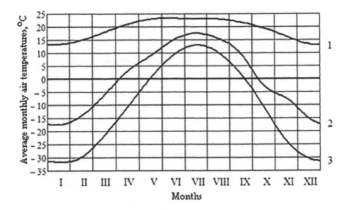

Figure 6.8 Change in average monthly air temperatures in various climatic regions HEP: I: Nam Chien; 2: Sayano-Shushensk; 3: Kankun.

usual combinations of loads and effects, a design year with the average long-term amplitude of fluctuations in average monthly temperatures is considered [15]. In unusual combinations of loads and effects, a year with the maximum amplitude of monthly average temperature fluctuations is considered as well as a year with a maximum monthly average temperature.

For a preliminary characterization of the temperature conditions for the design and construction of dams, a classification of temperature effects by severity was proposed in Ref. [177] (Table 6.3).

When performing temperature calculations of concrete blocks during the construction period, it is necessary to take into account daily changes in air temperature for the corresponding period of concrete time.

Temperature of the water in the reservoir determines the temperature of concrete near the upstream face dam and the dam as a whole and the foundation of the structure during the operational period. The nature of the change in water temperature over time and along the depth of the reservoir depends on the depth and flow rate of the reservoir, and climatic factors, including air temperature, wind speed.

When assessing the temperature regime of a reservoir, two zones are usually considered along the depth of the reservoir: the upper zone up to 40 m deep, within which there is an intense change in water temperature both in depth and in time, and the lower within which the water temperature change is insignificant.

Table 6.3 Classification of temperature effects by severity

The degree of severity of temperature effects	Average annual air temperature $T_{a,m}$, °C	Amplitude of fluctuations in average monthly temperatures A_{Ta}, °C
Favorable	12 and above	22 and less
Middle	6–12	28–30
Severe	Around 0	36–38
Particularly severe	−5 and below	44–46

In warm climates, the surface temperatures of the reservoir are close to the monthly average air temperatures. The difference in these temperatures does not exceed 1°C–3°C in one direction or another. The amplitude of water surface temperature fluctuations can reach 20°C–25°C.

Near the bottom of the reservoir, the water temperature is close to the minimum monthly average air temperature. The amplitude of fluctuations of this temperature does not exceed 1°C–2°C. Figure 6.9a shows a graph of the depth changes in the minimum and maximum temperatures of the water in the reservoir located in an area with a warm climate.

If the reservoir is located in an area with a severity climate when negative average monthly air temperatures occur in winter, the water surface temperatures in summer are close to average monthly air temperatures with deviations of 1°C–3°C. In the winter period after the formation of the ice cover, the water surface temperature is close to 0°C and can remain so for several months. In summer, the amplitude of temperature fluctuations in the surface of the water is close to the average monthly temperature of the warmest month. The water temperature at the bed of the reservoir in this case is close to 4°C, i.e., to a temperature that corresponds to the maximum density of water. The amplitude of fluctuations of this temperature is insignificant and does not exceed 1°C–2°C. Figure 6.9b shows the graphs of changes in depth of the minimum and maximum temperatures of the water in the reservoir located in an area with a severe climate.

To assess the temperature regime of reservoirs at the preliminary design stages, data from field observations of seasonal fluctuations in water temperature at operated analogous reservoirs that have parameters close to those of the designed reservoir and located in areas with climatic conditions close to design are usually used.

When performing estimated calculations of the thermal regime of nonflowing deep reservoirs at the preliminary design stages, a simplified approach based on the analysis of field observations of seasonal fluctuations in water temperature in operated

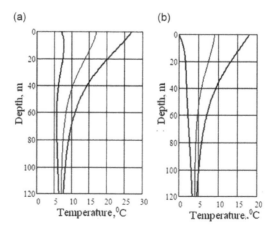

Figure 6.9 Change in depth of the minimum and maximum water temperatures in nonflowing deep reservoirs: (a) in an area with a warm climate; (b) in an area with a severity climate.

reservoirs can be applied. When performing the final calculations of the reservoir temperature, it is recommended to use the methodology described in Ref. [177].

Changes in the temperature of river water and groundwater are necessary when calculating the thermal regime of a concrete dam during the construction period in the event that construction water flows pass through unfinished structures. In addition, this information is used in the design of measures to regulate the temperature of concrete masonry during the construction period in particular in the design of pipe cooling of concrete blocks when natural river water is used as a cooler.

The nature of changes in the temperature of river water depends on the climatic conditions of the region where the river is located, the conditions of the river's water supply sources, and other factors. The maximum water temperatures in rivers are usually lower than the maximum monthly average air temperatures by 10°C–15°C. In areas with negative winter air temperatures, the temperature of river water in winter is close to 0°C.

Data on the natural temperatures of river water are obtained on the basis of an analysis of the results of hydrological surveys and hydrometeorological observations over a fairly long period (30–50 years).

During the operational period, the temperature of the water in the downstream of the dam may affect the temperature regime of the parts of the dam washed by water and in general the thermal regime of the "structure–foundation" system. The values and nature of the change in water temperature in the downstream depend on the temperature conditions in the reservoir in the area of the water inlets of the culverts of the HP.

Groundwater temperature can have an effect on the temperature of rocking concrete blocks and must be taken into account when performing temperature calculations of concrete dams during the construction period. Under natural conditions, data on the position of the surface and seasonal changes in groundwater temperature are obtained based on hydrogeological surveys.

After filling the reservoir together with a change in the hydrogeological regime of the dam foundation (occurrence of a filtration flow caused by the difference in levels of the upstream and downstream), the thermal regime of groundwater changes. At the same time, the temperature of groundwater after some time of operation of the structure becomes close to the temperature of the water in the lower layers of the reservoir.

Solar radiation or insolation, which is a flux of radiant energy from the sun to the Earth's surface, can have a significant effect on the surface temperature of concrete blocks during the construction period and on the temperature of the downstream face of the concrete dam during the operational period.

Solar radiation consists of *direct* and *diffuse* radiation. Direct radiation is the predominant component of solar radiation (about 3/4 of the total radiation) and is the radiant energy coming directly from the Sun in a cloudless sky. The diffused radiation coming from all points of the sky is formed due to the diffusing part of the solar radiation in the atmosphere. It is the diffused radiation that creates daylight and gives color to the sky.

In Ref. [12], data are presented for determining the total (direct and diffuse) solar radiation in a cloudless sky depending on geographic latitude, month of the designed year, and the position of the calculated surface that receives radiation.

Heat generation of concrete is one of the most important factors affecting the temperature regime of concrete dams during the construction period. Heat dissipation of concrete occurs due to complex physicochemical processes that occur during hardening of cement stone mainly due to the hydration of cement.

Heat generation of concrete is associated with self-heating of the constructed concrete masses as a result of which the temperature in the internal areas can increase significantly (sometimes by 40°C–50°C or more). As a result of subsequent cooling, an uneven temperature field is formed in the concrete mass, causing temperature stresses that can lead to cracking.

To assess the heat release, it uses the characteristic *specific heat generation of concrete*, which is the amount of heat that is released during hardening per unit mass of cement at the designed time after concrete preparation. In temperature calculations, another characteristic is widely used – the *intensity of specific heat of concrete* – which is the rate of change in time of specific heat of concrete.

Heat generation and its intensity depend on a large number of factors. These factors include cement consumption per $1\,m^3$ of concrete; chemical and mineralogical composition; the presence of pozzolana; fineness of cement grinding; age of concrete; initial temperature of the concrete mix; current temperature of concrete; and other factors.

The heat release process during hardening of concrete as well as the process of cement hydration takes place unevenly in time gradually fading.

Numerous experimental studies have shown that during the first 7–10 days of concrete hardening, up to 70%–80% of the total heat of cement hydration is released, after which the heat release process is significantly slowed down. Figure 6.10 shows a typical curve for the specific heat of concrete.

It is recommended to take the heat of concrete depending on the age of the concrete, type and mark of cement, in accordance with the data in Table 6.4.

Figure 6.11 shows the dependences of heat generation on the age of concrete for portland cement marks 300, 400, and 500 constructed in accordance with the data in Table 6.4.

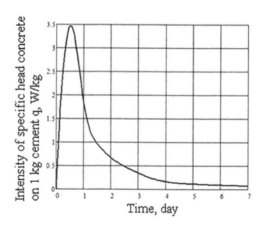

Figure 6.10 The intensity of specific heat concrete.

Table 6.4 Heat dissipation characteristics of concrete

Type of cement	Cement mark	Heat generation of concrete, kJ per I kg of cement aged concrete, days			
		3	7	28	90
Portland cement	300	210	250	295	300
	400	250	295	345	355
	500	295	335	385	400
Pozzolana portland cement	300	175	230	270	280
Slag portland cement	400	210	265	320	335

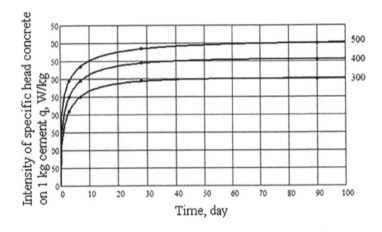

Figure 6.11 Dependence of concrete heat release on age.

At the final stages of the design of concrete dam, the data on the heat release of concrete are specified on the basis of special experiments in the process of research on the selection of concrete composition.

To carry out calculations of the temperature regime of concrete dams, the thermophysical characteristics of concrete and the rock foundation are needed. The values of these characteristics to a decisive degree affect the intensity of changes in the temperature field of the "structure–foundation" system both in time and in coordinates.

The characteristics used in performing temperature calculations include the following:

- Material density ρ, unit of density – kg/m^3;
- Specific heat c, numerically equal to the amount of heat that must be transferred to a unit mass of a substance to increase its temperature per unit; unit of measure – J/(kg°C);
- Thermal conductivity (the term "thermal conductivity coefficient" is often used in the technical literature) λ, numerically equal to the amount of heat flowing through a unit of surface area per unit time at a unit temperature drop per unit normal length to this surface; thermal conductivity unit – W/(m°C);

- Temperature conductivity (in the technical literature, the temperature conductivity coefficient is often used) characterizes the rate of temperature change in no equilibrium thermal processes; temperature conductivity is numerically equal to the ratio of thermal conductivity to volumetric heat capacity $C = c/\rho$

$$a = \frac{\lambda}{C} \text{ or } a = \frac{\lambda}{c \times \rho} \tag{6.21}$$

temperature conductivity unit $- m^2/s$.

In the general case, the indicated characteristics of hydraulic concrete depend on many factors. These factors include the size and mineralogical composition of the aggregate, the type and consumption of cement, the water/cement ratio, the setting conditions of concrete and its moisture regime, the age of the concrete, and temperature conditions.

When calculating the temperature regime of concrete dams, it is necessary to take into account the heat transfer conditions between the concrete surface and the external media (air or water). If the body surface temperature is higher than the ambient temperature, there is a heat flux directed from the body to the media; i.e., the body cools. When the ambient temperature is higher than the surface temperature of the body, the heat flux is directed from the media to the body; i.e., the body heats up.

As a thermophysical characteristic that characterizes the intensity of heat transfer in the contact zone between the concrete surface and the media, a physical quantity is used – *convection heat transfer coefficient* (or β_{cb}), which is the amount of heat passing per unit time through a unit surface of the body at a temperature difference between body surface and media equal to unit temperature. The unit of measure for convection heat transfer coefficient is $W/(m^2 {}^\circ C)$.

The value of the heat transfer coefficient by convection from the open surface of concrete into the air mainly depends on the wind speed: in the absence of wind, $\beta_{cb} = 3.4–4.5$ $W/(m^2 {}^\circ C)$; with an average wind speed, $\beta_{cb} = 11.5–17.5$ $W/(m^2 {}^\circ C)$; in strong winds, $\beta_{cb} \geq 30$ $W/(m^2 {}^\circ C)$ [15]. In Ref. [39], it is recommended to take the following values of the heat transfer coefficient by convection from the open surface of concrete: into the outside air, $\beta_{cb} = 24$ $W/(m^2 {}^\circ C)$; into the air inside hollow seams, shafts, and tents, $\beta_{cb} = 7–12$ $W/(m^2 {}^\circ C)$.

The value of the heat transfer coefficient by convection from the open surface of concrete into water depends on the speed of movement of water and is in the range $\beta_{cb} = 120–240$ $W/(m^2 {}^\circ C)$. Such high values of β_{cb} allow according to SR [12], to take $\beta_{cb} = \infty$ in practical calculations.

When calculating the temperature regime of high concrete dams, it is necessary to take into account the thermophysical characteristics of rock: density ρ, specific heat c, thermal conductivity λ, and temperature conductivity a. In the general case, these characteristics depend on the type of rock, their fissuring, humidity, temperature, and other factors. Typically, the values of these thermophysical characteristics are assumed to be constant and independent of these factors. Note that the degree of rocks fissuring, which largely determines the values of their physico-mechanical characteristics, significantly less affects the thermophysical characteristics of these soils.

Table 6.5 Thermophysical characteristics of rocks

Rock	ρ (kg/m^3)	c, kJ/(kg°C)	λ, W/(m°C)	$a \cdot 10^6$ (m^2/s)
Basalt	2,600–2,800	0.68–0.95	2.10–2.80	1.16–1.42
Gabbro	2,980	0.72	2.01	0.93
Granite	2,100–2,410	0.88–0.92	2.38–3	1–1.25
Diabase	2,970	0.70	2.22	1.07
Dolomite	2,510–2,550	0.72–0.96	3.02–3.33	1.36–1.58
Limestone	2,300–2,800	0.88–0.94	2.38–3.19	0.61–1.36
Quartzite	2,430	0.91	3.54	1.61
Sandstone	1,950–2,300	0.88	2.07–2.45	1.19
Ryolite	2,350	0.95	2.07	0.94

The thermophysical characteristics of the rock foundation of the dam should be determined on the basis of special studies that should be carried out when performing engineering–geological surveys.

At the preliminary design stages when performing temperature calculations of concrete dam, it can use the data on the thermophysical characteristics of some rocks shown in Table 6.5.

Calculations of the temperature regime of concrete dams are reduced to determining the temperature at any point in the structure at any time and are carried out by methods of the theory of thermal conductivity. In most cases, the temperature field changes over time – i.e., it is *nonstationary*; the temperature-invariable temperature field is called *stationary*.

When determining the temperature regime of the dam during the operational period in the general case, it is necessary to take into account the influence of water filtration in the foundation and solve the problem of heat and mass transfer for the selected designed area of the foundation. However with a sufficient degree of certainty for high concrete dams, temperature of the foundation of the structure can be taken equal to the temperature of the filtered water that comes from the lower layers of the reservoir.

FEM is an effective computational method for solving problems of the theory of thermal conductivity, which allows solving almost any stationary and nonstationary, two-dimensional, and spatial problems of the theory of thermal conductivity under any boundary conditions. In this case, homogeneous and inhomogeneous bodies, bodies with a time-varying internal heat release, and bodies with a time-varying configuration and boundary conditions can be considered. Due to the considerable complexity, such calculations should be performed at the final stages of dam design. Moreover, in order to obtain reliable calculation results, it is necessary to ensure sufficient accuracy of the input data.

6.2.4 Seismic effects

Concrete dams are earthquake-resistant structures. For example, during the devastating Sichuan earthquake in China (Wenchuan County) on May 12, 2008, the high arch dam of the Shaipai and the gravity dam of Bauzuzi did not receive a significant damage although the intensity of this earthquake significantly exceeded the designed value.

An earthquake is a dynamic oscillation that occurs in the Earth's crust due to its movements and deformations. In accordance with modern views, earthquakes are associated with the following circumstances.

It is believed that the outer shell of the Earth (lithosphere) consists of several giant tectonic plates. The thickness of such plates is about 80 km. Plates move relative to each other under the influence of convection flows arising at great depths under conditions of high temperatures and pressures. As a result of the interaction of the plates, stress concentration zones arise at their contacts increasing with a speed proportional to the speed of movement of the plates relative to each other. At the same time, there is a process of stress relaxation, which reduces the rate of their growth. If the stresses in the zones of their concentration do not have time to relax and reach the rock strength values, there is a fault (rupture). In such cases, it speaks of tectonic earthquakes with the greatest force.

The area around the fault is usually called the *center* or *hypocenter* of the earthquake. The projection of the hypocenter onto the Earth's surface is commonly called the *epicenter* of the earthquake.

Depending on the focal depth of H_{eq}, *deep-focus* earthquakes are distinguished at H_{eq} = 300–700 km, *intermediate* ones at H_{eq} = 60–300 km, *normal* ones at H_{eq} = 15–60 km, and *small-focus* at H_{eq} less than 15 km. The most destructive are small-focus earthquakes.

Rupture of rocks in the earthquake center is accompanied by the release of accumulated energy, most of which is spent on plastic deformation and destruction of rock, and the rest, a relatively small part of the energy, causes environmental oscillations.

The oscillations are accompanied by the propagation of seismic waves of two types: longitudinal and transverse. With the passage of longitudinal waves, time-varying compression-tensile strains arise, and the propagation of transverse waves is associated with slide strains. The propagation velocity of seismic waves depends on the density and elastic characteristics of the rocks. The propagation velocities of longitudinal V_p and transverse V_s waves in homogeneous isotropic elastic bodies can be found by the following formulas:

$$V_p = \sqrt{\frac{1 - v_{dyn}}{\left(1 + v_{dyn}\right) \times \left(1 - 2v_{dyn}\right)} \times \frac{E_{dyn}}{\rho}}, \, V_s = \sqrt{\frac{G_{dyn}}{\rho}} = \sqrt{\frac{E_{dyn}}{2(1 + v_{dyn}) \cdot \rho}} \qquad (6.22)$$

where

E_{dyn}, v_{dyn}, and G_{dyn} – dynamic elastic modulus, Poisson's ratio, and slide modulus, respectively;

ρ – the density of the material.

When longitudinal and transverse waves reach the surface of the Earth, they are reflected. The superposition of waves suitable to the Earth's surface on the reflected waves leads to interference of the waves. As a result, surface waves arise that are subdivided into *Love waves* and *Rayleigh* waves. *Love* waves are transverse horizontal oscillations in the direction perpendicular to the direction of wave propagation. With the passage of *Rayleigh* waves, particles move in a vertical plane describing an ellipse.

Oscillations of the Earth's surface during earthquakes at any point are an unsteady oscillatory process, which is characterized by rapidly varying accelerations as well as velocities and displacements that depend on them. The oscillation forms and peak values of these parameters depend on the geological structure of the rock mass and to a large extent on the distance of the considered point of the Earth's surface to the epicenter of the earthquake. Obviously, the peak values of the parameters of the oscillations of the Earth's surface have maximum values in the epicenter of the earthquake and decrease with distance from it.

To observe earthquakes in seismically hazardous areas, seismic stations are located where acceleration due to earthquake vibrations caused by earthquakes as well as during seismic exploration is recorded using a seismograph.

The most important characteristic of *magnitude* earthquakes is the decimal logarithm of the maximum amplitude (μ) of a seismic wave recorded by a standard-type seismograph at a distance of 100 km from the epicenter of the earthquake. The magnitude value is recalculated if the record was obtained on a seismograph of a different type or the record was made at a different distance from the epicenter of the earthquake. The use of magnitude as a characteristic of an earthquake was proposed by *Wadachi* (Japan) in 1931. Later in 1935, *C. Richter* (USA) improved the method for determining magnitude.

There is an empirical relationship between the total energy released in the center of the earthquake E, J, and magnitude M:

$$\lg E = 4.8 + 1.5\,M = K \tag{6.23}$$

For the energy assessment of the earthquake, the *Richter* scale is used. The value of the ball on this scale is taken equal to the magnitude value. The strongest earthquakes were observed in Colombia and Ecuador (1906), Japan (1933), and the Pacific Ocean 138 km from Honshu Island (2011). The magnitude of these earthquakes was 8.9.

The parameters of oscillations of the Earth's surface depend on the magnitude, the distance of the considered point of the Earth's surface to the source of the earthquake, and the geological structure of the rock mass.

To represent the seismic oscillation in the form of average peak horizontal acceleration, a graph was developed by *I. M. Idriss* [66], shown in Figure 6.12.

In addition to the energy characteristics of an earthquake in magnitude, the concept of *earthquake intensity* is widely used, which characterizes the manifestation of seismic oscillations at a particular point on the Earth's surface. Note that the intensity of the same earthquake at different points on the Earth's surface is different although the magnitude is the same [146].

Various scales are used to assess the intensity. In North America, the 12-ball *Mercalli* scale is widely used. In Japan, an *8-ball IMA scale* is used (minimum value 0 balls, maximum 7 points). Developed by *S. V. Medvedev* (Russia), *V. Schpoheuer* (Germany) and *V. Karnik* (Czech Republic), *12-point scale MSK-64* is used.

In Russia and Ukraine, the MSK-64 scale is currently used according to which the intensity of earthquakes I in balls can be determined on the basis of a qualitative description of the effects of earthquakes and a quantitative assessment of the parameters of the Earth's surface oscillations. A qualitative description of the effects of earthquakes is based on three criteria [192]: sensations of people and effect on surrounding objects; effects on buildings and structures of various types; and residual effects in soils and changes in the regime of groundwater (Table 6.6).

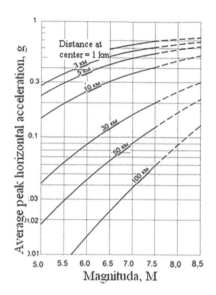

Figure 6.12 Dependence of the average peak horizontal acceleration on magnitude and distance to the earthquake center.

Table 6.6 Qualitative description of the effects of earthquakes on the MSK-64 scale

Intensity earthquakes I, balls	Brief description of the earthquake
1	Soil tremors are noted only by devices.
2	Soil tremors are felt in some cases by people.
3	An earthquake are felt by small people.
4	An earthquake is felt by many people. Glass rattling possible.
5	Swing hanging objects. Sleeping people wake up.
6	Light damage to buildings, thin cracks in the plaster. In some cases, in moist soils, cracks up to 1 cm wide are possible, landslides in mountainous areas. Changes in the flow rate of sources are observed.
7	In the plaster cracks and chipping pieces. Thin cracks in the walls. Waves on the surface of the water; raising sludge. Change in water level and flow rate of sources. Landslides in the sandy and gravel banks of rivers.
8	Large cracks in the walls, falling cornices, chimneys. Small landslides on steep slopes of embankments and excavations. Cracks in the soil up to several centimeters.
9	In some buildings, the collapse of walls, ceilings, roofs. Damage to artificial ponds; there are big waves on the surface of the water. On the plains of the flood. Frequent landslides, scree; the rocks are falling off. Cracks up to 10 cm wide and more in the ground.
10	Collapses in many buildings. Cracks up to 1 m wide in soils. Dangerous damage to dams and embankments. Large landslides on the banks of rivers and seas. Wide gaps parallel to the channels of watercourses. Water splashing out of their canals, rivers, lakes. The emergence of new lakes.
11	Disaster. Numerous cracks on the surface of the Earth, large landslides in the mountains. Serious damage to dams, bridges, railways. The definition of ball requires special studies.
12	Change of terrain. Severe damage or destruction of almost all ground and underground structures. The definition of ball requires special studies.

For a quantitative assessment of the parameters of oscillations of the Earth's surface, the MSK-64 scale contains the limits for the values of the following values corresponding to a particular ball I:

a_0 – the amplitude of acceleration of soil oscillations for periods of 0.1–0.5 seconds;
v_0 – the amplitude of the speed of soil oscillations for periods of 0.5–2 seconds;
x_0 – the amplitude of the displacements of the center of the pendulum of a standard seismograph with a natural oscillation period of 0.25 seconds and a logarithmic decrement of 0.5 (Table.6.7).

The values of the amplitude of acceleration of soil oscillations are used when calculating building structures for seismic effects. The data not shown in Table 6.7 for weak earthquakes with an intensity below $I = 5$ balls and for very strong earthquakes with an intensity above $I = 10$ balls can be obtained by nonlinear (parabolic) extrapolation.

One important characteristic is the *recurrence* (*frequency*) of earthquakes. When performing seismic resistance calculations of HS including concrete dams, it is necessary to take into account sufficiently strong earthquakes of rare recurrence. The value of the estimated recurrence period of earthquakes T_{ret} should be taken depending on the type of seismic effect and the class of structures equal to 500, 1,000, and 5,000 years [117,118].

The most complete earthquakes are characterized by *accelerogram*, which are the time-dependent seismic accelerations of the Earth's surface at this point in various directions.

Accelerograms of observed earthquakes are recorded at seismic stations using seismographs for fixed directions: north–south (*N-S*), west–east (*W-E*), and vertical (*V*). Accelerograms can be presented in the form of graphs, but now digitized accelerograms are usually used. The digitized accelerogram is a table of seismic acceleration values in this direction which are determined at fixed time points with a sufficiently small step, for example, in steps of 0.005, 0.01, and 0.02 seconds.

An important characteristic of the accelerogram is the *maximum peak acceleration* of the foundation a_p, which is taken as the seismic acceleration maximum in absolute value during the earthquake.

The nature of the accelerogram and the value of the maximum peak acceleration depend on the magnitude of the earthquake, the distance of the considered point of the Earth's surface from the center engineering–geological conditions, and other factors.

Table 6.7 Earth oscillation parameters on the MSK-64 scale

Intensity earthquakes I, balls	Ground oscillation characteristics		
	a_0 (cm/s^2)	v_0 (cm/s)	x_0 (mm)
5	12–25	1–2	0.5–1
6	25–50	2.1–4	1.1–2
7	50–100	4.1–8	2.1–4
8	100–200	8.1–16	4.1–8
9	200–400	16.1–32	8.1–16
10	400–800	32.1–64	16.1–32

The accelerogram corresponds to a *cyclegram* and a *seismogram*, which are records of the velocities and displacements of the Earth's surface in time at the point under consideration during an earthquake.

When performing seismic stability calculations of responsible structures, calculated accelerograms (and/or cyclegrams, seismograms) are used. The parameters of the calculated accelerograms corresponding to the calculated seismicity of the structure location site are established during seismological surveys and studies taking into account data on the speed, frequency, and resonance characteristics of the soils lying at the foundation of the structure. Moreover, according to the requirements of the SR [8], accelerograms recorded at the construction site are analogous to accelerograms obtained in areas similar to the area of the construction site under seismological conditions, and the synthesized accelerograms formed in accordance with the calculated parameters of seismic effects can be used.

Synthesized accelerograms are obtained by calculation methods based on statistical processing and analysis of a number of accelerograms and/or spectra of real earthquakes taking into account local seismological conditions.

Figure 6.13a, for example, shows the calculated accelerogram of horizontal oscillations of the Earth's surface, which was used when performing seismic resistance calculations of the main structures of the Dnestr HEP. The maximum peak acceleration of this accelerogram is $a_p = 0.26g$, where g is the gravitational acceleration. Figure 6.13b and c shows the cyclegram and seismogram corresponding to the calculated accelerogram.

In the seismic stability analysis of responsible structures, three-component accelerograms are used that describe the oscillations of the Earth's surface in three directions. In this case, it is necessary to consider the horizontal radial component (the direction of the site is the center of the earthquake), the horizontal tangential component (perpendicular to the radial component), and the vertical component of the accelerogram [266].

Based on observations of seismic activity, a general seismic zoning (GSZ) has been created, which is a division of the territory of a state or region into areas in which earthquakes of one or another intensity in balls are possible. In Ref. [8], maps are presented on which regions with the intensity of possible earthquakes in balls of the MSK-64 scale are plotted for various earthquake recurrences:

- Map A GSZ-97 for the earthquake recurrence period $T_{ret} = 500$ years;
- Map B GSZ P-97 for the earthquake recurrence period $T_{ret} = 1,000$ years;
- Map C GSZ-97 for the earthquake recurrence period $T_{ret} = 5,000$ years.

In addition, there is a list of cities and large settlements with an indication of the estimated seismicity maps GSZ-97 compiled for medium engineering–geological conditions.

According to Ref. [8], the seismicity in balls of the MSK-64 scale established by OCP-97 cards is considered as *standard seismicity* (indicated by I^{nor}).

When designing the HSs of I and II classes included in the head front, it is necessary to carry out a detailed seismic zoning in the framework of which a seismotectonic model of the area of the object is compiled. Such a model should include a map and characteristics of possible earthquake centers, information on the presence or absence

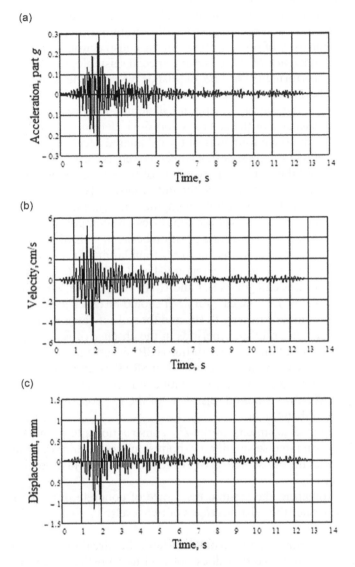

Figure 6.13 Accelerogram (a), cyclegram (b), and seismogram (c) used in seismic stability calculations of the main structures of the Dnestr HEP.

of active faults, and the possibility of slope displacements of large volumes. As a result
of a detailed seismic zoning, the *initial seismicity* of the area of the location of the objects is established in balls of the MSK-64 scale (indicated by I^{beg}).

For constructions of classes III and IV, the initial seismicity is assumed to be equal
to the standard seismicity, i.e., $I^{beg} = I^{nor}$.

With the same initial seismicity within the location of the HP, the intensity of the
seismic effect in the area of a particular structure may be different depending on the
engineering–geological conditions. In the case of rock foundations, the intensity of

seismic impact is lower, and in the case of clayey water-saturated soils, it is significantly higher. In addition, hydrogeological conditions, the nature of morphological changes in soils, the presence of possible discontinuous disturbances, and other factors affect the intensity of seismic effects in the construction zone.

To assess the influence of these factors, special seismological studies called *seismic microzoning* are performed. The task of seismic microzoning is to make a quantitative assessment of the influence of local conditions on the seismicity and nature of seismic oscillations within the site of a particular structure. As a result of seismic microzoning, the *design seismicity* of the construction site is determined in MSK-64 scale balls (denoted by I^{des}) and the parameters of the calculated accelerograms are set.

For class IV HSs at the final design stages as well as classes I–III HSs at the preliminary design stages, the estimated seismicity of the I^{des} construction site can be taken according to Table 6.8 depending on the initial seismicity and the soil category of the foundation determined by the results of engineering–geological surveys.

The foundation within the construction site can be composed of soils, which in their composition occupy an intermediate position between soils of categories I and II or II and III, for example, if layered soils lie at the foundation of the structure. In such cases in Ref. [8] in addition to the soil categories listed in Table 6.8, it is recommended to introduce intermediate categories I-II and II-III. In this case, the calculated seismicity of I^{des} for soils of intermediate category I-II should be taken as for soils of category II, for soils of intermediate category II-III, and for soils of category III.

In the construction and repair periods in the absence of water in the reservoir, the estimated seismicity of the I^{des} retaining HEP placement site can be reduced by 1 point. In accordance with the requirements of SR [8], seismic effects on concrete dams should be taken into account in areas with a standard seismicity of I^{nor} equal to six balls or more, with an estimated seismicity of the I^{des} construction site of seven balls or more on the MSK-64 scale.

With an estimated seismicity of the I^{des} construction site of more than 9 points as well as with an estimated seismicity of 9 points, in the presence of category III soils for seismic properties the construction of HS requires special justification and is allowed in exceptional cases.

When substantiating the seismic resistance of concrete dams like other HS, it is necessary to consider the seismic effects of two levels:

- MCE that the dam must withstand without the threat of destruction; at the same time, any dam damage that does not lead to a breakthrough of the head front, including damage that violates the normal operation of the object, can be allowed; the value of the recurrence period of permafrost MCE T_{ret}^{DLE} for concrete dams of I-III classes is taken equal $T_{ret}^{DLE} = 5{,}000$ years, and for IV class $- T_{ret}^{DLE} = 1{,}000$ years.

- DE which the dam must withstand without threatening the life and health of people while maintaining conditions that ensure the normal operation of the object; at the same time, residual displacements, deformations, cracks, and other damages that do not violate the normal operation of the object may be allowed; maintainability of the structure should be provided; the value of the recurrence period DE T_{ret}^{SLE} for concrete dams of all classes is taken to be $T_{ret}^{SLE} = 500$ years.

Table 6.8 Design seismicity of the construction site

Soil category	Soils	Designed seismicity of the construction site at the initial seismicity, points				
		6	7	8	9	10
I	Rock soils of all kinds (including permafrost in frozen and thawed conditions), not weathered and slightly weathered; coarse clastic soils of low moisture content from igneous rocks containing up to 30% of sand and clay aggregate; weathered and highly weathered rock and nonrock hard-frozen (permafrost) soils at a temperature of minus 2°C and lower during construction and operation on principle I (preservation of foundation soils in frozen state); shear wave velocity $V_s > 7{,}000$ m/s; the ratio of the longitudinal and transverse wave velocities $V_p/V_s = 1.7$–2.2, regardless of the degree of water saturation	-	-	7	8	9
II	Rock soils weathered and highly weathered, including permafrost, except those classified as category I; coarse soil, with the exception of those classified as category I; sands are gravelly, large and medium-sized, dense and medium-density, slightly moist and moist; sands are small and dusty, dense and of medium density, slightly moist; dusty clay soils with a fluidity index $J_L \leq 0.5$ with a porosity coefficient $e < 0.9$ for clay and loam and $e < 0.7$ for sandy loam; permafrost nonrock soils, plastically frozen or loose frozen, as well as hard frozen at temperatures above minus 2°C during construction and operation according to principle I; $V_s = 250 - 700$ m/s; $V_p/V_s = 1.7$–2.2 for nonsaturated soils; $V_p/V_s = 2.2$–3.5 for water-saturated soils	-	7	8	9	>9
III	Sands are friable regardless of the degree of moisture and size; sands are gravelly, large- and medium-sized, dense and medium-density, saturated; fine and dusty sands of dense and medium density moist and water-saturated; dusty clay soils with a fluidity index $J_L > 0.5$; dusty clay soils with a fluidity index $J_L \leq 0.5$ with a porosity coefficient $e \geq 0.9$ for clay and loam and $e \geq 0.7$ for sandy loam; permafrost nonrock soils during construction and operation according to principle II (assumption of thawing of soil of the base); $V_s < 250$ m/s; $V_p/V_s = 1.7$–3.5 for nonsaturated soils; $V_p/V_s > 3.5$ for water-saturated soils	7	8	9	>9	>9

For each level of seismic effects (MCE and DE), the values of the standard I^{nor}, initial I^{beg}, and calculated I^{des} of seismicity are set in accordance with the approaches described above.

It is necessary to know the calculated seismic impacts separately for MCE and DE. In the calculations of constructions of I and II classes at the final stages of design, the calculated effects are taken and designed accelerograms are determined during seismic microzoning and characterized by the calculated values of maximum peak accelerations a_p^{DLE} for MCE and a_p^{SLE} for DE. When making calculations of constructions of I and II classes at the preliminary stages of design and constructions of III and IV classes at the final stages of design, they are limited to determining the calculated values of the maximum peak accelerations a_p^{DLE} for MCE and a_p^{SLE} for DE.

The calculated values a_p^{DLE} in the calculations of concrete dams at the MCE are taken equal to for:

constructions of I and II classes

$$a_p^{\text{DLE}} = g \times A_{5,000} \tag{6.24}$$

constructions of class III

$$a_p^{\text{DLE}} = 0.93g \times A_{5,000} \tag{6.25}$$

constructions of class IV

$$a_p^{\text{DLE}} = g \times A_{1,000} \tag{6.26}$$

The calculated values a_p^{SLE} in the calculations of concrete dams on the DE are taken equal to for:

constructions of I and II classes

$$a_p^{\text{SLE}} = g \times A_{500} \tag{6.27}$$

constructions of III classes

$$a_p^{\text{SLE}} = 0.93g \times A_{500} \tag{6.28}$$

constructions of class IV

$$a_p^{\text{SLE}} = 0.8g \times A_{500} \tag{6.29}$$

In formulas (6.24)–(6.29), A_{500}, $A_{1,000}$, $A_{5,000}$ – the *calculated accelerations of the foundation in parts* g, are determined for an earthquake with the designed recurrence periods, respectively, $T_{\text{ret}} = 500\,\text{years}$, $T_{\text{ret}} = 1,000\,\text{years}$, and $T_{\text{ret}} = 5,000\,\text{years}$ depending on the initial seismicity, the site of the I^{beg} construction, and the soil category of the foundation for seismic properties according to Table 6.9.

If the seismic stability calculations of structures are made using the calculated accelerograms, the calculated values of the maximum peak accelerations of such accelerograms must be no less than those determined by formulas (6.24)–(6.29).

Table 6.9 Values of calculated foundation accelerations

Category soil	I^{beg}, ball									
	6		7		8		9		10	
	I^{des}, ball	A	I^{des}, ball	A	I^{des}, ball	A	I^{des}, ball	A	I^{des}, ball	A
I	-	-	-	-	7	0.12	8	0.24	9	0.48
I–II	-	-	7	0.08	8	0.16	9	0.32	-	-
II	-	-	7	0.10	8	0.20	9	0.40	-	-
II–III	7	0.06	8	0.13	9	0.25	-	-	-	-
III	7	0.08	8	0.16	9	0.32	-	-	-	-

When performing seismic resistance analysis of concrete dams, the following factors must be considered:

- Permissible residual deformation, cracks, and damage; this factor is taken into account by the coefficient k_f, the value of which for all HSs is equal to $k_f = 0.45$.
- Effect of dam height on inertial seismic loads; this factor is taken into account by the coefficient k_2, the value of which for concrete dams is taken equal to for dams of height:
  - up to 60 m $k_2 = 0.8$;
  - more than 100 m $k_2 = 1$;
  - ≥ 60 and ≤ 100– by linear interpolation.
- The influence of the damping properties of the structure; this factor is taken into account by the coefficient k_ψ, the value of which for concrete dams is equal to $k_\psi = 0.9$.

Given these factors, the calculated value of the seismic acceleration of the foundation a_s is:

$$a_s = k_f \times k_2 \times k_\psi \times a_p \qquad (6.30)$$

where
a_p – the calculated value of the maximum peak acceleration.

The calculated values of the seismic acceleration of the foundation a_s^{DLE} and a_s^{SLE} should be determined separately for the MCE and DE.

When calculating structures for seismic effects, the seismic coefficient K_s is often used, which is equal to the ratio of the calculated value of the seismic acceleration of the foundation a_s to the gravitational acceleration g, i.e.,

$$K_s = \frac{a_s}{g} \qquad (6.31)$$

The values of the seismicity coefficients K_s^{DLE} and K_s^{SLE} should also be determined separately for the MCE and DE, respectively.

In calculating the stability and strength of concrete dams, unusual combinations of loads and effects should be considered, in which in addition to the loads of the usual combinations it is necessary to take into account seismic loads separately for MCE and DE.

The number of seismic loads taken into account when making calculations of structures and their foundations includes the following loads:

- Seismic volume inertial loads arising due to an earthquake during oscillations of the structure and its foundation; the intensity of these loads according to Newton's second law is equal to the production of the density of the material and acceleration; the direction of volume inertial loads is opposite to the direction of acceleration;
- Seismic hydrodynamic pressure of water on the surface of the structure resulting from the development of inertial forces in the water of the reservoir and the interaction of the oscillating structure with the aquatic environment;
- Seismic pressure of sediment deposited before the construction arising due to the occurrence of volume inertial loads in the sediment mass and the interaction of the oscillating structure with this mass;
- Hydrodynamic pressure of seismic gravitational waves arising on the surface of the reservoir.

Height of the seismic waves Δh should be taken into account when designating the elevation of the dam crest above the calculated water horizon in the reservoir [8]. In the absence of zones of tectonic disturbances or residual deformations (movements) of the bed within the reservoir when the seismic wave is caused by a distant earthquake, its height Δh is determined by the formula:

$$\Delta h = K_s \times T_0 \sqrt{g \times h} \tag{6.32}$$

where
K_s – seismicity coefficient;
T_0 – the prevailing period of seismic oscillations of the reservoir bed, determined according to engineering–seismological studies and in their absence taken equal to $T_0 = 0.5$ seconds;
h – depth of the reservoir.

In the case of a deep reservoir ($h > 100$ m) and the possibility of seismotectonic deformations (movements) of the bed in it during earthquakes of intensity $J = 6$–9 balls, the height of the seismic wave is determined by the following formula:

$$\Delta h = 0.4 + 0.76 \times (J - 6) \tag{6.33}$$

At a known height of the seismic gravitational wave, its pressure on the dam is determined similarly to the determination of the pressure of wind waves.

In the general case, the direction of seismic actions can be arbitrary. Usually when calculating the strength of concrete dams, only horizontal seismic effect is taken into account, and when calculating the stability, it is inclined. Inclined seismic impact in accordance with the recommendations [8] is taken at an angle of 30° to the horizon.

To determine the seismic loads on concrete dams and other HS at different times, different theories of seismic resistance were used.

Static theory of earthquake resistance proposed by *F. Omori* and *Sano* in 1900 was recommended by the design standards of many countries to carry out calculations of buildings and structures for seismic effects until the mid-XX century. The essence of this theory was as follows: it was believed that the structure is undeformable and oscillates with an earthquake with the same parameters $a_{s, d}$, the foundation soil; i.e., the calculated values of seismic accelerations within the structure a_s were taken equal to the value of the calculated seismic acceleration of the foundation $a_{s, d}$. Volume inertial loads corresponding to seismic accelerations a_{sd} in the structure were considered as static. When making calculations in accordance with the static theory of seismic resistance, horizontal seismic effects were considered, the intensity of which was determined by one parameter – the value of the calculated seismic acceleration of the foundation a_s (or the seismicity coefficient K_s).

Elementary dynamic theory of earthquake resistance was proposed by *N. Mononobe* in 1921. This theory differs from the static one in that in determining inertial forces the dynamic coefficient was taken into account. This coefficient depended on the period of natural oscillations of the structure in the fundamental tone and on the period of oscillation of the base. When determining the dynamic coefficient, steady harmonic oscillations were considered. In 1927, *K.S. Zavriev* independently of *N. Mononobe* proposed using a dynamic coefficient that depends on the same parameters. However, the value of this coefficient turned out to be 2 times greater than that of *N. Mononobe* since *K.S. Zavriev* considered undamped harmonic oscillations starting at some moment time $t = 0$. With regard to HSs and in particular to concrete dams, the approaches of the elementary dynamic theory of seismic resistance have been fully developed by *Sh.G. Napetvaridze* [117]. In this theory, the seismic effect is taken into account in the form of one parameter – the calculated seismic acceleration of the foundation (or seismicity coefficient).

Linear spectral theory of seismic resistance is now widely used and recommended by design standards. This theory was proposed by *M.A. Bio* in 1934 [117]. In accordance with the linear spectral theory, seismic inertial load variables with respect to the height of the structure are proportional to the peak acceleration of the foundation and are determined depending on the forms and periods of own oscillations of the structure. The influence of water in the reservoir on the forms and periods of own oscillations of the structure can be taken into account.

Dynamic theory of seismic resistance allows us to fully take into account the main features of the calculated seismic effect, which is specified in the form of an accelerogram. According to the SR [8], it is recommended to use this theory when calculating responsibility structures. The dynamic theory of seismic resistance is based on the numerical integration of the equations of motion, which can be performed by two methods – decomposition in the forms of own oscillations or stepwise (in time) integration.

Wave theory of seismic resistance began to develop in the 60–70 years of the XX century at the Institute of Hydromechanics of the National Academy of Sciences of

Ukraine under the leadership of *L.I. Dyatlovitsky*. In accordance with this theory, the propagation of elastic waves arising from earthquakes at the foundation of the dam and the interaction of these waves with the structure and water of the reservoir are considered. At the same time for the "dam-foundation-reservoir" system, the dynamic problem of the mechanics of a solid deformable body within the computational domain occupied by the dam and the foundation and the fluid oscillation problem for the reservoir water are jointly solved. The calculation is made on a given seismogram characterizing the designed earthquake in the area of the structure. As a result, at any design time, dynamic stresses can be determined within the design area of the dam and foundation, which are used to assess the earthquake resistance of structures [69].

The stability and strength calculations of concrete gravity dams are often made according to a two-dimensional scheme in accordance with the static theory of earthquake resistance. During seismic oscillations in the dam, inertial loads s arise the value of which is determined in accordance with Newton's second law by the following formula:

$$s = -\rho_b \times a_{s,d} = -\rho_b \times a_s = -\rho_b \times g \times K_{s,d} = -\rho_b \times g \times K_s \qquad (6.34)$$

where

ρ_b – the density of concrete;
$a_s = a_{s,d}$ – the calculated acceleration of the dam and foundation;
$K_s = K_{s,d}$ – seismicity coefficient of the dam and foundation.

The minus sign in formula (6.34) shows that the vector of seismic inertial loads s is directed in the side opposite to the direction of the seismic acceleration vector a_{sd}.

During earthquakes, inertial forces also arise in the mass of water in the reservoir – hydrodynamic pressure. Based on the solution of the two-dimensional hydrodynamic problem for a semi-infinite rectangular computational domain occupied by a homogeneous compressible fluid on the vertical boundary of which oscillations of constant height are specified characterized by acceleration a_s or seismicity coefficient K_s, N. *Westergaard* obtained a formula for determining the hydrodynamic pressure [268]:

$$p_s = K_s \times \frac{8\rho_w \times g \times h_w}{\pi^2} \times \sum_{i=1}^{\infty} \frac{1}{(2i-1)^2 \times C_i} \times \sin\left[\frac{(2i-1)\times\pi}{2h_w} \times z\right] \qquad (6.35)$$

where

K_s, ρ_w, g – respectively, the seismicity coefficient, water density, and gravity acceleration;
h_w – depth of water in front of the dam;
z – immersion depth of the point at which the pressure is determined;
C_i – coefficient taking into account the compressibility of the liquid determined by the following formula:

$$C_i = \sqrt{1 - \frac{16\rho_w \times h_w^2}{(2i-1)^2 \times K_0 \times T_0^2}} \qquad (6.36)$$

where

K_0 – volumetric modulus of elasticity of water, the value of which can be taken equal to $K_0 = 2{,}000\,\text{MPa}$;

T_0 – the predominant period of seismic oscillations taken equal to $T_0 = 0.5$ seconds.

When calculating p_s values by formula (6.35), it will be enough limited to 5–6 members of the series (Figure 6.14).

At water depths less than 100 m without a large error, the value of C_i can be taken equal to 1. In this case, formula (6.35) takes the form:

$$p_s = K_s \times \frac{8\rho_w \times g \times h_w}{\pi^2} \times \sum_{i=1}^{\infty} \frac{1}{(2i-1)^2} \times \sin\left[\frac{(2i-1)\times\pi}{2h_w}\times z\right] \tag{6.37}$$

The ordinate of the seismic hydrodynamic pressure diagram at the bottom $p_{s,\,hw}$ is:

$$p_{s,hw} = 0.742 K_s \times \rho_w \times g \times h_w \tag{6.38}$$

Seismic hydrodynamic pressure force W_s is:

$$W_s = 0.543 K_s \times \rho_w \times g \times h_w^2 \tag{6.39}$$

This force is applied at a distance of $0.401h$ from the bottom.

To determine the seismic hydrodynamic pressure p_s on the vertical upstream face of a concrete dam with a sufficient degree of accuracy instead of formula (6.37), we can use the dependence:

$$p_s = K_s \times 6{,}168\rho_w \times g \times h_w \times a \times \left[1-\exp\left(-3.195\frac{z}{h_w}\right)\right] \tag{6.40}$$

Based on the decision of *N. Westergaard*, a simplified method for determining seismic hydrodynamic pressure is recommended by current design standards of Ukraine,

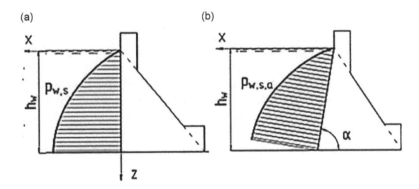

Figure 6.14 Seismic water pressure diagrams on the vertical (a) and inclined (b) upstream faces of a concrete dam according to the static theory of earthquake resistance.

Russia, the USA, and other countries for preliminary calculations of the seismic resistance of concrete dams.

In the case of an inclined upstream face of the dam, the seismic hydrodynamic pressure $p_{s,\alpha}$ is less than the pressure determined for the vertical upstream face of p_s. According to the recommendations of *Sh.G. Napetvaridze* [117], the ordinates of the diagram of seismic hydrodynamic pressure on the inclined upstream face of the dam can be found by the following formula:

$$p_{s,\alpha} = p_s \times \sin^2 \alpha \qquad (6.41)$$

where

α – the angle of inclination of the upstream face of the dam to the horizon;

p_s – hydrodynamic pressure on the vertical upstream face of the structure determined by formula (6.35).

Figure 6.14b shows a diagram of seismic hydrodynamic pressure on the inclined face of the dam. The ordinates of this diagram are calculated by formula (6.41).

The intensity and appearance of the sediment seismic diagram depend on the calculated earthquake and the thickness of the sediment layer. The presence of sediment affects the seismic hydrodynamic pressure on the dam. The rationale for the definition of seismic pressure of sediments on the upstream face of the structure described below is given in Ref. [36].

Sand and sand–gravel deposits are considered as a loose body and are characterized by density (taking into account water saturation) ρ_{ws} and conditional angle of internal friction φ_{wsr}. The value of φ_{wsr} should be determined by formula (6.16), into which instead of the value of v_{ws} we should substitute the value of the dynamic Poisson's ratio of the sediment soil v_{wsdyn}. At the preliminary design stages, $v_{wsdyn} = v_{ws}$ can be taken.

During earthquakes, sand or sand–gravel deposits are affected by the vertical volumetric forces of their own weight $(\rho_{ws} - \rho_w) \times g$ determined taking into account weighing in water and the horizontal volumetric seismic inertial forces $\rho_{ws} \times a_s = \rho_{ws} \times g \times K_s$. The angle of deviation from the vertical ε of the resultant force of self-weight of sediment suspended in the water and seismic inertial forces can be determined by the following formula:

$$\varepsilon = \text{arc tg} \frac{\rho_{ws} \times a_s}{(\rho_{ws} - \rho_w) \times g} \text{ or } \varepsilon = \text{arc tg} \frac{\rho_{ws} \times K_s}{\rho_{ws} - \rho_w} \qquad (6.42)$$

The diagram of the total pressure of sand or sand–gravel deposits caused by the action of volumetric forces of its own weight and volumetric seismic inertial forces is assumed triangular. The intensity of this pressure $p_{ws}\Sigma$ at a depth of h_{ws} is determined by the following formula:

$$p_{ws,\Sigma} = (\rho_{ws} - \rho_w) \times g \times h_{ws} \times \xi_{0,\Sigma} \qquad (6.43)$$

where

$\xi_{0,\Sigma}$ – the lateral pressure coefficient of sediment soil caused by the action of volumetric forces of its own weight and volumetric seismic inertial forces.

The actual seismic pressure of sand or sand–gravel sediments $p_{ws, s}$ is equal to $p_{wss} = p_{ws,\Sigma} - p_{ws}$, or:

$$p_{ws,s} = (\rho_{ws} - \rho_w) \times g \times h_{ws} \times (\xi_{0,\Sigma} - \xi_0)$$ (6.44)

In the case of a vertical upstream face of the dam, the expression for determining the value of $\xi_{0\Sigma}$ has the form:

$$\xi_{0,\Sigma} = \frac{\cos^2 (\phi_{ws,r} - \varepsilon)}{\left[\cos \varepsilon + \sqrt{\cos (\phi_{ws,r} - \varepsilon) \times \sin \phi_{ws,r} \times \cos \varepsilon} \right]^2}$$ (6.45)

and the value ξ_0 is determined by formula (6.14).

In the case of the upstream face of the dam inclined at an angle a to the horizontal, the expression for determining the value of $\xi_{0\Sigma}$ can be written as follows:

$$\xi_{0,\Sigma} = \frac{\sin^2 (\alpha + \phi_{ws,r} - \varepsilon)}{\cos \varepsilon \times \left[\sqrt{\sin (\alpha - \varepsilon) \times \sin \alpha} + \sqrt{\sin (\phi_{ws,r} - \varepsilon) \times \sin \phi_{ws,r}} \right]^2}$$ (6.46)

and the value ξ_0 is determined by formula (6.15).

Figure 6.15 shows the diagrams of seismic pressure p_{wss} defined by formula (6.44) of sand or sand–gravel deposits on the vertical and inclined upstream face of a concrete gravity dam [36].

In the presence of sandy or sandy–gravel sediments, the seismic hydrodynamic pressure on the vertical flat upstream face above the surface of the sediments should be determined by the formula of *N. Westergaard* (6.35) and by the inclined face – by formula (6.41). Below the sediment surface, the seismic hydrodynamic pressure on the flat upstream face should be taken constant in height and equal to the value at the mark

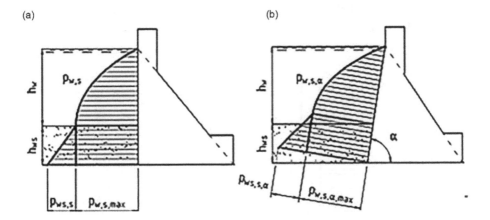

Figure 6.15 Diagrams of seismic pressure water and sand or sand–gravel sediment to the vertical (a) and inclined (b) upstream face of a concrete dam along the static theory of earthquake resistance.

of the sediment surface. In addition to the diagram of seismic pressure of sediments, a diagram of seismic hydrodynamic pressure on the vertical and inclined upstream face of a concrete dam is shown in Figure 6.15.

If small deposits in the form of clay, silt, and colloidal particles between which there are practically no bonds are deposited in front of the dam, they are considered as a heavy liquid. In this case, the seismic hydrodynamic pressure within the water layer of depth h_w above the sediment surface and the seismic pressure of the sediment within the sediment layer of thickness h_{ws} should be determined based on the solution of the corresponding hydrodynamic problem for a two-layer semi-infinite rectangular computational domain occupied by an inhomogeneous liquid, the dynamic characteristics of which change stepwise by surface sediment.

In Ref [62], the substantiation of the following approximate approach is given, which allows one to determine with sufficient accuracy the seismic pressure of water and the seismic pressure of sediments considered as a heavy liquid.

The seismic hydrodynamic pressure on the vertical upstream face of the dam within the water layer with depth h_w above the surface of sediment $p_{w,\,s}$ is determined by the *Westergaard* formula (6.35) into which instead of h_w value should be substituted h_1 value equal to

$$h_1 = h_w + \frac{\rho_{ws}}{\rho_w} \times h_{ws} \qquad (6.47)$$

where
  ρ_w and ρ_{ws} are, respectively, the density of water and the density of sediments saturated with water.

The seismic pressure considered as a heavy liquid of the sediment $p_{ws,\,sn}$ should also be determined by the *Westergaard* formula (Figure 6.16), in which the following changes should be made [268]:

•   Instead of the density of water ρ_w, substitute the density of sediments saturated with water ρ_{ws};

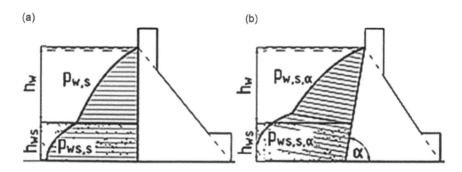

(a)   (b)

Figure 6.16 Diagrams of seismic pressure of water and sediment, considered as heavy liquid on the vertical (a) and inclined (b) upstream face of the concrete dam according to the static theory of earthquake resistance.

- Instead of the value of h_w, substitute the value of h_2 equal to:

$$h_2 = h_{ws} + \frac{p_w}{p_{ws}} \times h_w \qquad (6.48)$$

- Instead of the value z, substitute the value z_{ws1} equal to:

$$z_{ws1} = z_{ws} + h_2 - h_{ws} \qquad (6.49)$$

In formula (6.49), z_{ws} is the deepening under the surface of the sediment of the point at which the seismic pressure of the sediment is determined.

The seismic hydrodynamic pressure $p_{w,\,s,\,\alpha}$ and the seismic pressure of the considered sediment $p_{ws,\,s,\,\alpha}$ on the upstream face of the concrete dam inclined to the horizon at an angle with known values of $p_{w,\,s}$ and $p_{ws,\,s}$ are determined by formula (6.41).

6.3 Analysis of concrete gravity dams

6.3.1 Analysis of stresses by elementary methods

Calculations of gravity dams cut by flat temperature–sedimentary joints are usually made according to the two-dimensional problem considering a separate section or a conditionally cut section normal to the axis of the dam with a thickness of 1 m.

Calculation of gravity dams up to 60 m high is made by the method of materials resistance using the *Navier* hypothesis on the linear distribution of normal stresses over the calculated cross section. Gravity dams of greater height are also calculated at the preliminary design stages. The calculations are performed not on the full (see Section 6.2.1) but on the reduced composition of the loads and effects of the usual and unusual combinations. In this case, temperature effects are excluded from consideration, seismic loads are determined according to linear spectral theory based on a one-dimensional (cantilever) scheme, and the force effect of filtered water is taken into account only in the form of uplift forces at the concrete–rock contact.

In cases where the amplitude of seasonal fluctuations in the temperature of the outside air exceeds 17°C, the reduction in the width of the calculated sections of the dam body or its bottom due to the opening of construction joints at the downstream face under the influence of changes in air temperature should be taken into account.

Stresses are determined in horizontal design sections distributed over the height of the structure; sections are necessarily assigned in places of sharp changes in the dam profile, in places where concentrated forces are applied (ice pressure), in places of significant openings (Figure 6.17); the calculated section along the bottom of the dam is also required. Usually, at least five design sections are taken.

Normal vertical stresses on horizontal planes on the upstream σ_y^u and downstream σ_y^t faces of the dam are determined by the eccentric compression formula:

$$\sigma_y^u = -\frac{N}{b} + \frac{6M}{b^2}, \sigma_y^t = -\frac{N}{b} - \frac{6M}{b^2} \qquad (6.50)$$

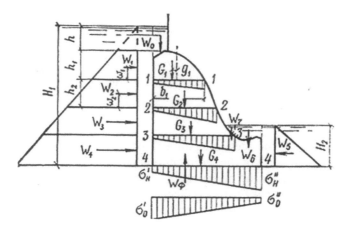

Figure 6.17 Scheme for calculating stresses in horizontal sections of the dam.

where

    b – width of the design section;

    N – the sum of the vertical forces acting above the section;

    M – the sum of the moments of all the forces acting above the section relative to the center of gravity of the section.

In formulas (6.50), the normal force N downward is considered positive, upward – negative. Clockwise bending moment M is considered positive, counterclockwise negative. Tensile stresses σ_y are considered positive, and compressive stresses are considered negative.

Figure 6.17 shows the stress plots σ_y in horizontal sections of dams.

Knowing the normal stresses σ_y^u and σ_y^t on horizontal planes, it is possible to determine the tangential stresses at the upstream τ_{xy}^u and downstream τ_{xy}^t faces of the dam as well as the normal stresses on vertical planes at the upstream σ_x^u and downstream σ_x^t faces.

Consider the equilibrium of an elementary triangle selected near the upstream face (Figure 6.18a). The width of such a triangle is dx and the height dy. If the coefficient of laying the upstream face of the dam $m_u = tg\alpha_u$, then:

$$dx = m_u \times dy \qquad (6.51)$$

where

    α_u – the angle of the upstream face of the dam with the vertical.

On the horizontal plane of the elementary triangle, normal σ_y^u and tangential τ_{xy}^u stresses act. The normal σ_x^u and tangential τ_{xy}^u stresses act on the vertical plane of this triangle. The hydrostatic pressure of water acts on the inclined plane of the triangle

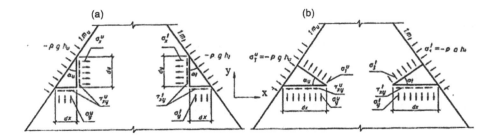

Figure 6.18 Scheme for determining stresses on the faces of the dam: (a) $\sigma_x{}^u$, $\sigma_y{}^u$, and $\tau_{xy}{}^u$ at the upstream face and $\sigma_x{}^t$, $\sigma_y{}^t$, and $\tau_{xy}{}^t$ at the downstream face; (b) $\sigma_1{}^u$ at the upstream face and $\sigma_3{}^t$ at the downstream face.

equal to $\rho_w \times g \times h_u$, where h_u is the water head at marks of the calculated section from the upstream side. Figure 6.18 shows the positive stresses σ_x, σ_y, and τ_{xy}.

From the equilibrium condition of the triangle, the tangential stress τ_{xy}^u at the upstream face of the dam is determined:

$$\tau_{xy}^u = \left(\sigma_y^u + \rho \times g \times h_u\right) \cdot m_u \tag{6.52}$$

and normal stress σ_x^u at the upstream face of the dam is:

$$\sigma_x^u = \sigma_y^u \times m_u^2 - \rho \times g \times h_u\left(1 - m_u^2\right) \tag{6.53}$$

At the vertical upstream face, a shear stress is $\tau_{xy}^u = 0$ and the normal stress σ_x^u is:

$$\sigma_x^u = -\rho \times g \times h_u \tag{6.54}$$

Similarly, the normal σ_x^t and tangential τ_{xy}^t stresses at the down face of the dam can be obtained:

$$\sigma_x^t = \sigma_y^t \times m_t^2 - \rho_w \times g \times h_t\left(1 - m_t^2\right) \tag{6.55}$$

$$\tau_{xy}^t = -\left(\sigma_y^t + \rho_w \times g \times h_t\right) \times m_t \tag{6.56}$$

In formulas (6.55) and (6.56), m_t represents the coefficient of laying the downstream face of the dam and h_t represents the water head the design section from the downstream.

Strength assessment of gravity dams is made according to the main stresses. When considering stresses in the vicinity of points lying on the faces of the dam (Figure 6.18b), it can make the following conclusion. One main area is located in the plane of the face, the second is perpendicular to the face, and the third is perpendicular to the longitudinal axis of the dam; i.e., it is located in the plane of the drawing.

The values of the main stresses on the faces of the dam can be determined from the equilibrium condition of elementary right-angled triangles selected near the faces (Figure 6.18b), and the legs of the triangles are the main planes.

On the upstream face of the elementary triangle, the main stress σ_3^u is equal to:

$$\sigma_3^u = -\rho_w \times g \times h_u \tag{6.57}$$

The main stress σ_1^u acts along the face of length $dx \times \cos\alpha_u$:

$$\sigma_1^u = \left(1 + m_u^2\right) \times \sigma_y^u + \rho_w \times g \cdot h_u \times m_u^2 \tag{6.58}$$

The main stresses σ_2^u at the planes perpendicular to the axis of the dam can be found by considering the conditions of plane deformation in accordance with Hooke's law:

$$\varepsilon_2^u = \frac{1}{E_b} \times \left[\sigma_2^u - v_b \times \left(\sigma_1^u + \sigma_3^u\right)\right] = 0 \tag{6.59}$$

where

$\varepsilon_2^u = 0$ – deformation in the direction of the longitudinal axis of the dam;
E_b and v_b – respectively, modulus of elasticity and Poisson's ratio of concrete.

From this expression follows:

$$\sigma_2^u = v_b \times \left(\sigma_1^u + \sigma_3^u\right) \tag{6.60}$$

If the upstream face of the dam is vertical, the expressions for determining the main stresses σ_1^u and σ_2^u take the form:

$$\sigma_1^u = \sigma_y^u, \sigma_2^u = v_b \times \left(\sigma_y^u - \rho_w \times g \times h_u\right) \tag{6.61}$$

Similarly, expressions can be obtained to determine the main stresses σ_1^t, σ_2^t, and σ_3^t, on the downstream face of the dam:

$$\sigma_1^t = -\rho_w \times g \times h_t \tag{6.62}$$

$$\sigma_3^t = \left(1 + m_t^2\right) \times \sigma_y^t + \rho_w \times g \times h_t \times m_t^2 \tag{6.63}$$

$$\sigma_2^t = v_b \times \left[\left(1 + m_t^2\right) \times \sigma_y^t - \rho_w \times g \times h_t \left(1 - m_t^2\right)\right] \tag{6.64}$$

The quantities included in these expressions m_t and h_t are the same as in formulas (6.55) and (6.56).

According to the received stresses, the strength of the dam is checked in accordance with the SR [14]. For all usual and all unusual combinations of loads and effects at all points, the condition for concrete compressive strength must be met:

$$\gamma_n \times \gamma_{\ell c} \times |\sigma_3| \leq \gamma_{cd} \times R_b \tag{6.65}$$

where

γ_n – reliability coefficient for the responsibility of the structure, taken depending on its class in the range from 1.1 to 1.25;

$\gamma_{\ell c}$ – coefficient of load combinations taken depending on the calculated combination of loads and the designed period (construction, operation, repair) ranging from 0.9 to 1;

γ_{cd} – coefficient of dam operating conditions taken depending on the calculated combination of loads and effects within from 0.9 to 1.1;

R_b – design concrete resistance to compression taken depending on the class of concrete.

In the old design standards BR 2.06.06–85 "Concrete and reinforced concrete dams", the formation of tensile stresses was allowed at the contact of the upstream face with the foundation and the depth of tensile stresses d_t was limited. In the code of rules SR 40.13330.2012 (the updated version of BR 2.06.06-85) [14], unfortunately this restriction was removed without comment. The authors of the innovation apparently believed that the problem of tensile stresses at the contact of the upstream face with the foundation can be solved by increasing the profile of the gravity dam. However, this is a fallacy. If tensile stresses on the upstream face can be avoided in this way, then tensile stresses are inevitable in the foundation under the upstream face. And it is to fight with them with the help of constructive measures and not by increasing the volume of concrete.

It is believed that it is useful to return to the "well-forgotten old" and to restore the restriction of the depth of tensile stresses at the contact of the upstream face of the dam with the foundation d_t, if the calculation is made according to the theory of elasticity under the assumption that the contact is resistant to tensile stresses. If the calculation is made taking into account the zero tensile strength of the contact, then the crack opening depth at the contact d_{cr} should be limited. The corresponding restrictions for d_t and d_{cr} are given in Tables 6.10 and 6.11 when calculating gravity dams for the full and reduced composition of loads and effects, respectively. From the point of view of the SSS of the dam body, the constraint conditions d_t and d_{cr} are equivalent.

It brings the forces acting in the dam to the normal force N and the moment of all forces M in the consideration of horizontal section of width b, and the eccentricity e. From the definition of normal stresses in the section (or at the contact) according to formula (6.50), it is easy to derive the formulas for determining d_t and d_{cr}:

$$d_t = \left(1 - \frac{b}{6e}\right) \cdot \frac{b}{2} \tag{6.66}$$

$$d_{cr} = 3e - \frac{b}{2} \tag{6.67}$$

in this case as can be seen d_t and d_{cr} are not directly dependent on the values of the normal force N and moment M.

According to the SR [14] when determining the stresses at the contact of a concrete dam by the materials' resistance, the main vector of all the forces acting on the dam, including uplift on the bottom (calculation scheme 1), is taken into account.

Figure 6.19 shows the results of such a calculation for a 105-m high-gravity dam under the action of hydrostatic pressure from the upstream and downstream, dead weight at γ_c = 2.4 t/m^3 (Figure 6.19a), and uplift (Figure 6.19b) specified in accordance with Section 6.2.2. Thus, according to calculation scheme 1, compression stress −2.5 t/m^3 was obtained at the upstream face at the contact (Figure 6.19c).

When solving the SSS of "dam-foundation" system, uplift is applied on the bottom of the dam which is subtracted from the stresses transmitted from the dam on bottom. Following this procedure first, it is important to calculate the contact stresses only from the forces transmitted by the dam (Figure 6.19e), then subtract the uplift from them (Figure 6.19b); as a result, a plot (Figure 6.19f) with a tensile stress of +52.8 t/m^2 and a length of the tension zone d_t = 4.04 m (calculation scheme 2) is obtained. Since the resistance to tensile contact of the dam with the foundation is usually close to zero, the contact should open.

Head of 100 m will be restored in the opened part of the contact (Figure 6.19g), which will propagate until it becomes numerically equal to the stress of 100 t/m^2 from the forces transmitted from the dam (Figure 6.19h). As a result, contact opened on the depth of d_{cr} = 8.37 m and the plot of real contact stress is shown in Figure 6.19i.

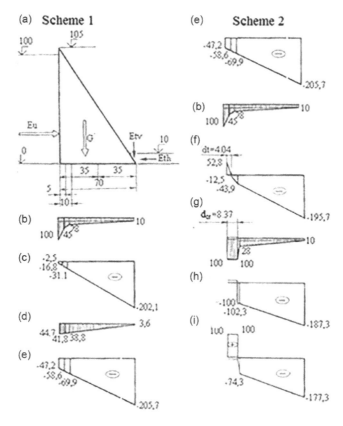

Figure 6.19 Schemes I (pos. a–c) and 2 (pos. a, e, b, f, g, and i) for calculating the stresses at the contact of a concrete dam 105 m high on a rock foundation.

In the calculation according to scheme 1, the uplift plot with a head of 100 m (Figure 6.19b) is transformed into the plot in Figure 6.19d with a significantly lower head of 44.7 m! According to scheme 1, the final stress plot (Figure 6.19c) is obtained by adding the stress plot only from the forces transmitted from the dam (Figure 6.19e) with "corrected" uplift plot (Figure 6.19d).

This explains the different results of calculations according to schemes 1 and 2.

Below are the formulas for determining the crack length d_{cr} at the contact of the dam with the foundation in the presence of contact tensile strength and in the absence of it [101].

If according to calculation scheme 2 at the contact with the upstream face, the tensile stress σ_c^u exceeds the contact tensile strength R_{tc} (Figure 6.19f), then a crack of length d_{cr} should be formed on which the piezometric pressure of the upstream h_u is restored (Figure 6.20d). On the continuous part of the contact, the stress at the upstream $(\sigma_c^u)'$ without taking into account the uplift is (Figure 6.20c):

$$\left(\sigma_c^u\right)' = -\frac{N_{n\pi}}{b-d_{cr}} + \frac{6N_{n\pi}\left(e_{n\pi}\right)_{oc}}{\left(b-d_c\right)^2} \tag{6.68}$$

where

the eccentricity $(e_{n\pi})_{oc}$ of the force $N_{n\pi}$ relative to the center of gravity of the continuous part of the contact O_c is:

$$\left(e_{n\pi}\right)_{oc} = \left(e_{n\pi}\right)_0 - \frac{d_{cr}}{2} = \frac{\left(M_{n\pi}\right)_0}{N_{n\pi}} - \frac{d_{cr}}{2} \tag{6.69}$$

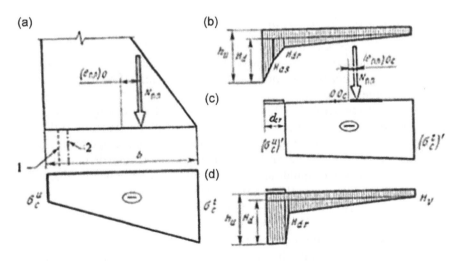

Figure 6.20 Scheme for determining the crack length d_{cr}: (a) plot of stresses from forces transmitted from the dam to a continuous contact and (c) on contact with a crack; (b) the plot of the uplift according to Ref. [14] at a continuous contact and (d) at contact with a crack; 1- and 2-axis cementation and drainage curtains, respectively.

At the end of the crack, the condition must be met:

$$\left(\sigma_c^u\right)' = h_u \times \gamma_w - R_{tc} \tag{6.70}$$

based on which the value of the crack length d_{cr} is determined:

$$d_{cr} = b - \frac{N_{n\pi}}{(h_u - R_{tc})} \cdot \left(\sqrt{1 + \frac{3(h_u - R_{tc}) \cdot \left[b - 2(e_{n\pi})_o\right]}{N_{n\pi}}} - 1\right) \tag{6.71}$$

where
  h_u – the piezometric pressure of the upstream.

With the tensile strength of the contact $R_{tc} = 0$, the crack length is:

$$d_{cr} = b - \frac{N_{n\pi}}{h_u} \cdot \left(\sqrt{1 + \frac{3h_u\left[b - 2(e_{n\pi})_o\right]}{N_{n\pi}}} - 1\right) \tag{6.72}$$

The set uplift on the dam bottom (Figure 6.20b) changes as shown in Figure 6.20d.

As a result, the plot of real t stress contact (see Figure 6.19i) is obtained by adding the plots shown in Figure 6.20a and d.

A comparison of the solution of the nonlinear SSS problem of the "dam-foundation" system with the calculation according to formula (6.72) shows that the crack lengths (contact openings) practically coincide.

6.3.2 Analysis of SSS by methods of elasticity theory

Dam analysis by elasticity theory methods are performed on the full composition of the loads and effects of the usual and unusual combinations for the construction, operation, and repair periods (Section 6.2.1). The loads and impacts taken into account in this have some features.

In the case of temperature effects, it is considered the change in the temperature state of the structure from the moment of the start of construction to the moment of complete cooling of the concrete masonry to mean annual temperatures. In this case, the initial temperature regime which is formed during the hardening of concrete, the temperature of closure of building joints, seasonal fluctuations in the temperature of the outdoor air, and water in the reservoir are taken into account.

Seismic effects are determined by the linear spectral theory using at least 10 forms of dam own oscillations. The periods and forms of own oscillations of a structure are determined from the solution of a two-dimensional dynamic problem.

The force effect of water on the surface of the dam and the foundation as well as the effect of filtered water in the body of the dam and the foundation is determined by an updated methodology (see Section 6.2.2).

In the first half of the XX century, the methodology for determining the SSS of gravity dams was based on the solutions to the plane problem of elasticity theory obtained by that time. In this case, the stresses in the dam from the dead weight and hydro-static pressure of the water were determined as in a semi-infinite acute-angle wedge (*M. Levy*'s decision). The influence of the dam vertex on its stress state was taken into account according to an approximate method proposed by *B. G. Galerkin*. The influence of the foundation was taken into account in accordance with a very cumbersome decision received by *F. Telke*.

At present in connection with the development of computer technology and numerical methods for solving boundary problems, the SSS of concrete dams is determined by numerical methods of the theory of elasticity (FEM, FDM, variation difference method, etc.) [64,101,111,175]. The most widely used theory of elasticity is FEM.

FEM allows us to take into account the totality of factors affecting the SSS dam: heterogeneity in the elastic properties of the dam associated with zoning of concrete; heterogeneity of the foundation and its anisotropy; the presence of fissures in the foundation; opening of construction joints; the presence of holes in the dam, etc.

In studies of the "dam-foundation" system SSS, the spatial model of the foundation is usually limited to a depth of (2–2.5) H, where H is the height of the dam. As a rule, such a depth of the foundation is sufficient to reproduce in the analysis of the nonuniformity of vertical displacements (settlements) of the dam, which affects the SSS of the dam and the foundation. However when comparing the calculated settlements of the dam with those obtained in nature measured using geodetic studies, it turns out that the measured settlements are significantly larger than the calculated. This is because the amount of settlements depends on the compressible thickness, and when creating a reservoir, its bed and foundation are compressed at a great depth under the dam.

Settlements of the structures of the Boguchan HEP were determined using geodetic observations on marks installed in the cementation adit at the contact of the upstream face of the concrete dam with the foundation relative to the benchmark 2 km away from the dam. The settlement of the benchmark itself is unknown. When comparing the measured settlements and the calculated vertical displacements of the dam obtained on the "dam-foundation" mathematical model, it turned out that they differ by two orders of magnitude.

A spatial finite-element model of the "dam-foundation-reservoir" system was created (Figure 6.21) with the help of which the dam settlements were determined with respect to the benchmark. The calculated and measured settlements became comparable with a significant increase in the depth of the compressible stratum up to 150 km.

In the process of research, an analysis was made of the effect of settlements of a reservoir bed on stresses and deformations of a concrete dam. It turned out that this influence is not significant, which allows not to take into account the influence of the reservoir when calculating the strength and stability of concrete dams and to limit the size of the calculated area of the foundation.

As knowing in rock foundation is under the upstream face of concrete dams, a decompaction zone forms, which affects the deformability of the foundation and the distribution of filtration forces. The formation of the decompaction zone can be taken into account as follows: contact finite elements having a real tensile strength are introduced into the foundation on the continuation of the upstream face; the SSS calculation of the dam section is made in three stages:

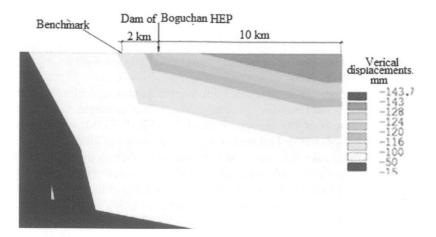

Figure 6.21 Settlements bed reservoir near concrete Boguchan HEP dam, mm.

- At the first stage, the stress field is determined at the foundation of the gravitational forces without taking into account the dam;
- At the second stage, the SSS of the dam is determined together with the foundation under the action of the loads acting on the dam;
- At the third stage, the obtained stress fields are added up (deformations at the foundation of the first stage are not taken into account).

At the designed depth, the horizontal tensile stresses arising from the loads on the dam will be neutralized by compressing natural stresses from the dead weight of the foundation (Figure 6.22).

Taking into account of nonlinear deformations when determining the decompaction zone, the standard values of the contact slide resistance parameters are increased by $\gamma_m = 1.25$ times compared to the designed values.

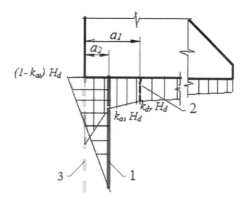

Figure 6.22 The decompaction zone (3) under the dam upstream face, I-pressure on the cementation curtain, 2-drainage curtain.

Since the permeability of the decompaction zone is significantly higher than that of the rock mass and cementation curtain, it is logical to apply a pressure to the cementation curtain, which is assumed to be linear with the ordinate equal to $(1 - k_{as})H_d$ at the dam bottom and zero at the end of the cementation curtain.

If underground circuit consists from only drainage, thus in instead of k_{as} it is must put k_{dr}, where k_{as} and k_{dr} are the head drop coefficients on the cementation curtain and drainage; $k_{dr} = 0.4$ and $k_{as} = 0.2$ during normal operation of the antifiltration and drainage devices and $k_{dr} = 0.5$ and $k_{as} = 0.3$ in case of a violation of the normal operation of the antifiltration and drainage devices [14].

In studies of the SSS of dams together with the foundation, it is recommended to take into account the presence of a decompaction zone in the foundation under the upstream face of the dam and the pressure on the grout according to the method described above.

Strength conditions of gravity dams calculated on the full composition of the loads and effects of the operational period are given in Table 6.10.

It is allowed to calculate a reduced composition of loads and effects of the usual and unusual dam combinations with a height of more than 60 m on the initial stages of design, and dams with a height of less than 60 m – at all stages of design; the relevant strength conditions are given in Table 6.11.

Table 6.10 Strength conditions of gravity dams calculated on the full composition of the loads and effects during operating period

---

At all points on the body of dams from all types usual and unusual combinations of loads and effects $\gamma_n \gamma_{lc} \sigma_3 \leq \gamma_{cd} R_b{}^a$

At the upstream face of the structure

Design features of dams and design cross sections	Usual load and effects combinations	Unusual load and effects combinations Not including seismic effects	Including seismic effects
A. Dams without extended seams			
Horizontal sections of the dam body without a waterproofing screen on the upstream face	Min $0.500a_1$ $d_t \leq 0.133b_d$ $d_{cr} \leq 0.181b_d$	$d_t \leq 0.167b_d$ $d_{cr} \leq 0.250b_d$	$d_t \leq 0.286b_d{}^b$ $d_{cr} \leq 0.668b_d{}^b$
The same with a waterproofing screen on the upstream face	$d_t \leq 0.167b_d$ $d_{cr} \leq 0.250b_d$	$d_t \leq 0.200b_d$ $d_t \leq 0.333b_d$	$d_t \leq 0.286b_d{}^b$ $d_{cr} \leq 0.668b_d{}^b$
Contact section without waterproofing of the contact with upstream face of dam	$d_t \leq 0.300a_2$ $d_{cr} \leq 0.750b$	$d_t \leq 0.083b$ $d_{cr} \leq 0.120b$	$d_t \leq 0.200b$ $d_{cr} \leq 0.333b$
The same with water-proofing of the contact with upstream face of the dam	$d_t \leq 0,083b$ $d_{cr} \leq 0.100b$	$d_t \leq 0,125b$ $d_{cr} \leq 0.166b$	$d_t \leq 0,200b$ $d_{cr} \leq 0.333b$

---

[a] When checking the strength at the down face, it is allowed to average the value σ_3 on part 4 m wide of the calculated horizontal section.

[b] In cases where d_t exceeds its limit value equal to $0.286b_d$, should:
  • at $0.286 < d_t < 0.320b_d$ – evaluate the strength of the structure in cross section under the condition $\gamma_n \gamma_{lc} \sigma_3 \leq \gamma_{cd} R_b$ with the determination of stress values σ_3 without taking into account the tensile strength of concrete at the upstream face of the dam;
  • at $d_t > 0.320b_d$, reinforce the upstream face of the structure considering the section of the dam body as reinforced concrete and ensuring the strength of the concrete in the compressed zone according to the condition $\gamma_n \gamma_{lc} \sigma_3 \leq \gamma_{cd} R_b$.

Table 6.11 Strength conditions of gravity dams calculated on the reduced composition of the loads and effects during operating period

At all points on the body of dams of all types usual and unusual combinations of loads and effects $\gamma_n \gamma_{lc} \sigma_3 \leq \gamma_{cd} R_b$

At the upstream face of the structure

Design features of dams and design cross sections	Usual load and effects combinations	Unusual load and effects combinations	
		Not including seismic effects	Including seismic effects
A. Dams without extended seams			
Horizontal sections of the dam body without waterproofing screen on the upstream face	$\sigma^u_y < 0^a$ $\sigma^u_y < 0.25\,\gamma_w H^u_d$	$d_t \leq 0.133 b_d$ $d_{cr} \leq 0.181 b_d$	$d_t \leq 0.286 b_d{}^a$ $d_{cr} \leq 0.668 b_d{}^a$
The same with a waterproofing screen on the upstream face	$d_t \leq 0.133 b_d$ $d_{cr} \leq 0.181 b_d$	$d_t \leq 0.167 b_d$ $d_{cr} \leq 0.250 b_d$	$d_t \leq 0.286 b_d{}^b$ $d_{cr} \leq 0.668 b_d{}^b$
Contact section without waterproofing of the contact with upstream face of dam	$\sigma^u_c < 0$	$d_t \leq 0.300 a_2$ $d_{cr} \leq 0.032 b^b$	$d_t \leq 0.200 b$ $d_t \leq 0.333 b$
Contact section with waterproofing of the contact with upstream face of dam	$d_t \leq 0.071 b$ $d_{cr} \leq 0.083 b$	$d_t \leq 0.083 b$ $d_{cr} \leq 0.100 b$	$d_t \leq 0.200 b$ $d_t \leq 0.333 b$

[a] If these conditions are not met, it should be guided by the note [b] Table 6.10.
[b] In determining the coefficient before b_d, it was assumed that $a_2 = 0.1 b$.

where

$\gamma_n, \gamma_{lc}, \gamma_{cd}$ – the coefficients adopted according to Section 6.6.1;

σ_3 – the maximum main compressive stress, MPa;

R_b – the design concrete compression resistance, MPa;

b – the width of the dam at the foundation, m;

b_d – the width of the calculated horizontal section, m;

b_h – the thickness of the vertex section with extended joints, m;

a_1 – the distance from the upstream face to the drainage of the dam body, m;

a_2 – the distance from the upstream face of the dam to the axis of the cementation curtain, m.

In both cases, the check of gravity dam's strength is carried out as rule for February and August.

6.3.3 Stability analysis

The forces acting on a concrete dam in various combinations can ultimately be reduced to the following resultants (Figure 6.23a): horizontal shear force T, vertical force V directed downward, and vertical uplift force on the bottom of the structure P directed upward.

Under the influence of these forces, the following types of disequilibrium of a structure located on a sufficiently solid foundation are possible:

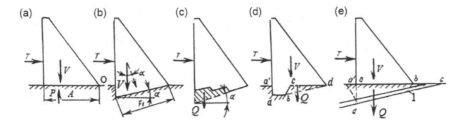

Figure 6.23 Schemes for calculating the slide stability of gravity dams: I-crack.

- The dam can be slided by force T at the foundation;
- The dam can be overturned by forces T and P around point O;
- The dam can be raised up by uplift P.

In accordance with possible types of disequilibrium of the dam, a check of its stability against sliding, overturning, and raising is carried out for the usual and unusual combinations of loads and effects.

Slide stability calculations. Depending on the type of connection of the dam with the foundation and on the features of its geological condition, the following schemes and design slide surfaces are distinguished:

- Flat slide along the horizontal bottom of a shallow dam taking place in the case of a homogeneous and sufficiently strong foundation (Figure 6.23a);
- Flat slide along the inclined bottom of the dam (Figure 6.23b); such a bottom is arranged if the dam is not ensured along the horizontal contact surface of the dam with the foundation;
- Slide along the stepped bottom of the dam (Figure 6.23c); such a bottom is arranged in the case that the inclined bottom but allows to reduce the volume of the cutting;
- Slide in the presence of a tooth (Figure 6.23d); such a slide can occur either along the acd line or along the abd line;
- Slide along dipping cracks in the foundation with reduced strength characteristics (Figure 6.23e).

In accordance with Ref. [11], the criterion for ensuring stability against slide is the condition:

$$\gamma_n \times \gamma_{lc} \times F \leq \gamma_{cd} \times R \qquad (6.73)$$

where
   γ_n and γ_{lc} – reliability coefficient for the responsibility of the structure and load combinations, respectively;
   γ_{cd} – coefficient of working conditions taken depending on the calculated slide surface; for slide surfaces passing through a concrete–rock contact $\gamma_{cd} = 0.95$; for slide surfaces passing through fissures in the foundation mass $\gamma_{cd} = 1$;

F and R – design values of, respectively, generalized slide forces and forces of ultimate resistance.

In the case of a horizontal slide surface (Figure 6.23a):

$$F = T \tag{6.74}$$

$$R = (V - P) \times tg\varphi + c \times A \tag{6.75}$$

where
   A – the slide surface area;
   $tg\phi$ and c – the design values of the slide characteristics taken depending on the rock of the foundation.

In the case of a slide surface inclined at an angle α (Figure 6.23b–d):

$$F = T \times \cos\alpha - V \times \sin\alpha \tag{6.76}$$

$$R = (V \times \cos\alpha + T \times \sin\alpha - P) \times tg\phi + c \times A \tag{6.77}$$

It should be borne in mind that the uplift P is directed perpendicular to the slide surface. For the cases in Figure 6.23c and d, the weight of the soil Q is added to the vertical force. In the case in Figure 6.23d when calculating the slide stability on part of the contact, the resistance of the thrust mass E_d of the rock foundation determined by formula (6.78) is taken into account.

When calculating the stability on the slide along the fissure (Figure 6.23e), the inclination of the fissure and the slide characteristics along it are taken into account. The values of the inclination angle and slide characteristics of the fissures are determined according to engineering–geological surveys in accordance with the detected fissure systems.

If the dam is deepened into the foundation, it is necessary to take into account the resistance force of the thrust array E_d of rock soils or backfill from nonrock soil from the downstream side. The calculated value of the force E_d in accordance with Ref. [11] is determined by the following formula:

$$E_d = \gamma_{c1} \times E_{p,d} \tag{6.78}$$

where
   γ_{c1} – the coefficient of working conditions taken depending on the ratio of the soil deformation modulus of the thrust array (backfill) E_s and E_f of the foundation: at $E_s/E_f \geq 0.8\, \gamma_{c1} = 0.7$; at $E_s/E_f \leq 0.1\, \gamma_{c1} = E_r/E_{p,d}$; at $0.8 > E_s/E_f > 0.1$ γ_{c1} determined by linear interpolation;
   E_r – the resting pressure determined by the following formula:

$$E_r = \frac{1}{r} \times \rho_f \times g \times h_f^2 \times \frac{v_f}{1 - v_f} \tag{6.79}$$

where

ρ_f and v_f – respectively, the density and Poisson's ratio of the soil of the thrust massif or backfill;

h_f – depth of the dam in the foundation;

$E_{p,d}$ – design value of passive resistance force.

Preliminary calculated values of the slide characteristics of rock soils can be taken according to Table 6.12.

In case of backfill, $E_{p,d} = E_r$. The values of the force $E_{p,d}$ for the persistent rock mass can be found by the following formula:

$$E_{p,d} = Q \times tg\left(\beta + \varphi_f\right) + \frac{c_f \times A_f \cdot \cos\varphi_f}{\cos\left(\beta + \varphi_f\right)}$$ (6.80)

where

Q – weight of the prism trust;

ϕ_f – friction angle corresponding to the value $tg\varphi_f$.

It should be noted that the resistance of the thrust array should be taken into account only if the dam is in close contact with the thrust array. The direction of force $E_{p,d}$ is assumed to be horizontal regardless of the inclination of the persistent face of the array.

Table 6.12 **The designed values of the slide characteristics of rocks**

Category	Rocks of the foundation	On the contact of concrete rock		On cracks in the foundation**	
		$tg\varphi$	C, MPa (kg/cm²)	$tg\varphi$	C, MPa (kg/cm²)
1	Rocks (massive, large block, layered, tiled, very weakly, and weakly fissured with $R_c \geq 50$ MPa) (500 kg/cm²)*	0.95	0.40 (4)	0.80–0.55	0.15–0.05 (1.5–0.5)
2	The same but medium fissured $R_c \geq 50$ MPa (500 kg/cm²)	0.85	0.3 (3)	0.80–0.55	0.15–0.05 (1.5–0.5)
3	The same but highly fissured $R_c = 15$–50 MPa (150–500 kg/cm²) and slightly fissured, weathered with $R_c = 5$–15 MPa (50–150 kg/cm²)	0.80	0.2 (2)	0.70–0.45	0.1–0.02 (1–0.2)
4	Half-rock with $R_c \leq 5$ MPa (50 kg/cm²)	0.75	0.15 (1.5)	0.65–0.45	0.05–0.02 (0.5–0.2)

* R_c – the standard value of the strength uniaxial compression.
** The values of $tg\varphi$ and C depend on the width of the fissure opening and the type of aggregate.

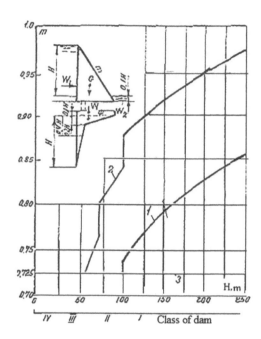

Figure 6.24 Graphs for preliminary determination of the incline of the downstream face of the gravity dam: H – height of the dam; W_1 and W_2 – hydrostatic pressure from the upstream and downstream; G – dead weight of the dam; W – uplift; m – incline the downstream face; 1 and 2 graphs $m = f(H)$ for a rock foundation with $tg\varphi = 0.75$, $C = 0.2\,\text{MPa}$ and $tg\varphi = 0.70$, $C = 0.1\,\text{MPa}$, respectively; 3-at $m \geq 0.725$ on the contact of the upstream face with the foundation, there are no tensile stresses.

Figure 6.24 shows graphs for the preliminary determination of the incline of the downstream face m of a gravity dam with a height H (up to 250 m) depending on the parameters of slide resistance $tg\varphi = 0.75$, $C = 0.2\,\text{MPa}$ (curve 1) and $tg\varphi = 0,70$, $C = 0.1\,\text{MPa}$ (curve 2), respectively.

Graphs plotted for a dam loaded:

- Hydrostatic pressure W1 from the upstream side at a mark coinciding with the dam crest and W2 from the downstream side with a pressure equal to 10% of the head H on the dam;
- Dead weight of the dam G;
- Uplift on the bottom of the dam W calculated from the condition of putting the head on the cementation curtain up to 40% of the head H and on the drainage up to 20% of the head H (Figure 6.24).

For the desired dam of construction class *I, II, III,* and *IV,* the coefficient stability of the dam on slide along the bottom equals 1.25, 1.20, 1.15, and 1.10, respectively; for dams with a downstream face $m \geq 0.725$, there are no tensile stresses at the contact of the upstream face with the foundation.

Calculations of stability against raising and overturning. Calculation of the stability of structures against raising is made only for low-pressure spillway dams in which the head is mainly supported by gates. The criterion for the stability of dams against raising is the condition:

$$\gamma_n \times \gamma_{\ell c} \times P \leq \gamma_{cd} \times V \tag{6.81}$$

where
> V – the sum of the vertical forces directed downward;
> P – uplift on the dam bottom directed upward.

When checking the stability of a gravity dam overturning, the condition must be met:

$$\gamma_{lc} \times M_t \leq \gamma_c \times M_r / \gamma_n \tag{6.82}$$

where
> γ_{lc}, γ_c, and γ_n – the coefficients of combinations of loads, working conditions, and reliability (see explanations to formula (6.65));
> M_t and M_r – the sums of the moments of forces tending to overturn and hold the dam, respectively, relative to the axis O_c, located in the middle of the shoulder plane BC (Figure 6.25).

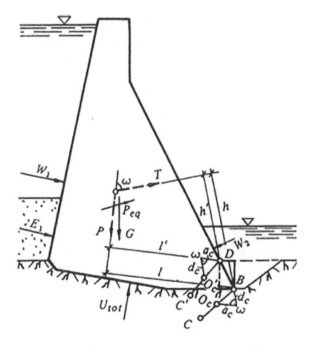

Figure 6.25 The scheme for calculating the gravity dam stability on overturning: O_c – the middle of the crumple area BC; O'_c – the middle of the crumple area of DC' – if there is an thrust.

Moments are determined from each force impact as a whole and not from its components. It is allowed to decompose the forces into horizontal and vertical components but it is necessary to attribute them to overturning and holding in accordance with the direction to which the moment of all the force belongs.

In the case shown in Figure 6.25, M_r should include moments from the dam weight G and water pressure from the downstream side W_2, whereas M_t should include moments from water pressure W_1 from the upstream, sediment pressure E_1, uplift U_{tot}, and seismic forces P_{eq}. The position of the axis O_c is found by the formulas:

$$a_c = P / \left(2 \times b \times R_{cs,\,m}\right) \tag{6.83}$$

$$d_c = \left[\left(0.5h - a_c \times \text{Cos}\omega\right)^2 + a_c\left(l - a_c\right)\right]^{1/2} - \left(0.5h - a_c \times \text{Cos}\omega\right) \tag{6.84}$$

where
  P – the result of holding forces;
  b – the width of the dam bottom;
  h – the shoulder of the force T, defined as the result of the overturning forces relative to point B;
  l – the shoulder of the force P relative to currents B;
  ω – the angle between the straight line a_c and d_c oriented normally to the forces P and T;
  $R_{cs,\,m}$ – the design value of the strength of the rock foundation on crumple.

At $R_{cs,\,m} > 20 \times \sigma$ (the σ-average normal stress at the bottom of the dam), it is allowed to calculate the dam stability according to the overturning scheme relative to point B. If there is a rock thrust from the downstream (in Figure 6.25 dashed line), the position O_c' should be found by formulas (6.83) and (6.84), while the distances a_c and d_c must be postponed from the point D of the intersection of the downstream face of the dam with the surface of the rock foundation.

The design values of the strength of the rock foundation on crumple are determined by the results of field experiments on stamps concreted to the rock foundation according to the formula:

$$R_{cs,\,m} = b \times \left(\sigma^2 + \tau_{\lim}{}^2\right) / 2 \times \left(l \times \sigma - h \times \tau_{lim}\right) \tag{6.85}$$

where
  $\sigma = PA_{pl}$, $\tau_{\lim} = T_{\lim}/A_{pl}$ – respectively, the average normal and ultimate shear stresses on the bottom of the concrete stamp at the moment of ultimate equilibrium;
  A_{pl} – the area of the stamp bottom;
  l and h – the shoulders of the forces P and T relative to the bottom edge of the stamp;
  b – the width of the stamp in the direction of slide.

The standard values of the strength of the rock foundation on crumple are determined by multiplying the calculated values on the safety factor for soil $\gamma_g = 1.25$.

Table 6.13 **Design values** $R_{cs,m,I}$

Rock soil category foundation [11]	I	2	3	4
The design values of the strength of the rock foundation on crumple $R_{cs,m}$ MPa (kg/cm²)	20 (200)	10 (100)	4 (40)	2 (20)

For dams of *I* and *II* classes under simple engineering–geological conditions at the preliminary stages of design and for dams of *III* and *IV* classes at all stages of design, the calculated values of the crumple strength $R_{cs,\ mI}$ can be accepted according to Table 6.13.

6.4 Analysis of lightweight dams

6.4.1 Gravity dams with extended joints

The arrangement of extended joints in gravity dams (Figure 6.26) can significantly reduce the filtration pressure on the dam bottom, which makes it possible to obtain concrete savings. When making calculations of gravity dams with extended seams, a separate section is considered.

Stresses in dams with extended seams can be determined by the elementary method and elasticity theory methods. Stresses in a dam with extended seams are determined by the elementary method in design sections – the position of which is assumed in the same way as in massive dams.

Normal stresses σ_y^u on the upstream and downstream σ_y^t faces of the dam in the designed horizontal section are determined by the eccentric compression formula:

$$\sigma_y^u = -\frac{N}{A} + \frac{M \times x_u}{J}, \sigma_y^t = -\frac{N}{A} - \frac{M \times x_t}{J} \tag{6.86}$$

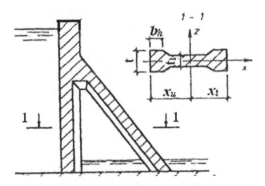

Figure 6.26 **Lightweight dam with extended joints.**

where

N and M – as before the sum of the vertical forces and bending moments of all forces, respectively, with respect to the center of gravity of the section;

A – the area of the designed cross section taking into account the cantilevered protrusions;

J – moment of inertia of the designed section relative to the z-axis passing through the center of gravity of the section (Figure 6.26);

x_u and x_t – the distance from the z-axis to the upstream and downstream faces of the dam, respectively.

As for massive dams, the forces N are M determined by the loads located above the designed cross section. For the section along the concrete–rock contact, uplift on the dam bottom is taken into account.

Knowing the stresses σ_y^u and σ_y^t and using formulas (6.52)–(6.55), stresses $\tau_{xy}^u, \sigma_x^u, \tau_{xy}^t, \sigma_x^t$ can be found and using formulas (6.56)–(6.59), (6.62)–(6.64) are the main stresses $\sigma_1^u, \sigma_2^u, \sigma_3^u, \sigma_1^t, \sigma_2^t, \sigma_3^t$ at the dam faces.

Stresses in dams with extended joints by elasticity theory methods are determined in the same way as for massive dams.

To ensure the strength of the gravity dam with extended joints, the condition (Equation 6.65) for compressed concrete zones and the conditions given in Table 6.14 must be observed [14].

In Table 6.14, t is the section size in the direction of the dam axis; t_1 is the wall thickness of the section within the expanded joints; b is the thickness of the section top along the end section; η is the coefficient equal to

$$\eta = 4(t_1 / t - 1/2)^2 \tag{6.87}$$

The device of the longitudinal cavity at the foundation of the gravity dam (Figure 6.27) significantly reduces the uplift on the dam bottom [119]. This allows us to reduce the volume of concrete in the dam and thereby get a more economical solution.

Stresses in a dam with a longitudinal cavity by the elementary method are determined according to the following approximate scheme. The upper part of the dam is calculated in the usual way as a massive dam. The lower part is considered as a massive frame with rigid knots on which the forces of its own weight water pressure from the upstream side as well as the loads transferred by the upper part of the dam act. After determining the forces in the sections of the considered frame, the stresses in these sections are found.

The stability calculations of gravity dams with longitudinal cavities are made similarly to the calculations of massive dams. This takes into account significantly less uplift on the dam bottom as well as a decrease in adhesion at the contact of the dam with the foundation due to the device of the longitudinal cavity.

6.4.2 Anchored dams

The idea of anchoring the gravity dam for the first time was implemented with the strengthening of the Scherfa dam in Algeria in 1934 due to insufficient stability and its increase by 3 m high. In order to eliminate tensile stresses on the upstream face, cables

Table 6.14 Strength conditions in design sections of gravity dams with extended joints

Horizontal design sections	Usual load combinations	Unusual load combinations	
		Not including seismic effects	Including seismic effects
In calculations by the elementary method			
Dam sections	$\sigma_y^u \leq -0.25\rho \times g \times h_u$	$d_t \leq 0.133\eta \times b$	$d_t \leq 0.286\eta \times b$
Contact section	$\sigma_y^u \leq 0$	$d_t \leq 0.3\eta \times a_2$	$d_t \leq 0.2\eta \times B$
In calculations by method of elastic theory			
Dam sections	$d_t \leq \min\begin{cases}0.5\eta \times a_1 \\ 0.5\eta \times b_h \\ 0.133\eta \times b\end{cases}$	$d_t \leq \min\begin{cases}0.167\eta \times b + 0.667\left(1-\frac{t_1}{t}\right) \times b_h \\ 0.167b\end{cases}$	$d_t \leq \min\begin{cases}0.286\eta \times b + 0.667 \times \left(1-\frac{t_1}{t}\right) \times b_h \\ 0.286b\end{cases}$
Contact section	$d_t \leq 0.3\eta \times a_2$	$d_t \leq 0.083\eta \times B + 0.667\left(1-\frac{t_1}{t}\right) \times a_2$	$d_t \leq 0.2\eta \times B + 0.667\left(1-\frac{t_1}{t}\right) \times a_2$

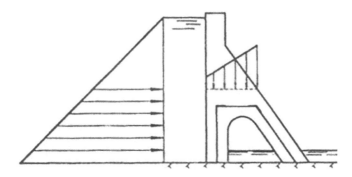

Figure 6.27 Gravity dam with longitudinal cavity.

were used that were laid in special wells in the dam body and foundation sandstones and cemented to a depth of 22–24 m. With jacks on the crest of the dam, the cables were tensioned by a force that was generally equal to the weight of one-third of the dam.

When designing an anchored dam, the following three conditions must be met:

1. Ensure the stability of the dam sliding;
2. Ensure the stability of the dam overturning;
3. Ensure the absence of tensile stresses on the upstream face.

The anchoring effect is shown on the example of a gravity dam with a height of 50 m with a vertical upstream face and incline of downstream face $m = 0.7$, at a reservoir level on dam crest mark (for simplicity). The specific gravity of the concrete is $\gamma = 2.5$ t/m³, and the uplift on the dam bottom is taken according to a linear plot with an ordinate equal to 0.5 of the head (Figure 6.28a).

When the reservoir is filled, the stresses at the contact of the upstream face are equal to zero $\sigma_u' = 0$, while those at the contact of downstream face are compressive $\sigma_d'' = 1$ MPa (Figure 6.28b). When the reservoir is empty (i.e., during the construction or repair period), the stresses on the contact of the upper face are compressive $\sigma_u' = 1.25$ MPa, while those on the contact of the downstream face $\sigma_d'' = 0$ (Figure 6.28b).

If the incline of the downstream face is reduced to $m = 0.5$, then, when the reservoir is full the tension $\sigma_u' = -0.9$ MPa will appear at the contact of the upstream face and the compressive stresses will increase to $\sigma_d'' = 2.1$ MPa at the contact of downstream face (Figure 6.28c). When the reservoir is empty, the stresses at the contact of the upstream face are compressive $\sigma_u' = 1.5$ MPa, whereas those at the contact of the downstream face $\sigma_d'' = 0$ (Figure 6.28c).

To satisfy the condition of the absence of tensile stresses at the contact of the upstream face, a compressive force of 625 tons must be applied. Then when the reservoir is filled at the contact of the upstream face $\sigma_u' = 0$, and at the contact of the downstream face, the compressive stresses are equal to $\sigma_d'' = 1.5$ MPa (Figure 6.28d). When the reservoir is empty, the compressive stresses at the contact of the upstream face will increase to $\sigma_u' = 2.2$ MPa, whereas small tensile stresses $\sigma_d'' = -0.5$ MPa will appear at the contact of the downstream face (Figure 6.28d), which are not dangerous.

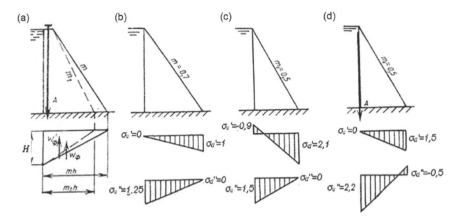

Figure 6.28 Scheme for calculating anchored dam.

The creation of prestressing in the region of the upstream face makes it possible to reduce the incline of the downstream face of the gravity dam and thereby obtain a more economical solution. The dam is prestressed by means of anchors embedded in the rock foundation, which are pulled by special jacks. These jacks are usually located on the crest of the dam.

The calculations of the stress state, total strength, and stability of anchored dams are made in the same way as the corresponding calculations of massive gravity dams. When designing anchored dams, special attention is paid to ensuring the reliability of anchoring, which is achieved by observing the following conditions.

The first condition ensuring the strength of the anchor itself is written as follows (Figure 6.29):

$$\gamma_n \times \gamma_{\ell c} \times N_A \leq \gamma_c \times A_a \times R_s \tag{6.88a}$$

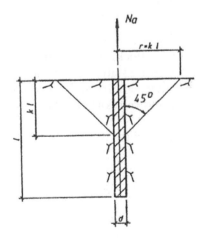

Figure 6.29 The scheme for determining the bearing capacity of the anchor.

where

$\gamma_n, \gamma_{tc}, \gamma_c$ – coefficients whose values are taken to be the same as in Equation (6.65);
N – force in the anchor;
R_s – the design resistance of the anchor material;
A_s – anchor cross-sectional area.

The second condition that determines the length of the anchor embedment into the rock foundation has the form:

$$\gamma_n \times \gamma_{tc} \times N_A \le \gamma_c \times \pi \times d \times \ell \times R_g \qquad (6.88b)$$

where

d and ℓ – respectively, the diameter of the well and the length of the embedded part of the strand;
R_g – adhesion of the cement mortar to the rock taken equal to half of the designed tensile strength of the mortar; the length of the anchor in the foundation ℓ is close to the value $25d$.

The third condition that determines the resistance of the rock to crumple in the volume of the cone is written as (Equation 6.89):

$$\gamma_n \times \gamma_{\ell c} \times N_A \le \gamma_c \left(\frac{1}{3} \rho_g \times g \times k^3 \times \ell^3 + \frac{\pi \times k^2 \times \ell^2 \times c_g}{\cos 45°} \right) \qquad (6.89)$$

where

ρ_g – rock density;
c_g – specific adhesion of the rock foundation;
k – coefficient characterizing the depth of a possible zone of rock crumple and thereby taking into account the behavior of weak rocks.

6.5 Analysis of arch dams

6.5.1 General information

Arch dam is a shell of variable thickness and curvature, based on the bottom and sides of the gorge. Arched dams are statically indefinable systems; therefore, seasonal fluctuations in air and water temperature in the reservoir cause temperature stresses in it. In some cases, temperature stresses can be commensurate with stresses caused by hydrostatic pressure of water.

Temperature stresses can lead to the opening of joints, to the formation and opening of cracks, which can significantly affect the nature of the static operation of arch dams. In order to reduce the influence of temperature stresses on the strength of arch dams, various measures are used; for example, heat shielding walls are arranged from the bottom side.

The influence of dead weight and filter uplift on the bottom on the SSS of thin arch dams is usually small. However, in arch-gravity dams, the dead weight of the structure and the filter uplift on the bottom play a significant role.

Very significant for arch dams are seismic impacts. In this case, the least favorable stress state arises in the dam from seismic impacts directed across the river valley.

For arch dams, stresses resulting from shrinkage and swelling of concrete can be significant. These stresses depending on the method of concrete laying are usually defined as stresses caused by a fictitious increase and a decrease in temperature in the corresponding zones of the dam.

Stresses in arch dams due to self-weight forces should be determined taking into account the sequence of construction and the conditions for monolithic intercolumn and intersectional construction joints. Monolithization is carried out during the construction of the dam as the temperature-shrinkage deformation of the construction period attenuates.

In calculating the SSS of arch dams, it is necessary to take into account the influence of the bed deflection of the reservoir and the so-called collapse of the banks which arise as a result of the hydrostatic pressure of the water on the bed and banks of the reservoir. It is believed that this effect is relatively small. For gravity dams, this influence has been investigated (see Section 6.3.3), and it is really not significant.

The SSS studies of arch dams are carried out in order to check its strength as well as the stability and strength of bank slopes (abutments) perceiving the loads from the dam.

Determining the SSS of such a complex design is associated with significant mathematical difficulties. Therefore, simplified calculation methods have been developed in each of which there are limitations on the applicability of the calculation scheme in accordance with which the operation of the arch dam is investigated.

6.5.2 Independent arches method

The method of independent arches is that the arch dam in height is divided by horizontal sections into a series of arcs working independently of each other. Each of these arcs is affected by hydrostatic pressure, sediment pressure, temperature effects, seismic effects, and wave pressure.

The simplest stresses in such arches are determined under the assumption of articulated support of the heels. In this case, the stresses σ in the arches can be found by the boiler formula:

$$\sigma = \frac{p \times r_n}{d},$$

(6.90)

where
  p – the hydrostatic pressure of water;
  r_n – outer radius of the arch;
  d – thickness of the arches.

The boiler formula is very approximate. It does not allow us to take into account the conditions for closing the heels of arches, the flexibility of the foundation, and a number of other factors. Therefore, the strength of arch dams calculated by the method of independent pivotally supported arches is estimated using low permissible compressive stresses. The values of such permissible stresses are usually taken equal to 2.5 MPa for thin upper arches, whereas for the middle and lower arches, the permissible stresses are reduced in height to 0.8–1 MPa.

In calculations of pinched independent arches, ductility of the foundation is usually taken into account by the *Vogt–Telke* method. A circular arch of radius r_o of constant thickness d with a central angle of $2a_o$ is considered. The arch material is concrete characterized by elastic modulus E and Poisson's ratio v. The arch is loaded with a uniformly distributed radial load p (Figure 6.30).

In accordance with the decision of construction mechanics, bending moments M and normal N and transverse Q forces in any section of the arch making an angle α with its axis, can be found by the following formulas:

$$
\left.
\begin{aligned}
M &= -p \times r_o^2 \times \left(1 + \frac{d}{2r_o}\right) \times A \times \left(\frac{\sin \alpha_o}{\alpha_o} - \cos \alpha\right) \\
N &= -p \times r_o \times \left(1 + \frac{d}{2r_o}\right) \times (1 - A \times \cos \alpha) \\
Q &= -p \times r_o \times \left(1 + \frac{d}{2r_o}\right) \times A \times \sin \alpha
\end{aligned}
\right\}
\tag{6.91}
$$

where

$$
A = \frac{2\sin \alpha_o}{\left(12 \times \dfrac{r_o^2}{d^2} + 1\right) \times \left(\alpha_o + \dfrac{1}{2}\sin 2\alpha_o\right) - 24 \times \dfrac{r_o^2}{d^2} \times \dfrac{\sin \alpha_o^2}{\alpha_o}}
\tag{6.92}
$$

The arch deflection in the key f_0 is determined by the following formula:

$$
f_o = \frac{1 - v^2}{E} \times p \times \frac{r_o^2}{d} \times \left(1 + \frac{d}{2r_o}\right) \times \frac{(\alpha_o - \sin \alpha_o) \cdot (1 - \cos \alpha_o)}{\alpha_o + \dfrac{1}{2}\sin 2\alpha_o - \dfrac{2\sin \alpha_o^2}{\alpha_o \times \left(1 + \dfrac{d^2}{12 r_o^2}\right)}}
\tag{6.93}
$$

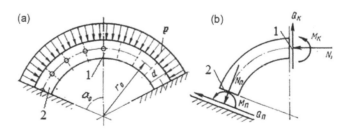

Figure 6.30 Scheme for calculating an independent arch: (a) plane of arch, (b) positive directions of forces and moments: l-key, 2-heels.

From the known forces N and moments M, the normal stresses σ on the arch faces in any section can be found by the eccentric compression formula.

In thick arches with relatively small central angles, zones of significant tensile stresses can arise. Such zones are usually located near arches heel on the upper face and in the area of the key on the dam down face (Figure 6.31). Tensile stresses in these areas may exceed concrete tensile strength. In this case, cracks occur and parts of the cross sections with cracks are turned off from the arch. As a result, a so-called secondary arch is formed which perceives the existing loads.

The secondary arch is an arch of variable thickness. The stiffness of such an arch in the crack zones is significantly lower than the stiffness of the original arch without cracks. The outline of the faces of the secondary arch is calculated assuming that the concrete does not work in tension.

Let a bending moment M and a compressive force N act in a certain section. The eccentricity e_o of the force N is

$$e_o = \left| \frac{M}{N} \right| \tag{6.94}$$

The condition for the appearance of a crack in the considered section is the inequality:

$$e_o > \frac{d}{6} \tag{6.95}$$

where
  d – the thickness of the original arc without cracks.

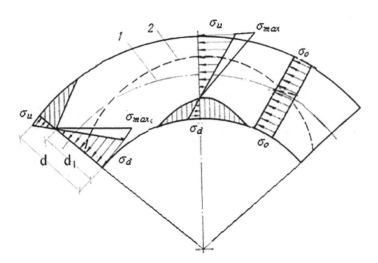

Figure 6.31 Scheme of the secondary arch (zones with tensile stresses are shown by vertical hatching): 1-axis of the arch; 2-pressure curve; σ_u and σ_d – stress on the upper and down faces; σ_{max} – stress in the secondary arch; σ_o – average stress.

The thickness of the section decreases to d_1, the value of which is determined by the following formula:

$$d_1 = 3(0.5d - e_o) \tag{6.95a}$$

The stress diagram in the section of the secondary arch is assumed to be triangular (Figure 6.31). The maximum compressive stress σ_{max} is equal to

$$\sigma_{max} = 2\frac{N}{d_1} \tag{6.96}$$

Thus, the appearance of cracks in the arch elements of dams does not lead to their destruction, if the maximum compressive stresses in the resulting secondary arches do not exceed the compressive strength of concrete. The formation of secondary arches explains the performance of arch dams with significant even nondesign overloads.

Analysis of the operation of the secondary arches allows us to choose the momentless outline of the axis of the arches, corresponding to the axis of pressure. These considerations explain the adoption in some cases of a noncircular outline of the axis of the arcs.

6.5.3 Arches–central console method

When calculating arch dams by the arches–center console method, the dam is considered as spatial structure consisting of a system of arches, filling the entire volume of the dam, but not connected to each other, and one central console in the place of the highest dam height (Figure 6.32). The movement of the points of the center console and the corresponding key points on the axes of the arches under the influence of hydrostatic pressure of water are considered equal.

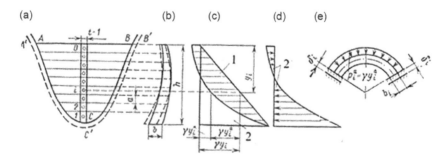

Figure 6.32 The scheme for calculating arch dams using the arches–central console method: (a) the allocation of arches and the center console, (b) transverse cross section of the center console, (c) the distribution of hydrostatic load between the arches and the console, d) the diagram of the load on the console, (e) diagram of the load on the ith arch; I and 2 – parts of hydrostatic load on the arches and console, respectively; h – dam height; b_i – thickness of the i-th arch; y_i –depth of the i-th arch; γ – specific weight of water; δ_1 – Telke fictitious recess.

Usually in the method of arches and the center console of all possible movements of the common points of the arches and the center console, only radial movements are considered. From the equality of these movements, it follows that the hydrostatic pressure of the water is distributed in some way between the arches and the center console. It is assumed that part of the hydrostatic pressure perceived by each arch is constant along its length. This assumption allows us to take into account in some way the influence of other consoles, except the central one.

The arch–center console method can be used to calculate thin and thick arch dams located in symmetrical or close to symmetrical gorges.

The equation of bending of the center console can be written as the equation of bending of a beam of variable section in the form:

$$\frac{d^2}{dy^2}\left[D(y)\times\frac{d^2u(y)}{dy^2}\right]=q(y) \tag{6.97}$$

where

u(y) – horizontal displacement of console points varying in height y;
D(y) – cylindrical stiffness variable in height:

$$D=\frac{E\times d^3(y)}{12(1-v^2)} \tag{6.98}$$

where

d(y) – the thickness of the center console varying in height equal to the thickness of the arches;
E and v – the elastic modulus and Poisson's ratio of a concrete dam, respectively;
q(y) – the horizontal load unevenly distributed in height transmitted to the center console.

It should be noted that the load q(y) acting on the center console can be represented as the difference between the external load p(y) and the load $p_o(y)$ perceived by the arches. The load $p_o(y)$ is proportional to the deflection of the arches in the key. From the equality of these deflections to the movements of the center console, it follows:

$$p_o(y)=k(y)\cdot u(y) \tag{6.99}$$

where

k(y) – the height variable coefficient of elastic resistance of the arches determined by the following formula:

$$k(y)=\frac{1}{f_o(y)} \tag{6.100}$$

where

$f_o(y)$ – radial deflection in the key of the arch located at a height y and loaded with a uniformly distributed radial load of unit intensity.

The value $f_o(y)$ is determined by the usual methods of structural mechanics depending on the conditions of support of the arches. For rigidly jammed arches (or taking into account the ductility of the foundation according to the *Vogt–Telke* method for fictitiously elongated arches), the value $f_o(y)$ can be found by formula (6.93), in which $p = 1$ should be taken.

Given (Equation 6.99), the expression for $q(y)$ can be written as:

$$q(y) = p(y) - k(y) \times u(y) \tag{6.101}$$

Substituting this expression in Equation (6.97), we obtain the main differential equation of the arches–center console method:

$$\frac{d^2}{dy^2}\left[D(y) \times \frac{d^2u(y)}{dy}\right] + k(y) \times u(y) = p(y) \tag{6.102}$$

Equation (6.102) is an equation for the bending of a beam of variable cross section located on an elastic Winkler base with a variable coefficient of elastic resistance.

Integration of Equation (6.102) should be performed subject to the following boundary conditions:

on the crest of the dam at $y = H$:

$$\left.\frac{d^2u}{dy^2}\right|_{y=H} = 0, \frac{d}{dy}\left[D(y) \times \frac{d^2u}{dy^2}\right]_{y=H} = 0 \tag{6.103}$$

at the base of the dam at $y = 0$:

a. For a rigidly clamped center console (or taking into account the ductility of the foundation according to the *Vogt–Telke* method for a fictitiously elongated console):

$$\left.u\right|_{y=0} = 0, \left.\frac{du}{dy}\right|_{y=0} = 0 \tag{6.104}$$

b. For an elastically clamped center console (taking into account the ductility of the foundation according to *Vogt*):

$$\left.u\right|_{y=0} = \frac{k_5 \times M_o}{E_o \times d_o} + \frac{k_3 \times Q_o}{E_o}, \left.\frac{du}{dy}\right|_{y=0} = \frac{k_1 \times M_o}{E_o \times d_o{}^2} + \frac{k_5 \times Q_o}{E_o \times d_o} \tag{6.105}$$

where

M_o, Q_o, N_o – moment, transverse, and normal forces in the calculated reference section;

E_o – foundation deformation modulus;

d_o – the thickness of the arch dam in the section under consideration;

k_1, k_2, k_3, k_5 – the *Vogt* coefficients.

The solution of Equation (6.102) in quadratures for arbitrary functions $D(y)$, $k(y)$, $p(y)$ is associated with significant mathematical difficulties. Therefore, the integration of Equation (6.102) under the boundary conditions (Equation 6.103), (Equation 6.104), or (Equation 6.105) is usually performed by the numerical method, for example, the FEM, the successive approximation method, the variation method, etc. [58].

As a result, a system of n linear algebraic equations with respect to the desired values of the displacements of the points of the center console u_i ($i = 1, 2,...n$) is obtained.

According to the horizontal movements of the points of the center console u_i, in accordance with (6.101), a part of the external load perceived by the arches $p_{ai} = k_i \times u_i$ is determined.

From the obtained p_{ai} values, the forces in the sections of each arch element are determined. The values of the forces depend on the conditions of support of the heels of the arches. So for rigidly jammed arches (or when taking into account the ductility of the foundation according to the *Vogt–Telke* method), the moments M and the forces N and Q in the sections of the arches are determined by formulas (6.91). From the forces found, it is not difficult to determine the normal stresses on the faces of the arches.

Stresses in horizontal sections of the center console are composed of stresses from its own weight and stresses caused by part of the hydrostatic pressure of water perceived by the center console. It is believed that stresses from the dead weight of the dam occur only in consoles.

Stresses in the sections of the center console caused by hydrostatic pressure of water are determined as follows. The well-known deflection function $u(y)$ determines the bending moment M:

$$M = D \times \frac{d^2 u}{dx^2} \tag{6.106}$$

or in finite differences:

$$M_i = \frac{D_i}{h^2} \times \left(u_{i+1} - 2u_i + u_{i-1} \right) \tag{6.107}$$

The bending moments M_i in the calculated sections determine the stresses σ_i on the faces of the center console using the well-known formula:

$$\sigma_i = \pm \frac{6 \times M_i}{d_i^2} \tag{6.108}$$

6.5.4 Other methods

In the method of arches–consoles, the spatial work of the arch dam as a cross system of horizontal arches and vertical consoles is studied (Figure 6.33). From the condition of equality of radial displacements of arches and horizontal displacements of consoles at the points of their intersection, the hydrostatic pressure of water is distributed between arches and consoles.

There are varieties of this method when at the intersection points of arches and consoles, the rotation angles and vertical displacements are also equalized.

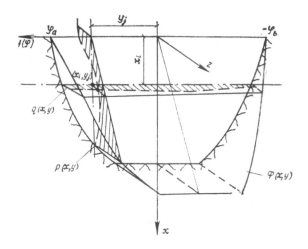

Figure 6.33 Separation of hydrostatic load on arches and consoles.

The arches–console method is a rather laborious method and requires the use of specially designed computer calculation programs.

Despite the certain approximation of this method, it allows us to assess the stress state of the arch dam more reliably than the methods described above. The method can be used to calculate thin and thick arched dams in symmetric and asymmetric gorges.

In the shell theory method, the arch dam is schematized as a thin or thick shell, which relies accordingly on the bed and banks of the gorge. The problem is solved on a computer using any numerical method (e.g., FDM or FEM). Based on the forces found in the solution in the sections of the shell, the stresses at the calculated points of the dam are determined. The method of the theory of shells makes it possible to accurately determine the SSS of thin and thick arch dams in gorges of arbitrary shape.

The method of elasticity theory allows us to study the operation of the system "arch dam-foundation" as a continuous inhomogeneous elastic three-dimensional body. In this case, the entire set of existing loads, including the water pressure on the banks of the gorge, can be taken into account. The solution to the problem is performed on a computer usually FEM. The method of elasticity theory allows us to get more reliable information about the SSS arch and arch-gravity dams of any kind, as well as their foundation and bank of the gorge.

Experimental methods based on studies of physical models made of various materials make it possible not only to study the stress state of arch dams, but also to determine safety factors and schemes for the possible destruction of the structure. These methods make it possible to evaluate the operability and reliability of the dam design when designing arch dams in complex topographic and engineering–geological conditions.

In connection with the development of numerical methods and the introduction of computer programs in practice, the role of experimental methods is reduced due to the cost and speed of research.

When investigated the reliability of the Nam Chien arch dam, erected in Vietnam in 2012, design studies of the FEM of the SSS of the dam were carried out in the construction and operational periods. Taking into account the phased construction of the dam and filling the reservoir, calculations of the stability of bank abutments, as well as studies of the limit state of the dam according to SIX scenarios, were made [105,236].

The height of the dam is 113 m without plug and 135 m with plug, and the length along the ridge is 273.3 m (arched part). The thickness of the arch dam on the plug is 17.4 m, while that on the crest 6 m. The shape coefficient of the site is 2.42, and the shapeliness coefficient of the dam is 0.154. At the foundation of the dam, antifiltration and drainage measures were performed: reinforcing cementation, cementation, and drainage curtains (Figure 6.34).

On the basis of engineering–geological and topographic data, a three-dimensional finite-element model of the system "arch dam–rock foundation" was created. The dimensions of the mathematical model in the plane were 540 × 570 m, and the height of the model is 620 m; from above, the model is limited by the natural surface of the relief. In the foundation model, zones of preservation of bedrocks (zones IB, IIA, IIB) and explosive disturbances of the fourth and fifth order (Figure 6.35) are identified.

To reproduce the real contact strength of the dam with the foundation close to zero, the contact was approximated by special contact finite elements. In addition, two planes simulating virtual slide planes were reproduced in the right-bank abutment; in the base on the extension of the upper face, contact elements were introduced that simulated the decompression zone and bound together with the slide planes a design block for which stability calculations were made according to the method described in Section 6.5.7.1. In the studies, six stages of the dam construction were reproduced (Figure 6.36); at the seventh stage of the calculation, the water pressure at the NHL and the temperature effect were applied.

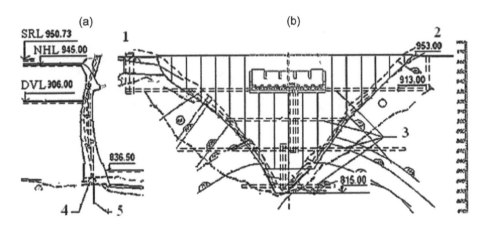

Figure 6.34 Nam Chien arch dam (Vietnam): (a) section along the center console, (b) a view from the downstream; I – left bank, 2 – right bank, 3 – drainage galleries, 4 – cement curtain, 5 – drainage.

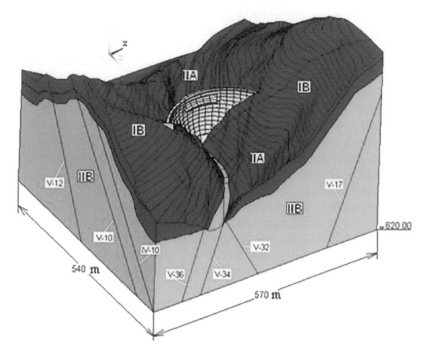

Figure 6.35 Mathematical geomechanical model of the system "arch–rock foundation". Axonometric, view from the downstream.

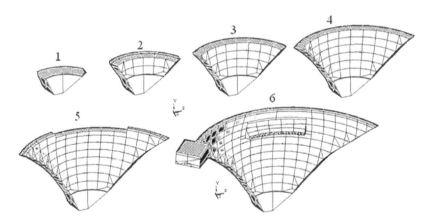

Figure 6.36 Six stages of construction Nam Chien arch dam.

After substantiating the strength and stability of the "dam-foundation" system in the operational stage, studies of the system in the first ultimate state were performed [106,236].

In practice, two methods (scenarios) are used to bring the structure to the first limit state (loss of bearing capacity or destruction):

1. Increase of load on the dam;
2. Reduction of the strength properties of the "dam-foundation" system.

The first method was used in testing physical models of the "dam-foundation" system, when the dead weight of the dam and the pressure on the upper face were proportionally increased. In some cases, only the pressure on the upper face of the dam increased. While during the test the direction of the main vector of forces acting on the dam changed, it was believed that in this way a well-known catastrophe was simulated on the arch dam Vaiont (Italy), when due to a huge landslide in the reservoir a wave of water 100–150 m high passed over the dam. In this case, the safety factor is defined as the ratio of ultimate load to operational.

In the second method, the safety factor is defined as the ratio of the real strength parameters of the "dam-foundation" system to the virtual ones at which the system goes into a limiting state.

Both methods were used in the studies to bring the system "arch dam–rock base" to the first limit state in six scenarios.

Scenario 1: a virtual decrease of the parameters slide resistance along the planes restricting the design block in the bank abutment of the arch dam. The studies reproduced the scheme for calculating the stability of bank abutment arch dams developed in Ref. [101] and regulated by the SR [11]. Detailed research results for this scenario are presented in Section 6.5.6.3. With a $K = 7$-fold decrease in the parameters of slide resistance along the designed slide planes 1 and 2, the bearing capacity of the dam turned out to be far from exhausted; in other words, the kinematic scheme of the loss of stability of the dam incorporated in the stability calculation turned out to be unrealistic.

Scenario 2: a virtual decrease in the parameters of slide resistance along the contact of the dam with the foundation, i.e., checking the possibility of dam slide in contact with the foundation. With relatively gentle bank abutments, a slide of the arch dam along the contact with the foundation in the upstream–downstream direction is possible. In the studies, a virtual decrease in the parameters of slide resistance along the contact of the arch dam with the foundation by $K = 2$, 4, and 7 times was considered. At the same time, the strength of the dam concrete was not reproduced in the studies, and the obtained compressive stresses were estimated by the standard concrete compressive strength $R_{bn} = 25.80$ MPa, whereas tensile stresses by the allowable tensile stress $R_{btn} = 2.16$ MPa. At $K = 7$, the stresses in the dam did not exceed the compressive and tension strength of concrete, while the dam moved on 8.9 mm in contact with the foundation; i.e., for the arch dam under consideration, such a limiting state is not relevant.

Scenario 3: a virtual decrease in the strength of a concrete dam jammed in contact with the foundation. In this scenario, it was assumed that the ultimate state (loss of the bearing capacity of the arch dam) occurs due to a virtual decrease in the tension and compression strength of concrete by $K = 2$–4 times. For this purpose, special solid 65 three-dimensional finite elements were used with which the destruction of concrete as a brittle material was simulated.

At the standard concrete strength ($K = 1$), there formed 21 cracks on the upstream face; new cracks began to form in the dam as the concrete strength decreased; and at $K = 4$, the number of cracks increased significantly to 108. With a subsequent virtual decrease in concrete strength, the solution to the problem diverged, which testified to the transformation of the dam into kinematic variable system. Thus, the dam safety factor was determined equal to at least four.

Scenario 4: a virtual decrease in the strength of the dam concrete, taking into account the ductility of the rock foundation. The strength of the rock foundation was not modeled in the studies, and its stress state was evaluated from the point of view of the compression strength of the foundation. The studies were carried out with standard parameters of concrete compression and tension strength and with a virtual decrease of $K = 1.5$ and 1.7 times. An analysis of crack formation in the dam suggests that at $K = 1.7$, the bearing capacity of the dam was far from exhausted. With the subsequent virtual decrease of the concrete strength parameter, it was not possible to obtain an exact solution to the problem due to its poor convergence. The actual dam safety factor is higher than 1.7. This is evidenced by a safety factor of four in the previous scenario 3 and compressive stresses at the foundation far from the compression strength of the rock mass dam.

Scenario 5: a virtual increase in the load on the upstream face of the dam. In this scenario, the pressure on the upstream face of the arch dam is increased by changing the volumetric weight of the water at a constant reservoir water level. The uplift on the bottom of the arch dam and the dead weight of the dam remained constant; sediment pressure on the upstream face of the dam was not taken into account. The studies were performed with a virtual increase in the load on the upstream face of the arch dam by $K = 2$, 2.5, and 3 times. Compressive stresses in the central part of the upstream face increased from $-10\,MPa$ at $K = 2$ to $-20.5\,MPa$ at $K = 3$; there are no tensile stresses on the upstream face of the dam. Opening of the contact of the dam with the foundation occurred under the plug and the upstream face in the lower part of the banks. With a subsequent decrease in concrete strength, an exact solution to the problem was not obtained due to its poor convergence. Safety factor of dam is assumed to be three.

Scenario 6: a virtual increase in the load on the dam upstream face (due to the overflow of water through the dam, similar to the accident at the Vaiont Dam). The studies were performed with a virtual increase in the load on the upstream face of the arch dam at $K = 1.13$, 1.95, and 3.38 times. As the load on the dam increased, the vertical displacements changed sign: at $K = 1$, displacements were directed downward, and at $K = 1.95$ and 3.88, those were directed upward; the maximum displacement of 55 mm was obtained on the upstream face. The deflections of the dam increased significantly as the load on the dam increased, and at $K = 3.38$, it amounted to 240 mm in the key on the crest. Deformed state of the dam is presented in Figure 6.37; isochromes of the main displacement vector at a scale of 500:1 are shown. It can be seen that the dam has moved to the upstream in contact with the foundation, including the right-bank pier.

The arch dam project was completed at the Hydroproject (Kharkov), and design studies at the IIGH.

6.5.5 Assessment of arch dams strength

In accordance with SR [14], the main compressive stresses σ_3 must satisfy the condition [156]:

$$\gamma_n \times \gamma_{lc} \times |\sigma_3| \leq \gamma_{cd} \times \gamma^c_{cda,1} \times R_b, \tag{6.109}$$

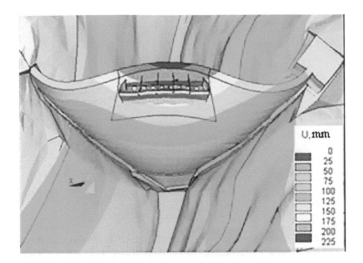

Figure 6.37 Nam Chien arch (Vietnam). The deformed state of the arch dam in scenario 6 at $K = 3.38$. Displacement vector U, mm. The scale of displacements is 500:1.

the main tensile stresses σ_1 must satisfy the condition:

$$\gamma_n \times \gamma_{lc} \times \sigma_1 \leq \gamma_{cd} \times \gamma^t_{cda,1} \times R_{bt} \tag{6.110}$$

where

R_b and R_{b1} – respectively, the design resistance of concrete to compression and
tension;

γ_n, γ_{lc}, γ_{cd} – coefficients taken in accordance with the explanations to formula
(6.65);

$\gamma^c_{cda,1}$, $\gamma^t_{cda,1}$ – coefficients of working conditions of arched dams, taken equal to
$\gamma^c_{cda,1} = 0.9, \gamma^t_{cda,1} = 2.4$.

6.5.6 Analysis of arch dams stability

6.5.6.1 Analysis of bank abutments stability

The first study of the stability of an arch dam was carried out during the design of the
Luixieh dam in China, while it was believed that the stability of the dam determines
the bank abutment under the influence of the dam forces [235]. The calculation was
made as follows: from the dam, an arch element 1 m high and a cantilever element
resting on the heel of the arch conditionally stood out (Figure 6.38). Calculated slide
planes were drawn through the upper heel point at an angle a to the heel and having
a dip angle λ equal to the tilt angle of the foundation pit. The calculation took into
account: normal N_a and transverse Q_a forces in the heel arch, the weight of the console
G_a, and the weight of the sliding rock mass G_{sk}.

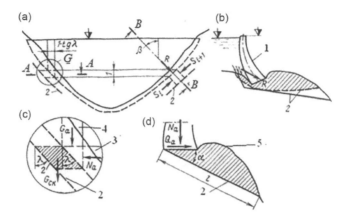

Figure 6.38 Scheme for calculating the stability of arch dam in flat sections: (a) view from the downstream, (b) section along B-B, (c) node G, (d) section along A-A; 1 – dam, 2 – plane of slide, 3 – arc element, 4 – cantilever element, 5 – horizontal terrain.

In the presence of a system of fissures or a weak interlayer, the position of the slide plane was determined by their position. Otherwise, the position of the dangerous plane was determined by selection as giving the smallest safety factor. The coastal stability coefficient was defined as the ratio of the holding forces on the slide plane to the slider forces directed to the downstream.

When designing the Inguri arch dam in the 1970s of the last century when substantiating its stability, the limiting state of the design blocks in spatial conditions was analyzed, isolated in the bank abutments in accordance with the geological structure [101]. Three planes bounding the calculated rock block were selected according to the geological structure; according to the fissures or fracture systems, the fourth surface was the surface of the shore and the foundation pit of the arch dam (Figure 6.39a). A slide of the design block was assumed along two planes 1 and 2, and a detachment along plane 3 was caused by the zone of decompaction of the foundation under the upper face of the dam [164].

The limit state of the rock block was investigated [216,228,235], arising from a virtual decrease in the parameters of slide resistance along slide planes bounding the rock block, under the assumption that the forces from the arch dam do not change as the limit state of the block is reached. It was believed that the rock block goes to the limit state when the slide and holding forces are equal along the two slides 1 and 2 (Figure 6.39b) – in the particular case of slide along one slide plane and separation along two others (Figure 6.39c).

The block stability coefficient was determined as the ratio of the real slide parameters $tg\varphi$ and C to virtual $tg\varphi_s$ and C_s, at which the block was in the limiting state. All rock blocks possible according to the conditions of the geological structure were analyzed, and the stability coefficient of bank abutment was determined as the smallest of the considered ones. It was believed that the bank abutment stability coefficient determined the stability of the arch dam [256].

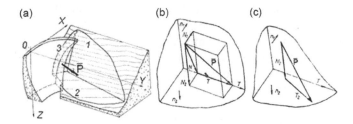

Figure 6.39 Schemes for calculating the stability of an arch dam in spatial conditions: (a) design block, (b) slide along two planes I and 2, and (c) slide along one plane 2.

From the calculation of the SSS of the arch dam, the main force vector $\vec{P}$ [101] is transmitted from the arch dam and the filtration forces along the planes 1–3; then, from the equilibrium of forces Equation (6.111) written in vector form, the normal forces $\vec{N}_1$ and $\vec{N}_2$ were determined on the slide planes 1 and 2 and the slide force $\vec{T}$ on line of the planes 1 and 2 intersection:

$$\vec{N}_1 + \vec{N}_2 + \vec{T} = \vec{P} \tag{6.111}$$

In the limiting state with virtual slide parameters $tg\varphi_s$ and C_s, the resistance forces and the slide force T are equal to:

$$N_1 tg\varphi_{1s} + N_2 tg\varphi_{2s} + C_{1s}A_1 + C_{2s}A_2 = T \tag{6.112}$$

where
A_1 and A_2 – the areas of the slide planes 1 and 2.

Whereas the safety factor k is equal to:

$$K = tg\varphi_1 / tg\varphi_{1s} = tg\varphi_2 / tg\varphi_{2s} = C_1 / C_{1s} = C_2 / C_{2s} \tag{6.113}$$

it is to get that:

$$K = \left[N_1 tg\varphi_1 + N_2 tg\varphi_2 + C_1 A_1 + C_2 A_2 \right] / T \tag{6.114}$$

If it was found from solution (6.111) that the normal force is $N_1 < 0$ (or $N_2 < 0$), the rock block was sliding along plane 2 (or 1) with a separation from plane 1 (or 2, Figure 6.39c). The safety factor K in this case is equal to:

$$K = \left[N_{20} tg\varphi_2 + C_2 A_2 \right] / T_{20} \tag{6.115}$$

where
N_{20} and T_{20} – the projections of the main force vector from the arch dam P on the normal and on plane 2 (or 1, if $N_2 < 0$).

Analysis of the calculation of stability of bank abutment arch dam shows that for $N_1 > 0$ and $N_2 > 0$, the tendency of the rock block to slide along the incidence lines of the slide planes 1 and 2 from the dead weight of the block and the vertical forces transmitted from the arch dam is not taken into account.

In Ref. [101], this drawback was eliminated and a slide block scheme was developed with its separation into two compartments by plane 12 passing through the intersection line of planes 1 and 2 (Figure 6.40). In Figure 6.40, the following parameters are marked:

G_1 and G_2 – own weights of compartments;
P_1 and P_2 – forces transmitted from the arch dam to the compartments;
U_1, U_2, and U_{12} – filtration pressure of water on the planes 1, 2, and 12;
T_1, T_{12}, and T_2 – parts of the force T, referred to planes 1, 2, and 12.

From the calculation of the SSS of the arch dam were determined two main vectors of force $\vec{P}_1$ and $\vec{P}_2$, transferred from the arch dam to the first and second compartments, and the equations of equilibrium of forces acting on each compartment of the calculation unit separately were written:

$$\vec{N}_1 + \vec{N}_{12} + \left(\vec{T}_1 + \vec{T}_{12}\right) = \vec{P}_1 \tag{6.116}$$

$$\vec{N}_2 + \vec{N}_{12} + \left(\vec{T}_2 + \vec{T}_{12}\right) = \vec{P}_2 \tag{6.117}$$

where:

$$T_i = N_i tg\varphi_i + C_i A_i \tag{6.118}$$

$$K = tg\varphi_i / tg\varphi_{is} = C_i / C_{is} \left(i = 1, 2, 12\right) \tag{6.119}$$

In the system of six equations (three for each compartment), there were seven unknowns: $\vec{N}_1$, $\vec{N}_2$, $\vec{N}_{12}$, γ_1, γ_2, γ_{12}, and K (γ_1, γ_2, and γ_{12} – the angles between the line of

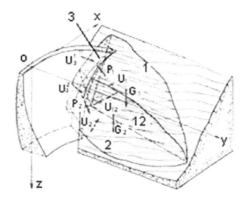

Figure 6.40 Scheme for calculating the stability of an arch dam when the calculated rock block is divided by plane into two compartments: 1 and 2 – slide planes; 3 – separation plane.

dip plane i and the direction of slide along plane i ($i = 1, 2, 12$). The missing equation was deduced from consideration of the slide kinematics on three slide planes 1, 2, and 12 (Figure 6.41):

$$tg\gamma_{12} = \left(tg\gamma_1 \sin\xi_{2-12} - tg\gamma_2 \sin\xi_{1-12}\right) / \sin\xi_{1-2} \qquad (6.120)$$

where

ξ_{1-2}, ξ_{1-12}, and ξ_{2-12} – the angles between planes 1 and 2, between planes 1 and 12, and between planes 2 and 12, respectively;

Δ_{T1} and Δ_{T2} – displacements along planes 1 and 2, respectively;

Δ_{T12} – mutual displacements of the compartments along the plane 12.

A system of seven equations with seven unknowns is nonlinear, and its solution is presented in Ref. [101].

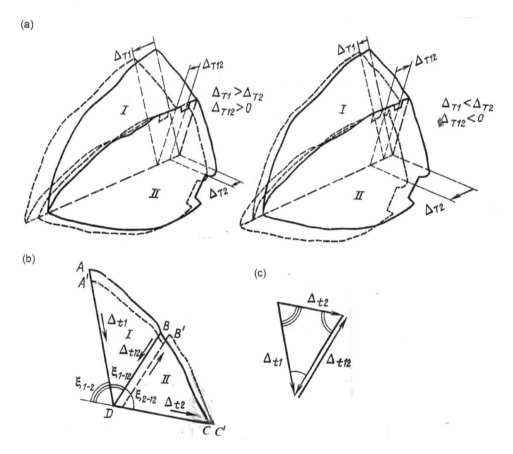

Figure 6.41 The scheme of virtual displacements of two compartments of rock block in the ultimate state: (a) two options for displacements of compartments in space, (b) displacements compartments in a plane perpendicular to line T, (c) kinematic relationship of displacements in a plane perpendicular to line T.

In accordance with SR [14], the stability coefficient of the bank abutment of the arch dam must satisfy the condition:

$$K \geq \gamma_n \times \gamma_{lc} / \gamma_{cd} \times \gamma_{cda2} \tag{6.121}$$

where

γ_n and γ_{lc} – reliability coefficient for responsibility and combinations of loads given in the explanation to formula (6.65);

γ_{cd} – the coefficient of working conditions equal to 0.75 in the stability calculations of the arch dam bank abutments;

γ_{cda2} – the coefficient of working conditions, taken 1 in the calculations excluding seismic effects and 1.1 in the calculations taking into account seismic effects.

6.5.6.2 Analysis of arch dam stability in wide site

It was believed that in extreme equilibrium under the influence of operational loads, the dam can rotate together with some volume of the rock base around an axis passing through the center of arch pressure in the plane of any radial section of the dam coinciding with the intersection joint (Figure 6.42).

The stability safety coefficient K was calculated as the ratio of the sum of the moments of the reactive forces ΣM_p developing on the slide surface with the calculated parameters of slide resistance to the sum of the moments of the active forces ΣM_a acting in normal operation on the considered part of the dam:

$$K = \Sigma M_p / \Sigma M_a \tag{6.122}$$

The active forces acting in normal operation relied on (Figure 6.43):

- Hydrostatic pressure on dam E, wave pressure, and ice pressure;
- Component of the dead weight of the dam parallel to the bottom plane $T(G'')$, if the bottom has a slope.

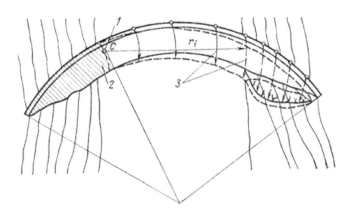

Figure 6.42 Kinematic scheme of the shoulder rotation of the arch dam with the base: I – instantaneous axis of rotation, 2 – fixed part of the dam, 3 – directional surfaces of rotation.

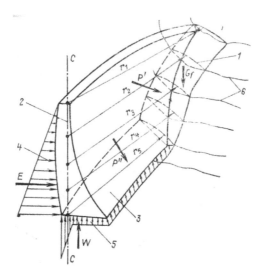

Figure 6.43 Scheme of loads and forces acting on the shoulder of the dam: 1 – surface of rotation, 2 – instantaneous axis of rotation, 3 – bottom of the dam, 4 – hydrostatic pressure, 5 – uplift, 6 – horizontal terrain.

The moment of active forces was defined as:

$$\Sigma M_a = E \times r_E + T(G'') \times r_G'' \tag{6.123}$$

where
   r_E and r_G'' – the radii of the points of the application E and G''.

The reactive forces on the slide surface are the components of the main vectors P' and P'' normal to the slide surface, acting on the bottom of the dam and on the bank, multiplied by the parameters $tg\varphi'$ and $tg\varphi''$, respectively, and the cohesion force equal to the product of the slide surface areas A' and A'' to parameters C' and C'', respectively; the normal to the slide surface component of the weight of the rock mass G_r is multiplied by $tg\varphi'$, i.e.,

$$\Sigma M_p = \left[N(P') \times tg\varphi' + C \times A\right] \times r_A' + \left[N(P'') \times tg\varphi'' + C'' \times A''\right] \times r_A''$$
$$- W \times r_W \times tg\varphi + N(G_r) \times r_G \times tg\varphi' \tag{6.124}$$

In accordance with SR [14], the stability coefficient of an arch dam in a wide range must satisfy the condition:

$$K \geq \gamma_n \times \gamma_{lc} / \gamma_{cd} \times \gamma_{cda2} \tag{6.125}$$

where
   γ_n and γ_{lc} – reliability coefficient for responsibility and combinations of loads given in the explanation to formula (6.65);

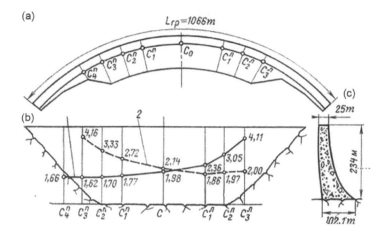

Figure 6.44 Stability coefficients K of the arch-gravity dam of the Sayano-Shushensk
HEP: (a) plane, (b) view from the downstream, (c) cross section of the
dam; I and 2 – rotation of the left and right shoulders, respectively.

γ_{cd} – coefficient of working conditions equal to 1.1 in analysis of the stability of an
arch dam in a wide range;

γ_{cda2} – coefficient of working conditions taken 1 in the analysis excluding seismic
effects and 1.1 in the analysis taking into account seismic effects.

The method was used to justify the overall stability of the arch–gravity dam of the
Sayano-Shushensk HEP. Figure 6.44 shows graphs of the stability safety coefficients
obtained for a number of positions of the instantaneous axis of rotation when turning
the left and right bank of the dam shoulder. In the calculation, the following slide resistance parameters of the rock foundation were taken: $tg\varphi = 1$ and $C = 0.5\,\mathrm{MPa}$. The
smallest safety factor for the shift of 1.62 was obtained for the left-bank shoulder.

A common drawback of the two calculations of the stability of arch dams is the
incomplete consideration of the joint work of the arch dam and bank abutment, since
the forces from the dam remain unchanged as the virtual limit state sets in.

6.5.6.3 Example of arch dam stability study

Studies of the bearing capacity of the arch dam Nam Chien on the volumetric mathematical geomechanical model of the system "arch dam–rock foundation" [105,106,236]
take this circumstance into account.

In the model (see Figure 6.35) in the left-bank abutment using contact elements,
three fissures were reproduced in accordance with the geological structure (two of
them are calculated slide planes, and one is the separation plane) that bound the calculation block (Figure 6.45).

As a result of stability calculations of the three design blocks according to the
method described above for Block No. 1, the lowest stability coefficient was obtained
equal to $k = 2.32$, which determined the degree of stability of the arch dam.

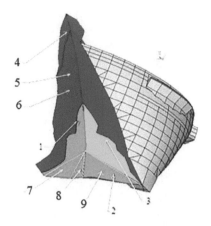

Figure 6.45 **Three design blocks in the left-bank abutment of the arch dam: I and 2 – slide planes; 3 – separation plane (de-compaction zone); 4 – zone IB; 5 – zone IA; 6 – zone IIB; 7 – Block No. 3; 8 – Block No. 2; 9 – Block No. I.**

The SSS studies of the "arch dam–rock foundation" system according to scenario 1 were carried out with a virtual decrease in the parameters of slide resistance along slide planes 1 and 2 by $K = 1, 3, 5$, and 7 times (see Section 6.5.4). Normal forces N_1 and N_2 were calculated from normal stresses on slide planes 1 and 2, and slide forces T were calculated from tangential stresses oriented along the intersection line of planes 1 and 2.

As the slide resistance parameters decreased, the dam moved toward the left bank. The virtual maximum horizontal displacements across the site amounted to –31, –30.6, –30.2, and –35.8 mm at $K = 1, 3, 5$, and 7, respectively, while the vertical displacements of the dam remained unchanged. The maximum deflections (displacements along the stream) of the arch dam changed insignificantly within 58.8 … 62.7 mm.

The largest main tensile stresses S1 on the dam downstream face near the right bank and on the crest did not exceed +2 MPa (Figure 6.46), which is less than the allowable concrete tensile stress of 2.49 MPa. The main compressive stresses S3 on the upstream face reached –10 … –15 MPa at the right-bank abutment (Table 6.15), which is significantly less than the concrete compressive strength of 25.80 MPa.

With a 7-fold decrease in the parameters of slide resistance along the calculated slide planes 1 and 2, the bearing capacity of the dam was far from exhausted. An analysis of the displacements of the arch dam with a change in K shows:

- Displacement of the upper part of the arch dam along slide plane 1 was accompanied by a shortening of the span of the arch dam;
- Trust forces in the dam at the upper (951 m) and lower (880 m) marks increased;
- Lower part of the dam prevented displacement (Figure 6.46).

Such a pattern of displacements of the arch dam indicated that the slide of the design block along plane 1 is unlikely from the kinematic point of view, since the ratio

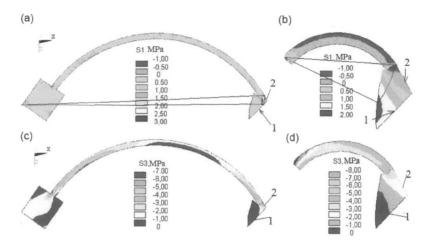

Figure 6.46 The main stresses in the arches at $K = 7$: (a) S1 at mark 961 m, (b) S3 at mark 961 m, (c) S1 at mark 880 m, and (d) S3 at mark 880 m; 1 – plane of slide 1; 2 – surface rupture.

Table 6.15 The main results of the SSS investigations "arch dam–rock foundation" system

Parameters		$K = 1$	$K = 3$	$K = 5$	$K = 7$
The main stresses in the dam,	S1	2	2	2	2
MPa	S3	−10	−11	−15	−15
Slide displacements along slide	1	7.50	17	37.50	59.50
planes, mm	2	12.16	18.26	37.50	59.50
Normal force on the slide	N1	-	7,153	7,465	7,763
plane, MN	N2	-	2,498	2,387	1,858
Slide force, MN	T	-	3,561	2,096	1,647
The results of calculating the	N1, MN	4,355			
stability of the design block	N2, MN	2,174			
according to Section 6.5.6.1	T, MN	5,497			
	K	2.32			

between the resistance forces and the slide forces along the design slide planes 1 and 2 increased (Table 6.15).

It is noteworthy that as the slide resistance parameters decreased, the block movement along the T line was 59.5 mm but the SSS of the arch dam changed insignificantly. In all likelihood, the obtained displacement values and the displacement trajectory of the block in the left-bank abutment turned out to be not "dangerous" for the arch dam, and with other trajectories, a stronger influence on the SSS of the dam is possible.

The stability coefficient of the arch dam $K = 2.32$ calculated on the analysis of the stability of rock blocks in the bank abutment according to the limit state (see Section 6.5.6.1) should be considered as the lowest estimate of the degree of stability [236].

6.6 Substantiation of reliability and safety

6.6.1 Normalization of reliability and safety

Reliability and safety of HS in Russia is ensured in accordance with the federal laws (FL) "On the Safety of HS" [1], "On Technical Regulation" [2], and "Technical Regulations on the Safety of Buildings and Structures" [3] by mandatory application at all stages of design, construction, and operation of BR and SR included in the list [4, 5], approved by the Governments of the Russian Federation (DGRF).

In the SR, "HS. The main provisions" [9] are established:

- Class of HS depending on socioeconomic responsibility, operating conditions, and installed capacity of HEP (table B.2 of mandatory annex B);
- Service life of the main HS (clause 8.10 of the SR);
- Permissible level of risk of an accident at the head front structures.

At all stages of the HS life cycle, namely, during the design, construction, operation, conservation, and liquidation, a safety declaration [5] is drawn up, which is the main document containing information on the compliance of the HS to safety criteria.

Safety criteria are the limit values of quantitative and qualitative indicators of the state of the HS, corresponding to the acceptable level of risk of HS accident. Safety criteria are presented in a deterministic form, since the conditions of strength and stability in the current regulatory documents for design are also expressed in a deterministic form using a system of safety factors that indirectly provide an acceptable level of risk of a HS accident.

The most important component of the system for ensuring the reliability and safety of HS is a complex of design justifications for design decisions made in accordance with the system of SR [8–16].

Assessment of the technical state of the HS under construction and in operation is made by quantitative and qualitative diagnostic indicators using their criteria values of the first and second level:

K1 – the first (warning) level of the value of the diagnostic indicator, upon reaching which the stability, mechanical and filtration strength of the HS and its foundations still correspond to the conditions of normal operation.

K2 – the second (limit) level of the value of the diagnostic indicator, beyond which the state of the HS becomes pre-emergency, when further operation of the HS in the design regime is unacceptable.

An operational assessment of the technical state of HS and their engineering safety is made by comparing the measured (or calculated on the basis of field measurements) quantitative and qualitative diagnostic indicators with their criteria values taking into account the predicted interval of their change.

For operating HS, the following technical states are distinguished:

- Workable (normal), when the values of diagnostic indicators do not exceed K1;
- Partially operational (potentially dangerous), when the value of at least one diagnostic indicator reaches K1 or exceeds the range of its values predicted for

this combination of loads, but does not exceed K2. This state of the HS in which its further temporary operation does not yet lead to the threat of an immediate breakthrough of the head front and the HS in this state can be exploited subject to the implementation of appropriate measures to improve engineering safety;

- Inoperative (pre-emergency), when the value of at least one diagnostic indicator exceeded K2; in this case, continued operation in the design regime is unacceptable without special permission of the state supervisory authority; before restoring, the required level of safety restrictions should be introduced on the operation mode of the HS.

If the criteria values of K1 exceeded, it is necessary to verify the reliability of the measurement and calculation results as well as the validity of the accepted values of K1. If necessary, an expert commission is created with the involvement of design and research organizations to clarify the assessment of the state of the HS and its safety level.

Quantitative criteria values of K1 and K2 diagnostic indicators are established on the basis of designed estimates of the reaction of the structure to the usual and unusual combination of loads and effects, respectively. During the operation of HS when adjusting the composition of diagnostic indicators and their criteria values, the data of field observations obtained for the entire period of construction and operation of HS are used.

Below as an example, the composition of the diagnostic indicators for the concrete dam and the "dam-foundation" system of the Boguchan HEP is given.

1. Quantitative diagnostic indicators of the SSS system "concrete dam-foundation".

    1.1. Horizontal displacements along the stream of dam points.
    1.2. The total vertical displacement of the dam from the initial observation cycle.
    1.3. Increments in the vertical movements of the dam as the reservoir fills.
    1.4. Irregularity of settlements within one section of a concrete dam and adjacent sections.
    1.5. The state of the contact joint "concrete–rock" at a distance of 3 m from the upstream face.

2. Quantitative diagnostic indicators of the filtration state of the "dam-foundation" system.

    2.1. The water levels in the piezometers located at the foundation of the concrete dam in front of the grouting curtain, between the grouting and the drainage curtains, as well as behind the drain.
    2.2. The total discharge of water filtered through a concrete dam and its foundation.

3. Diagnostic indicators of loads and effects on a concrete dam.

    3.1. Hydrostatic pressure of water from the upstream and downstream faces.
    3.2. Uplift on the bottom of the dam.
    3.3. Ambient temperature (air, water).

4. Qualitative diagnostic indicators of the state of the concrete dam.

    4.1. Deformation, wear and corrosion of concrete, reinforced concrete, and metal elements of structures.

4.2. The presence of emptiness and cavities in the foundation and body of the dam.

4.3. The presence and development of cracks and other damage on the faces of the dam, in the zones of interfacing with Earth structures.

4.4. Leaks in the galleries of structures, traces of leaching of concrete.

4.5. Clogging, overgrowing, freezing of drainage devices.

4.6. Exceeding the duration of the drainage galleries in the drained state.

4.7. Mechanical damage to spillway elements.

4.8. The performance of instrumental control systems.

To ensure the reliability and safety of concrete dams and their foundations according to the conditions of strength and stability, the inequality

$$F \leq R \tag{6.126}$$

where

F – value of the generalized force effect (force, moment, stress);

R – value of the generalized bearing capacity.

The values of F are determined by the results of analysis and studies of the SSS of the "structure–foundation" system, and the values of R depend on the characteristics of construction materials, foundation soils, and other factors.

The generalized force action F depends on the loads acting on the structure, which are random variables. For example, hydrostatic pressure (one of the main loads on the dam) is a random variable. This pressure depends on the water level before the construction, which in turn is a function of the discharge of water in the river. Since water discharge in a river is usually considered a random variable, the hydrostatic pressure dependent on it is also a random variable.

The generalized bearing capacity R is also a random variable, because it depends on the physical–mechanical characteristics of the dam concrete and foundation soils, which are the random variables. The limit strength of concrete is a random variable. In the experimental determination of the values of this quantity, samples are tested according to a strictly defined program. The obtained values of the limit strength of the tested samples are different although close to each other; i.e., limiting the strength of the material is a random variable.

In most cases, there is no upper limit for the generalized force effect F and a lower limit for the generalized bearing capacity R. Therefore, the absolute requirement that conditions (Equation 6.126) be satisfied is not valid. It can only say that this condition must be satisfied during the life of the dam or structure with a probability close to unity. It follows that engineering calculations can be considered probabilistic calculations.

Existing traditional deterministic methods for calculating structures contain elements of a probabilistic approach in a more or less veiled form.

When designing HS, structures, and their elements until the middle of the XX century, the **calculation method for permissible stresses** was used. In accordance with this method, the strength condition was

$$\sigma_{max} \leq [\sigma] \tag{6.127}$$

where

σ_{max} – maximum design stress at a point in the structure of the highest expected load;

$[\sigma]$ – permissible stress adopted in accordance with the requirements of design standards depending on the material, type of construction, and type of stress state.

The use of inequality (Equation 6.127) of deterministic values of the maximum stress caused by the highest expected load and the allowable stress, significantly lower than the average value of the limit strength of the material, made it possible to ensure a sufficiently high reliability of the structure during operation.

The calculation method for permissible stresses with some additions is currently used in the design of some elements and structures of HS at the preliminary stages of design.

As science and technology developed, the calculation method for permissible stresses was replaced by the **calculation method for the safety factor**. The condition of strength or stability in the calculations by the method of safety factor was written as

$$F_n \leq \frac{R_n}{k} \tag{6.128}$$

where

F_n – normative load;
R_n – normative bearing capacity;
k – safety factor.

All quantities included in expression (Equation 6.128) are deterministic. However, the safety factor, the value of which is significantly greater than unity, allows taking into account the variability of external loads and bearing capacity in a veiled form, thereby ensuring a sufficiently high reliability of structures during operation.

A further step in the development of structural analysis methods is the development of the **limiting state method**, the application of which is regulated by the current SR for the design of HS [11,14,108]. This method also called the *semiprobabilistic method* is deterministic in form. However, it allows us to take into account the variability of loads and effects, the variability of the bearing capacity, and other probabilistic factors.

A natural development of the method for calculating structures by limiting states is the **probabilistic method**, which has been intensively developed in recent years [34,35]. This method fully takes into account the probabilistic nature of the loads and effects, the properties of materials and structures, and their operating conditions.

In accordance with the current SR for the design of HS [11,14], to assess their reliability and safety calculations should be performed using the limiting state method.

The fundamental concept of this method is the *concept of limiting states*. The limiting states are understood as such states of the structure upon reaching which the elements of the structure, the structure as a whole, and its foundation cease to satisfy the specified operational requirements, or the requirements during the performance of work. There are two groups of limiting states.

The first group of limiting states includes states upon reaching which the structure or its foundation becomes completely unsuitable for operation due to exhaustion of strength or loss of stability. To assess the possibility of the limiting states of the first group, calculations are made of the total strength and stability of the "structure–foundation" system, the total filtration strength of the foundations and soil structures, the strength of individual elements of the structure, the destruction of which can lead to the termination of operation of the structure; calculations of structural movements, on which the strength or stability of the structure as a whole, etc.

The second group of limiting states includes states upon reaching which the structure becomes unsuitable for normal operation. To assess the possibility of the limit states of the second group, calculations of the local strength of the foundations, calculations of displacements and deformations, calculations of the formation or opening of cracks and construction joints, calculations of local filtration strength, and strength calculations of individual elements of the structure are not considered when performing calculations to assess the possibility of limit states of the first group.

When carrying out calculations of HS, their structures, and foundations, the following conditions must be observed to ensure that limit states do not occur

$$\gamma_{lc} \times F \leq \frac{R}{\gamma_n} \tag{6.129}$$

where
> F – the design value of the generalized force effect (force, moment, stress), strain, or other parameter by which the onset of the limiting state is estimated;
>
> R – value of the generalized bearing capacity, deformation, or other parameter (when calculating the first group of limiting states, the calculated value; when calculating the second group of limiting states, the normative value), established by the design standards of certain types of HS, determined taking into account the safety factors for the material γ_m or soil γ_g and the operating conditions γ_c;
>
> γ_{lc} – load combination coefficient;
>
> γ_n – reliability coefficient for the responsibility of the structure.

The fulfillment of inequality (Equation 6.129) guarantees the prevention of the limiting state of the structure. To obtain an economical technical solution, it is necessary that for the least favorable design case, the value of the right-hand side of expression (Equation 6.129) does not exceed the value of the left-hand side by more than 10%.

The possibility of the onset of the HS-limiting state is determined using the determinate values F and R included in expression (Equation 6.129), although by their nature these values are random. The influence of the variability of these quantities is taken into account by the corresponding standard coefficients.

When determining the calculated values of the generalized force, deformations, and displacements included in the expression (Equation 6.129), the design values of the loads Q are determined by the following formula:

$$Q = Q_n \times \gamma_f \tag{6.130}$$

where

Q_n – the standard value of loads and effects, determined by the SR of individual types of HS, their structures, and foundations (usually, the value of Q_n is close to the average value);

γ_f – load reliability coefficient taking into account possible load deviations in an unfavorable direction.

The values of the reliability coefficients for the load γ_f in the calculations of the HS for the first group of limit states are taken in accordance with Table 6.16, whereas in the calculations for the second group of limit states, $\gamma_f = 1$.

For loads, the normative values of which are established on the basis of statistical processing of a long-term series of observations, experimental studies, actual measurements, and determination taking into account the dynamic coefficient, the value of the reliability coefficient for the load is taken to be unity.

It should be noted that according to Ref. [9], when calculating the total strength and stability of concrete dams on the usual combination of loads and impacts, the value of the reliability coefficient for the load is taken to be equal to unity. These loads include dead weight, temperature and humidity effects, soil loads determined at the calculated values of soil characteristics, and dynamic loads.

The design material resistance R, which is included in expression (Equation 6.131), is determined by the formula:

$$R = \frac{R_n}{\gamma_m} \tag{6.131}$$

where

R_n – standard resistance of the material;

γ_m – material safety factor.

The values of the standard resistance of the material are determined by the results of statistical processing of test data of the samples. In this case, the arithmetic mean

Table 6.16 Reliability factors for load γ_f

Loads and effects	Reliability coefficient values for load γ_f
Water pressure on the surface of the structure and foundation; filtration pressure, wave pressure, pore pressure	1
Net weight of the structure (without soil weight)	1.05 (0.95)
Ground weight	1.1 (0.9)
Lateral soil pressure	1.2 (0.8)
Sediment pressure	1.2
Ice loads	1.1
Temperature and humidity effects	1.1
Seismic effects	1

Note: The values of γ_f indicated in brackets refer to cases where the use of lower values of the coefficients leads to unprofitable loading of the structure.

value of the material resistance and the standard deviation are found. For a random value of the resistance of the material, the values of this resistance of any security can be calculated. According to the current design standards, the security (one-sided probability) of the standard resistance values of all materials is taken equal to 0.95.

The SR for the design of concrete and reinforced concrete structures of HS [15] provides the standard and designed values of the resistance of HS of various classes to axial compression and tension (see also Section 5.1). The standard and design resistance of the reinforcement, depending on its class and working conditions, is also given there (see also Section 5.7).

The reliability coefficient for the material allows us to take into account a possible decrease in the design resistance of the material compared to the standard. Such a decrease is due to the variability of the material properties, as well as to factors such as the test procedure, type of stress state, etc. When calculating structures according to the first group of limiting states, the values of material safety factors are always greater than unity. When calculating structures for the second group of limiting states, the value of the reliability coefficient for the material is taken to be unity.

In calculations of soil structures, as well as concrete structures together with the foundation, soil characteristics such as friction coefficient, specific adhesion, deformation modulus, compaction coefficient, filtration coefficient, etc. are used. The designed values of these characteristics R_g are determined by the formula

$$R_g = \frac{R_{gn}}{\gamma_g} \tag{6.132}$$

where

    R_{gn} – standard values of soil characteristics;
    γ_g – soil safety factor.

According to the current design SR, the standard values of soil characteristics are determined on the basis of statistical processing of soil test data as the arithmetic mean value in a series of tests.

The reliability coefficient for soil has the same meaning as the coefficient of reliability for material. The values of the reliability coefficient for soil γ_g are accepted according to the SR [11] in the range from 1.05 to 1.25.

The loads and impacts taken into account in the calculations of HS are taken in the most unfavorable, but possible combinations separately for periods of construction, normal operation, and repair (see Section 6.2.1). Such combinations are established on the basis of an analysis of the working conditions of the structure at all stages of construction and operation. Distinguish between the usual and unusual combinations of loads and effects. The usual combinations of loads and impacts include permanent, temporary, and long-term and short-term loads and effects with an annual probability of exceeding more than 0.01. Unusual combinations of loads and effects are formed from the usual combinations with the addition of one of the unusual loads of rare repeatability.

The combination of loads is taken into account when making calculations of HS using the load combination coefficient γ_{lc}. The value of the load combination coefficient is taken depending on the group of limiting states and the period of construction work (normal operation, construction, and repair) considered in the calculations, as well as the estimated combination of loads and effects.

In the calculations for the first group of limiting states for the usual combination of loads and effects, the value of the load combination coefficient γ_{lc} is taken depending on the calculation period:

- During normal operation – $\gamma_{lc} = 1$;
- During period of construction and repair – $\gamma_{lc} = 0.95$.

In the calculations for the first group of limit states for an unusual combination of loads and impacts, the value of the load combination coefficient γ_{lc} is taken depending on the annual probability of exceeding the special load:

- At an unusual load, including seismic at the level of the DE, the annual probability of exceeding 0.01 or less – $\gamma_{lc} = 0.95$;
- At an unusual load except for the seismic annual probability of exceeding 0.001 or less $\gamma_{lc} = 0.9$;
- With a seismic load at the level of the MCE – $\gamma_{lc} = 0.85$.

In the calculations for the second group of limit states, the value of the load combination coefficient is taken to be $\gamma_{lc} = 1$.

To account for various types of structures, constructions, foundations, types of building materials, the approximation of accepted design schemes, the type of limiting state, and other factors, the coefficient of working conditions γ_c is used. The value of this coefficient ranges from 0.75 to 1.15 and is adopted according to the design standards of individual types of structures.

The reliability coefficient for the responsibility of the structure γ_n takes into account the class of the structure and the significance of the consequences when certain limit states occur. When calculating structures according to the limiting states of the first group, the value of this coefficient is taken to be $\gamma_n = 1.25$ for class I buildings, $\gamma_n = 1.20$ for class II buildings, $\gamma_n = 1.15$ for class III buildings, and $\gamma_n = 1.10$ for class IV structures. When calculating the structures according to the limiting conditions of the second group, $\gamma_n = 1$ is taken. When calculating the stability of natural slopes, the coefficient value is taken for the class of the nearby designed structure.

It is important to note that the condition (Equation 6.129), ensuring the prevention of the onset of limiting states, must be observed at all stages of the construction and operation of HS including at the end of their assigned service life. Usually, the designated service life of the main HS should be assigned no less than the estimated service life. Estimated service life of structures is regulated by the SR for design and are taken equal: for structures of I and II classes –100 years; for constructions of III and IV classes – 50 years.

6.6.2 Comparison of reliability criteria for Russian SC and USA standards

The most widespread in the world practice of designing HS are the standards of American organizations: Federal Energy Regularly Commission (FERC), Bureau of Reclamation (BR), and US Army Corps of Engineers (US ACE).

If the FERC is state-owned and the FERC president is approved by the USA president [199], then the other two organizations are not state-owned. It should also be noted that the FERC's task is to oversee the safety of energy facilities, but not to design them; BR [142] and US ACE [195] are design organizations.

6.6.2.1 Comparison on dam strength

It is necessary to write the expressions for the generalized strength coefficients of concrete dam:

$$\gamma_{str} = \gamma_n \gamma_{lc} / \gamma_c \gamma_\tau \gamma_\eta \tag{6.133}$$

where

γ_n, γ_{lc}, and γ_c – the above-mentioned coefficients;

γ_τ – coefficient taking into account the change in concrete strength with age (see Table 5.5);

γ_η – coefficient taking into account the difference in concrete strength in control samples and construction (see the note to Table 5.5).

And a generalized stability coefficient:

$$\gamma_{stab} = \gamma_n \gamma_{lc} / \gamma_c \tag{6.134}$$

Tables 6.17 and 6.18 show the numerical values of the general coefficients of the concrete strength of the dam γ_{str} and the stability of the dam γ_{stab} for the usual combinations of loads and effects with seismic and without seismic effects and construction and repair case when calculating the limiting states of the first group [103].

When using the solutions of the linear theory of elasticity instead of the condition of tensile strength in horizontal sections of the dam and at the contact of the dam with the foundation under upstream face, a restriction of the length of the tensile zone d_t is introduced; these restrictions are given in Tables 6.10, 6.11, and 6.14.

Table 6.17 **The values of the generalized strength factors γ_{str} and $\gamma_m \gamma_{str}$ according to the Russian SR (construction of class I, concrete age by loading time 0.5 years)**

Combination load and impact		γ_{str} on		$\gamma_m \gamma_{str}$ on	
		Compression	Tensile	Compression	Tensile
I		2	3	4	5
Usual		1.25	1.39	1.62	2.08
Unusual without seismic		1.02	1.12	1.33	1.68
Unusual with seismic	DE	1.09	1.08	1.42	1.62
	MCE	0.97	0.97	1.26	1.45
Construction and repair		1.19	1.32	1.55	1.98

Note: $\gamma_m \gamma_{str}$ – the generalized coefficient of strength in relation to the standard strength of concrete R_{bn} and R_{btn}.

Table 6.18 Values of the generalized stability coefficients γ_{stab} and $\gamma_g\gamma_{stab}$ according to Russian SC (class I construction)

The combination of loads and effects	γ_{stab} on		$\gamma_g\gamma_{stab}$ on	
	Fissure in the massif foundation	Contact concrete–rock and in the massif	Fissure in the massif foundation	Contact concrete–rock and in the massif
1	2	3	4	5
Usual	1.25	1.31	1.56	1.64
Unusual without seismic	1.12	1.18	1.40	1.48
Unusual with DE	1.18	1.25	1.48	1.56
seismic MCE	1.07	1.12	1.34	1.40
Construction and repair	1.19	1.25	1.49	1.56

Note: $\gamma_g\gamma_{stab}$ – generalized coefficient of stability in relation to the standard values of the parameters of slide resistance of the rock foundation.

Calculations of local strength are made according to the limiting states of the second group (with the coefficients $\gamma_n = \gamma_{lc} = \gamma_c = 1$).

The stresses at the foundation, which can be determined by the theory of elasticity, must satisfy the following conditions:

on cracks in the foundation and on contact with the dam:

$$\left.\begin{aligned} \sigma_i &< R_{tn,j}, \\ \theta_j &= -\sigma_i tg\phi_{n,j} + C_{n,j} > 1 \end{aligned}\right\} \tag{6.135}$$

along the massif of the foundation (with a rectilinear envelope of circles of the Mohr):

$$\sigma_3 < R_{tn,m} \tag{6.136}$$

$$\theta_m = 2\left[\left(-\sigma_1 tg\ \varphi_{n,m} + C_{n,m}\right)\times\left(-\sigma_3 tg\varphi_{n,m} + C_{n,m}\right)\right]^{1/2}\Big/\left(\sigma_1 - \sigma_3\right) > 1 \tag{6.137}$$

where

σ_j – the normal stress in the fissure or at the contact ("-" compression);

σ_1 and σ_3 – the main stresses in the foundation mass ("-" compression, $\sigma_1 > \sigma_3$);

θ_J and θ_m – the coefficients of local strength along the fissure (contact) and in the massif, respectively;

$R_{tn,j}$ and $R_{tn,m}$ – the standard values of tension strength along the fissure (contact) and in the massif, respectively;

$tg\varphi_{n,j}, C_{n,j}$ and $tg\varphi_{n,m}, C_{n,m}$ – the standard values of the parameters of slide resistance along the fissure (contact) and in the massif, respectively.

When using nonlinear methods for studying the "dam-foundation" system SSS, it is allowed not to satisfy the first strength conditions (Equation 6.135 and Equation 6.136), while the depth and position of the decompaction zone in the foundation under the upstream face should be analyzed and possible increases in drainage discharges should be assessed.

Comparison of the values normalized in Russia and the USA, strength and stability factors lead at first glance to the conclusion that Russian SRs are more "bold" than American ones.

Indeed, the values of strength coefficients for concrete regulated by the Russian SR at an age of 0.5 years for γ_{str} compression are in the range from 1.25 (the usual combination of loads) to 0.97 (unusual combination for the MCE seismic) and in the γ_{str} tensile range within 1.39–0.97, respectively (see columns 2 and 3 in Table 6.17), whereas according to American standards (Table 6.20):

- US ACE [195] – γ_{str} values are in the range of 3.3–1.1 per compression for the same load combinations (column 3): at usual combination tension is not allowed, whereas at unusual combination $\gamma_{str} = 1.1$ (column 4);
- BR [142] – in the range of 3 for compression and tension (columns 5 and 6);
- FERC-2002 [199] – in the range of 3–1.3 for compression (column 7), tensile stresses are not limited by themselves, but the secondary system is checked for compression and slide after excluding the tensile zone (column 8).

The compressive and tensile strength of concrete used in the American Standards does not contain material safety factors γ_m. When recalculating the concrete strength coefficients at the age of 0.5 years for the standard compressive and tensile strengths, the $\gamma_m \gamma_{str}$ values for Russian SR increase at the usual combination of loads and effects to 1.62 compressive and 2.08 tensile, and at the unusual combination with the MCE seismic up to 1.26 for compression and 1.45 for tension (see columns 4 and 5 in Table 6.17), but remain smaller than by American standards (see Table 6.20). The most noticeable difference is the usual combination of loads when checking the compressive strength of structures.

Table 6.20 shows the strength coefficients γ_{str} according to the Russian SR for compression and tension for concrete at the age of 1 year by the time of loading, based on the design concrete resistance to compression R_b and tensile R_{bt} (columns 9 and 10). Comparison with American standards shows that the difference in strength coefficients is huge, especially at the top of the table (when checking the compressive strength for the usual combination of loads, the difference is 163%–189%). Unstable compression strength coefficients less than 1 appear for all unusual combinations of loads and impacts.

In American standards, strengths are assigned relative to the average compressive strength f_c' of a 15 × 15 × 15 cm cube.

In Russian practice, the selection of concrete composition and its quality control is also carried out according to the average compressive strength R of a cube measuring 15 × 15 × 15 cm. Then, the transition is made from cube strength to prismatic strength R_{pr} and then to standard R_{bn} and design R_b concrete compressive strength; unfortunately, this transition was not given in Ref. [9,15]. Table 6.20 shows this transition for four classes of concrete [103].

At one time, the average compressive strength in kg/cm^2 of cubes 15 × 15 × 15 cm R (column 3) determined the concrete mark M (column 1 in Table 6.19). Then, concrete classes appeared (column 2), and according to Ref. [15], the concrete class in terms of

Table 6.19 The compressive and tension strength of concrete (MPa)

M	B	R	R_{pr}	R_{bn}	R_b	R_{bt}	$f_t^{'}$	$R_{bt}/f_t^{'}$
1	2	3	4	5	6	7	8	9
150	10	13.1	9.8	7.5	6	0.57	2.46	4.32
150	12.5	16.4	12.3	9.5	7.5	0.66	2.86	4.33
200	15	19.6	14.7	11	8.5	0.75	3.23	4.31
250	20	26.2	19.6	15	11.5	0.90	3.91	4.34

compressive strength corresponded to the value of the prismatic concrete compressive strength R_{pr} in MPa at the age of 0.5 years (column 4). The prismatic strength R_{pr} (compressive strength of 15×30 cm cylinders) was 75% of R:

$$R_{pr} = 0.75\, R \tag{6.138}$$

standard compression resistance R_{bn} (column 5) is determined depending on the prismatic strength R_{pr} according to the following formula:

$$R_{bn} = (1 - \alpha\, c_v) R_{pr} = 0.5868\, R \tag{6.139}$$

where
  $p = 0.90$ – security for massive structures;
  $c_v = 0.17$ coefficient of variation;
  $\alpha = 1.28$ coefficient;
  R_b – design concrete compressive strength (column 6) at reliability coefficient for the material $\gamma_m = 1.3$ is determined using the following formula:

$$R_b = R_{bn}/g_m = R_{bn}/1.3 = 0.4514\, R = R/2.215 \tag{6.140}$$

- Average compressive strength R (column 3) of cubes $15 \times 15 \times 15$ cm (R corresponds to the compressive strength $f_c^{'}$ according to American standards, column 8);
- Prismatic compressive concrete strength R_{pr} (column 4);
- Standard compressive strength R_{bn} (column 5);
- Design compressive strength R_b (column 6);
- Design tension strength R_{bt} (column 7);
- Ratio $R_{bt}/f_t^{'}$

In Table 6.20 in column 10, the values of compressive strengths coefficient γ_{str} according to Russian SR are calculated with respect to the average compressive strength R, reflecting the real ratio with American standards. Comparison of the γ_{str} compressive strength coefficients shows that for the usual combination of loads, their values according to the Russian SR (2.77) and American standards (3–3.3) are close. But Russian SR are somewhat bolder (Table 6.20). At unusual combinations without a seismic and with the OBE seismic, as well as for the construction period, Russian SR are more conservative: $\gamma_{str} = 2.26$–2.64 against $\gamma_{str} = 2$ according to American standards. At unusual combinations of loads with the MCE seismic and PMF flood, Russian SR are

Table 6.20 Comparison of strength coefficients in γ_{str} dam by US Army Corps of Engineers – 1995 [195], Bureau of Reclamation [142] and FERC-2002 [199] and Russian SR [11–14]

Combination of loads		Strength coefficient for								
US Army Corps of Engineers – 1995, Bureau of Reclamation, FERC-2002	Russian SR	US Army Corps – 1995		Bureau of Reclamation		FERC-2002		Russian SR		
		Compression[a]	Tension[b]	Compression[a]	Tension[b]	Compression[a]	Tension[b]	Compression $R_b{}^{c}$	R^{d}	Tension[e]
1	2	3	4	5	6	7	8	9	10	11
Usual	Usual	3.3	Not allowed	3	3	3	Checking the secondary system for compression and slide after exceptions from the work of the tensile zone	1.14	2.77	1.32
Unusual	Unusual without seismic	2	1.7	2	2	2		0.93	2.64	1.07
	Unusual with seismic DE							0.99	2.26	1.03
	Construction and repair							1.08	2.41	1.26
Extreme	Unusual at PMF	1.1	1.1	1	1	1.3		0.93	2.26	1.07
	Unusual with seismic MDE							0.88	2.15	0.92
	Construction + seismic DE							-	-	-

a With respect to the compressive strength f'_c of the cube 15 × 15 × 15 cm.
b Relative to tension strength f'_t.
c Relative to the design concrete compressive strength R_b at the age of 1 year.
d With respect to the average compressive strength R of a cube of 15 × 15 × 15 cm.
e Relative to the design concrete tension strength of R_{bt} at the age of 1 year.

even more conservative: γ_{str} = 2.15–2.26 against γ_{str} = 1.1–1.3 according to American standards.

Comparison of tension strength γ_{str} by Russian SR and American standards given in Table 6.20, without comment, is illegal. The fact is that the values of the designed tension strengths R_{bt} are normalized in Ref. [15] for the concrete class, and the tension strength f_t according to American standards is calculated depending on the compressive strength f_c according to the following formula:

$$f_t' = 0.4436(f_c')2/3 \tag{6.141}$$

the tension strength f_t' is 4.3 times greater than the design tension strength R_{bt}. It does not seem possible to explain why such a big difference in tension strengths in Russia and the USA but if the tension strengths γ_{str} in tension strengths in Russian SR (column 11 in Table 6.20) increase by 4.3 times, then Russian SR will be much more conservative in comparison with American standards.

In fairness, it should be noted that the tension strength coefficients γ_{str} as shown in Table 6.20 according to the American standards and the Russian SR are largely conditional since as a rule the tension strength of concrete is ignored and the compressive or slide strength of the secondary system is analyzed after exceptions from the work of the tensile zone.

6.6.2.2 Comparison on dam stability

Comparison of the γ_{stab} stability coefficients according to the Russian SR and American standards shows that (Table 6.21):

1. For the usual combination of γ_{stab} loads according to American standards (columns 3–6) is significantly higher than according to the Russian SR (columns 8 and 9), with the exception of the γ_{stab} according to FERC-2002 when using residual slide strength (column 7);
2. The difference in γ_{stab} somewhat decreases with special load combinations during the construction period without seismic and with the OBE seismic, and according to FERC-2002 when using residual slide strength γ_{stab} is less than in the Russian SR;
3. With unusual load combinations at PMF and with the MCE seismic American standards are bolder than the Russian SR, according to BR for contact slide and FERC-2002 when using residual slide strength γ_{stab} are generally 1.

The significant difference in γ_{stab} according to American standards and Russian standards for the usual combination of loads is explained by the fact that American standards use peak slide strength, while Russian SR regulate the use of residual strength; therefore, the γ_{stab} coefficients according to FERC-2002 when using residual slide strength and the Russian SR are quite close. Nevertheless, it should be noted that American standards are more stringent in the usual combination of loads and effects.

The same tendency is observed with unusual combinations of loads without the seismic and with the DE seismic and in the construction period (with unusual combinations according to American terminology).

Table 6.21 Comparison of g_{stab} dam stability coefficients by US Army Corps of Engineers – 1995 [195], Bureau of Reclamation [142], FERC-2002 [199], and Russian SC [11–14].

Combination of loads		Stability coefficients by						
US Army Corps of Engineers – 1995, Bureau of Reclamation, FERC-2002	Russian SR	US Army Corps – 1995[1]	Bureau of Reclamation		FERC-2002 for using of slide strength		Russian SR[2]	
			Slide on fissure	Slide on contact	Peak	Residual	Slide on fissure	Slide on contact
1	2	3	4	5	6	7	8	9
Usual	Usual	2	4	3	3	1.5	1.56	1.64
Unusual	Unusual without seismic	1.7	2.7	2	2	1.3	1.40	1.48
	Unusual with seismic DE						1.48	1.56
	Construction and repair						1.49	1.56
Extreme	Unusual at PMF	1.3	1.3	1	1.3	1	1.40	1.48
	Unusual with seismic MCE						1.34	1.40
	Construction + seismic DE						-	-
After seismic	-	1.3	-	-	2	1.1	-	-

1 The values of the slide stability coefficients are given for the case of reliable information on the parameters of the foundation slide resistance.
2 The values of the slide stability coefficients for standard values of slide resistance parameters.

Despite the different procedure for determining the parameters of slide resistance, with unusual load combinations at PMF and with the MCE seismic (with extreme combinations – according to American terminology) as already indicated, American standards are bolder than Russian SR. Perhaps this is due to lower requirements for stability during the short-term loads and effects.

Unlike Russian SR, the American standards impose rather stringent requirements on the stability of structures after seismic effect, and it is prescribed to take into account irreversible disturbances that occurred during an earthquake.

6.6.2.3 Comparison on limitation of extension zone and the depth

As for the restriction of the depth of the extension zone d_t in the horizontal sections of the dam and at the contact of the dam with the foundation in the region of the upstream face, Russian SR are less different from American standards (Table 6.22).

Comparing the standards of US ACE-1995 with Russian SR, it can be noted that (Table 6.22):

1. For the usual combination of loads in Russian SR, a small tensile zone is allowed, while according to American standard, tensile zone is not allowed;
2. With unusual combinations of loads without the seismic and with the DE seismic, the restrictions on the depth of the extension zone d_t are quite close;
3. With unusual combinations of loads with PMF and the seismic MCE, Russian SR are more conservative;
4. The absence of a limitation on the depth of the extension zone d_t according to BR and FERC-2002 explicitly means that the compressive or slide strength of the secondary system is analyzed after exclusion from the work of the extension zone. This requirement corresponds to the Russian condition (Table 6.10) on the limitation of σ_3 and the main compressive stress in the section, determined without taking into account of working concrete on tension.

The approach for determining the depth of the extension zone d_t adopted in US ACE-1995 differs from the calculation procedure used in BR.

According to US ACE, the resultants in horizontal sections of the dam and in contact with the foundation are determined taking into account that all the acting forces, including uplift, and restrictions are placed on its position. It is prescribed to act in the same way in Russian SR. However, it is obvious that with close to zero tension strength of construction joints in the dam and contact with the foundation, a crack should form. The conditional technique outlined above is explained by the desire of the developers of standards and norms to make life easier for designers at the time without forcing them to carry out calculations of secondary systems for compression after excluding the tension zone from work.

Otherwise, a crack is formed, the length of which is determined by iterations: at each iteration, in the section where the constraint condition d_t is not fulfilled, the pressure of the upstream is set and stresses σ_z are recalculated for the remaining section width, reduced by the length of the crack. Iterations are repeated until there are only compressive stresses on the contact.

Table 6.22 Comparison of restrictions on the position of the resultant N_R and the length of the extension zone d_t according to US Army Corps of Engineers – 1995 [195], Bureau of Reclamation [142], FERC-2002 [199], and Russian SC [11–14]

Combination of loads		Position restrictions N_R and d_t by						
US Army Corps of Engineers – 1995, Bureau of Reclamation, FERC-2002	Russian SR	US Army Corps of Engineers – 1995			Bureau of Reclamation	FERC-2002	Russian SR	
		N_R in the middle section	d_t/b	e/b			d_t/b	e/b
1	2	3	4	5	6	7	8	9
Usual	Usual	1/3	0	0.167	Not limited	Not limited	0.03[a]	0.177
Unusual	Unusual without seismic	1/2	0.167	0.25			0.083	0.2
	Unusual with seismic DE						0.2	0.278
	Construction and repair						-	-
Extreme	Unusual at PMF	1	0.333	0.5			0.083	0.2
	Unusual with seismic MDE						0.2	0.278
	Construction + seismic DE						-	-

Notes: N_R – the vertical component of the resultant of all forces acting in the horizontal section of the dam or at the contact of the dam with the foundation, including uplift; d_t – the length of the extension zone from the upstream side in the horizontal section of the dam or at the contact of the dam with the foundation; b – the width of the horizontal section of the dam or the contact of the dam with the foundation; e – eccentricity N_R.

a When calculating d_t/b for the usual combination, $a_1 = 0.1b$ and $a_2 = 0.1b$ were taken.

Above are formulas (6.71) and (6.72) for determining d_{cr} during work by tension contact and at zero tension strength.

According to the Russian SR, the verification of the depth of the extension zone is carried out according to the second group of limiting states, when the coefficients $\gamma_n = \gamma_{lc} = \gamma_c = 1$; therefore unlike BR, there is no safety factor in formulas (6.71) and (6.72).

6.6.2.4 Comparison on foundation strength

The criteria for compressive strength used by US ACE-1995, FERC-2002, and BR for a rock base (Table 6.23) are similar to the criteria for concrete (see Table 6.20); tensile stresses at the base are not limited.

A completely different approach to assessing the strength of a rock foundation is regulated in Russian SC [11]:

- As already indicated, the local strength should be checked for the second group of limit states at the coefficients $\gamma_n = \gamma_{lc} = \gamma_c = 1$;
- Tensile normal stress at the contact of the dam with the foundation or on the fissure should not exceed the corresponding standard tension strength (the first condition in Equation (6.135));
- Triaxial tension is not allowed at the foundation (the first condition in Equation (6.135));
- Compressive stresses in the foundation are not explicitly limited, but must satisfy the Mohr–Coulomb criterion (second conditions in Equations (6.135) and (6.137)).

Table 6.23 Comparison of compressive strengths at the foundation according to US Army Corps of Engineers – 1995 [195], Bureau of Reclamation [142], FERC-2002 [199], and Russian SC [11–14]

Combination of loads		Coefficient compressive strengths			
US Army Corps of Engineers – 1995, FERC-2002, Bureau of Reclamation	Russian SR	US Army Corps – 1995	Bureau of Reclamation	FERC-2002[a]	Russian SR
Usual	Usual	$\sigma_3 \leq [\sigma]$	4	3	Local strength test (6.135) and (6.136) and (6.137)
Unusual	Unusual without seismic	$\sigma_3 \leq [\sigma]$	2.7	2	
	Unusual with seismic DE				
	Construction and repair				
Extreme	Unusual at PMF	$\sigma_3 \leq 1.33[\sigma]$	1.3	1	
	Unusual with seismic MDE				
	Construction + seismic DE				

Notes: $[\sigma]$ – permissible compression stress
[a] If the concrete compressive strength f_c is greater than the foundation; otherwise, compressive stresses in concrete are limited; this condition guarantees the dam against capsizing.

In the SR [14], dams are regulated to check the total strength in the first group of limiting states. In other words, at all points of the body of the dams with the usual and all unusual combinations, there are unusual combinations of loads and effects should the condition $\gamma_n \gamma_{lc} \sigma_3 \leq \gamma_{cd} R_b$ (see Table 6.10) be met, with the coefficients $\gamma_n, \gamma_{lc}, \gamma_c, \gamma_\tau,$ and γ_η not equal to 1.

In fact, this condition is a test of local strength since the stress cannot characterize the total strength of the dam; in addition, the limitation of the stress at a point cannot be a harbinger of the limiting state of the first group.

In the SC [11], it is regulated to check the local strength of the rock foundation according to Equations (6.135) and (6.136) and (6.137) but strength coefficients are bolder for the rock foundation than for the dam; in other words, its components in the "dam-foundation" system are not equally reliable. Moreover, more stringent requirements are imposed on dam concrete–artificial material with adjustable physical–mechanical properties than on rock foundation material created by nature, the ideas about the physical–mechanical properties of which are not exhaustive, which is not logical and unjustified.

A direct comparison of the numerical values of the strength and stability coefficients regulated in the Russian SC and American standards, which at first glance shows a more "bold" approach to justifying the reliability of HS in Russia, is wrong. It is necessary to take into account the methods for determining the strength characteristics of materials, the practice of accounting for loads and effects, the methods used for calculating studies of the SSS of HS, and their foundations.

Given the existing experience in studies of the SSS and stability of the "concrete dam–rock foundation" system, it should be noted:

> for gravity dams on a rock foundation above 80–100 m, the determining reliability criterion is the strength condition of tensile stress at the contact of the dam upper face with the foundation, and also the condition for limiting the depth of the zone of tensile stress at the contact; the strength condition of compressive stress in the region of the downstream face (even with the formation of the secondary system) as well as the condition of stability of the dam are automatically fulfilled with safety larger, the higher the dam.

The comparable analysis of strength criteria according to Russian SR and USA standards indicates the cumbersomeness of the system of divided reliability coefficients and the conventionality of the two groups of limit states adopted in Russian SC.

The physics of the phenomena accompanying the limiting states remains unknown and is not studied.

Transition to the limiting states happened more in form than in essence. The transition period has dragged on, SR are reissued, their names change to SP [8–16], so far no progress has been seen.

Chapter 7

Gravity dams

"When in all his height
The ridge standed
His feast table toast
The crown of his outfit"

Boris Pasternak (1890–1960),
Russian poet

7.1 Types of gravity dams and conditions of their work

Gravity dams are structures whose stability under the influence of all forces and effects is ensured mainly by their weight.

Reliability of concrete gravity dams is based first of all on the high compressive strength of the concrete; due to the limited tensile strength of concrete, gravity dams are designed in such a way as to exclude tensile stresses in the dam body under normal operating conditions. Limited tensile stresses are allowed under unusual conditions (for example, with seismic effects), as well as with short-term temperature effects.

Massive gravity dams are simple in design. Their height approached 300 m, reaching 285 m on the Grande Dixence Dam in Switzerland (see Figure 1.19) built back in 1962. There is every reason to further increase the height of gravity dams.

Due to the simplicity of the design, and consequently simpler construction conditions compared to other types of concrete dams, the possibility of construction in a variety, including complex natural conditions (hydrological, climatic, topographic, geological, and seismic), gravity massive dams are the most common type of concrete dams.

Lightweight types of concrete dams (gravity with extended seams and buttresses), characterized by a decrease in concrete volume but with a more complex structure and construction technology in recent decades of the XX century were practically stopped building because of low efficiency [119].

Massive gravity dams are designed in wide sites (at $\ell/h \geq 10$, where ℓ is the width of the gorge at the level of the dam crest, h is the height of the dam); in relatively narrow sites (at $\ell/h \leq 5$) and sites at $5 < \ell/h < 10$, gravity dams can be considered in parallel with arch and arch-gravity [217].

The technology of erecting gravity massive dams is the most important factor determining to a large extent their efficiency and competitiveness.

In modern technologies to ensure the fastest exothermic heat dissipation, equalize the temperature of concrete, prevent crack formation during the construction period, and generally achieve the monolithic of the dam, concrete is laid in blocks with various dam cutting schemes (see Figure 1.6):

- with columnar cutting into blocks when the blocks are laid in the "pillars" with the formation of longitudinal joints, which after cooling of the concrete are cemented or, when the joints are made wide (up to 1.5 m), concreted;
- with sectional cutting when the entire dam section between the transverse joints is concreted with one block 0.7–1.5 m high;
- blocks layers 0.3–0.4 m high in large sizes from RCC and from a rigid concrete mixture compacted by vibratory rollers.

Development and implementation in the last quarter of the XX century technologies for erecting massive gravity dams from RCC ensured high-speed construction and reduction of labor costs, cost, and construction time. The efficiency and competitiveness of gravity massive dams has dramatically increased compared to even dams made of soil materials. Given that the advantages of such dams are provided primarily by erection technology, their design and technology must be inextricably linked.

In a relatively short period, massive dams using RCC technology have become most widespread. In 2010, the number of operated and constructed gravity dams in the world from RCC reached 470 and in 2015 – 600 [257]. The height of the gravity massive dam from RCC Miel-1 built in Colombia in 2002 reached 188 m. In 2009, the highest gravity dams from RCC: Longtan, 216.5 m high, and Guangzhao, 201 m high, in China and in 2015, the Gibe-III dam, 245 m high, in Ethiopia were constructed. Of the 102 gravity dams built in 2015 over 60 m high, about 75% were built from RCC [222].

Gravity dams by purpose are divided into the following:

- spillways, weir, and deep openings;
- deaf;
- station, within which the intakes and turbine conduits of HEP are located.

Deaf sites are usually performed by mating spillway and station dams with the banks.

By height, gravity dams are divided into the following:

- low (up to 40 m high);
- medium (from 40 to 100 m);
- high (from 100 to 150 m);
- super high (more than 150 m).

Massive gravity dams are designed as follows:

- triangular profile close to the "classic";
- with a vertical upstream face or close to it;
- with concrete apron;
- flattened with a symmetrical triangular profile.

In gravity dams, transverse joints of various designs are used:

- solid flat dividing the dam into separate independently working sections;
- not continuous notch-seams;
- extruded, hinged, or monolithic joints, due to which the dam works in a spatial pattern in relatively narrow sites.

Depending on the height of the dams, the socio-economic responsibility of the HP, and the consequences of the accident, the dams are divided into four classes (see Section 2.1).

To monitor compliance with the design criteria and conditions of the "dam-foundation" system and evaluate its safety during construction and operation at dams of classes I, II, and III, constant monitoring is provided, as a rule with an automated diagnostic monitoring system. Monitoring is necessary for timely adoption of measures in case of deviation of the dam state from the design one. The volume, composition, and types of control CME and information-diagnostic system (IDS) are determined by the specific conditions and parameters of the structure [68,153,186].

Improving the designs of massive dams is aimed at improving their SSS and, above all, the upstream face and the contact zone of the foundation, to ensure the durability of concrete in the outer zones especially in harsh climates, and to increase seismic resistance and technology improvement.

7.2 Rolled concrete dams

Unlike ordinary concrete, RCC is laid from a rigid concrete mixture with a low cement content and a high content of pozzolana (fly ash) in layers with a height of mainly 0.3–0.4 m and compacted by vibratory rollers. The physico-mechanical characteristics of RCC depend not only on the composition but also on the technology of erection and vibration rolling. As construction practice has shown, dams made of RCC have significant advantages which are achieved by [263]

- the use of rigid low-cement concrete mixtures with a relatively low heat release, which allows concreting dams with large block layers with the removal of exothermic heat due to surface cooling without complicated measures to regulate the temperature regime;
- increased crack resistance of concrete masonry;
- high early strength of RCC providing the possibility of movement of vibratory rollers and other mechanisms on the surface of the laid concrete;
- use of simple technological schemes, high performance equipment, minimizing complex, and time-consuming auxiliary operations (formwork, preparing blocks for concreting, and thermal control measures), which allow using flow methods, increasing the intensity of concrete work and reducing the time and cost of construction.

The design of gravity dams from RCC does not have fundamental differences from the design of dams from ordinary concrete; however, the specific features of RCC must be taken into account. The main difference between such dams is a large number of horizontal (slightly inclined) joints between the layers of RCC. The physico-mechanical properties of concrete in the joint zone may differ from those of RCC, which may affect the monolithic of the dam and filtering conditions.

Depending on the binder content (cement and pozzolan), the following types of dams from RCC are usually distinguished [93,162]:

- from lean concrete with a low binder content (less than $100 \, \text{kg/m}^3$);
- from concrete with an average binder content ($100–149 \, \text{kg/m}^3$);
- from concrete with a high binder content ($150–300 \, \text{kg/m}^3$).

Japanese dams made of RCC are characterized by average binder content (120–130 kg/m^3) but with higher cement consumption.

Of the dams constructed from RCC over the past 10 years, dams with a high binder content account for more than 50%, while high density, strength, and water tightness of concrete including horizontal joints have been achieved.

RCC dams are built in different climatic zones including in harsh climatic conditions, for example, Upper Stillwater in the USA, 91 m high, Bureya, 140 m high, in Russia, and Beni Haroun in Algeria, 118 m high, and humid tropical climate, for example, Pangue in Chile with a height of 113 m and Miel-1 in Colombia with a height of 188 m.

The parameters of the constructed and under construction high dams of RCC are given in Table 7.1.

In most RCC dams, the upstream faces and abutment to the rock foundation are formed by laying ordinary concrete, which complicates the technology. Screens are often made on the upstream face of a number of dams from rolled concrete (see Section 7.3.5). The device of the external faces and especially the upstream face of ordinary concrete allows for waterproof and frost resistance in harsh climatic conditions; within ordinary concrete drainage, seals of expansion joints, installation of CME, and if necessary, reinforcement are arranged. The implementation of the downstream faces of the dam and spillways has been widely used, which simplifies the formation of the downstream face and weirs and provides a favorable aesthetic perception as well as accessibility for examinations. Such spillways used at relatively low specific discharges (usually less than 25 m^2/s per meter) recently used at higher specific discharges provide efficient dissipation of the energy of the discharge stream.

The trend of recent years is an increase in the relative volume of RCC in the dam body with a decrease in the volume of ordinary concrete in combination with the use of an increased binder content while maintaining a rigid concrete mixture.

To reduce the use of ordinary concrete, in China and other countries, laying of vibrated RCC enriched with cement mortar GEVR as well as RCC enriched with cement mortar (GE RCC) is successfully applied on many dams of RCC with a high binder content and a narrow strip up to 1 m wide at the faces of the dam and rock foundation. RCC in the body of the dam is laid in inclined layers, which allows for the monolithic of the dam. For the first time, such a technology was applied on the Jiangya dam, 131 m high.

The realization of the potential advantages of RCC dams depends to a large extent on the harmonious combination of dam design and construction technology.

In dams made of RCC, in order to ensure the most favorable conditions for their construction, intake and water conduits of HEP strive to extend beyond the dam profile and reduce the number of galleries, mines, intersectional transverse joints.

At large HPs where the construction of a concrete dam determines the general construction time, optimization of layout and structural solutions with the provision of effective technology of RCC can reduce construction time and cost by up to 25% or more [88,92].

When performing calculations of dams from RCC, the possibility of internal anisotropy caused by the presence of a large number of horizontal or inclined joints in the dam and, accordingly, a possible decrease in strength, slide, and filtration parameters along these joints should be taken into account.

Table 7.1 High RCC dams

No	Name	Country	Years of construction (years of laying RCC)	Height (m)	Length (m)	Laying of faces Upstream/in lower part	Downstream	Concrete volume (RCC) thousand m³	Content cement + pozzolana (kg/m³)
1	Tamagawa	Japan	1980–1990 (1983–1986)	100	441	Vertical	0.81	1,150 (772)	91 + 39
2	Sakaigawa	Japan	1985–1993 (1988–1991)	115	226	Vert./0.8	0.78	718 (373)	91 + 39
3	Yantan	China	1985–1998 (1989–1992)	110	525	Vertical	0.80	905 (629)	55 + 104
4	Capanda	Angola	1987–2004 (1989–1992)	110	1203	Vertical	0.70	1,154 (757)	70 + 100
5	Tridomil	Mexico	1987–2004 (1991–1992)	100	250	Vert./0.24	0.80	681 (362)	148 + 47
6	Miyegase	Japan	1988–2001 (1991–1995)	155	400	0.20/0.60	0.63	2,001 (1,537)	91 + 39
7	Satsunaigawa	Japan	1987–1999 (1991–1995)	114	300	Vert./0.4	0.80	750 (250)	42 + 78
8	Urayama	Japan	1990–1999 (1992–1995)	156	372	Vert./0.65	0.80	1,860 (1,294)	91 + 39
9	Pangue	Chile	1992–1996 (1995–1996)	113	410	Vertical	0.80	740 (670)	80 + 100
10	Gassan	Japan	1988–2001 (1994–1998)	123	393	Vertical	0.80	1,160 (731)	91 + 39
11	Jiangya	China	1995–1999 (1996–1999)	131	370	Vertical	0.80	1,386 (1,100)	87 + 107
12	Beni Haroun	Algeria	1993–2000 (1998–2000)	118	714	Vertical	0.80	1,900 (1,690)	82 + 143
13	Porce 2	Colombia	1994–2001 (1996–2001)	123	425	0.10	0.75	1,445 (1,305)	132 + 88
14	Dachaoshan	China	1997–2002 (1998–2001)	111	460	Vert./0.20	0.70	1,287 (757)	94 + 94
15	Miel-1	Colombia	1997–2002 (2000–2002)	188	345	Vertical	0.75	1,730 (1,669)	85 to 160 + 35
16	Bureya	Russia	1976–2004 (1992–2002)	136	714	Vertical	0.70	3,500 (1,200)	95 to 110 + 25
17	Takizawa	Japan	1999–2007 (2001–2003)	140	424	0.15/0.70	0.72	1,800 (810)	84 + 36
18	Ralko	Chile	1999–2004 (2002–2003)	155	360	Vertical	0.80	1,640 (1,596)	133 + 57
19	Sidi Said	Morocco	2001–2005 (2003–2004)	120	600	Vertical	0.60	660 (600)	65 + 15
20	Cindere	Turkey	1994–2005 (2003–2005)	107	280	0.70	0.70	1,680 (1,500)	50 + 20
21	Suofengying	China	2000–2008 (2003–2005)	122	165	Vert.0.25	0.70	739 (421)	64 + 95
22	Nagai	Japan	2000–2010 (2003–2006)	126	381	Vert./0.50	0.73	1,200 (703)	91 + 39
23	Baise	China	2001–2007 (2003–2006)	130	734	Vertical	0.80	2,672 (1,995)	80 + 132
24	Longtan	China	2001–2009 (2004–2008)	216.5	849	Vert./0.25	0.70	7,458 (4,952)	99 + 121
25	Guangzhao	China	2004–2008 (2006–2008)	201	412	Vert./0.25	0.75	2,870 (820)	61 + 91
26	Koudiat	Algeria	2002–2008 (2006–2008)	121	500	0.65	0.65	1,850 (1,650)	77 + 87
27	Labrena II	Spain	2005–2008 (2007–2008)	120	685	0.05	0.75	1,600 (1,400)	81 + 104
28	Yeywa	Myanmar	2001–2010 (2007–2008)	134	680	Vertical	0.80	2,800 (2,430)	75 + 145
29	Jinanqiao	China	2005–2010 (2006–2008)	160	640	Vert./0.3	0.75	3,600 (2,400)	72 + 108
30	Ban Ve	Vietnam	2004–2011 (2007–2010)	135	480	Vert./0.3	0.85	1,750 (1,520)	80 + 120

No	Name	Country	Years of construction (years of laying RCC)	Height (m)	Length (m)	Laying of faces Upstream/in lower part	Laying of faces Downstream	Concrete volume (RCC) thousand m³	Content cement + pozzolana (kg/m³)
31	Ban Chat	Vietnam	2006–2010 (2007–2009)	130	425	Vertical	0.80	1,600 (1,200)	70 + 150
32	Guandi	China	2007–2013 (2010–2012)	168	516	Vert./0.30	0.70	4,710 (2,970)	-
33	Cine	Turkey	1995–2011 (2004–2010)	137	300	0.10/0.20	0.85	1,650 (1,560)	85 + 105
34	Dong Nai 4	Vietnam	2004–2012 (2009–2011)	128	481	Vertical	0.80	1,370 (1,288)	85 + 95
35	Son La	Vietnam	2004–2012 (2008–2010)	140	900	Vertical	0.80	4,800 (2,700)	60 + 160
36	Gibe III	Ethiopia	2008–2015 (2011–2014)	249	610	0.10/0.25	0.65	6,300 (5,900)	70–120
37	Guanyinyan	China	2008–2016 (2011–2015)	168	1250	Vert./0.1	0.75	9,364 (6,473)	-
38	Ayvali	Turkey	2011–2016 (2013–2016)	178	405	Vertical	0.75	1,900 (1,650)	50 + 115
39	Grand Ethiopian Renaissance	Ethiopia	2010–2018 (2013–2017)	180	1790	Vertical	0.75	10,500 (10,100)	75 + 120
40	Las Cruces	Mexico	2013–2018 (2014–2017)	185	815	Vert./0.10	0.80	2,400 (2,000)	-

7.3 Constructions of basic elements of dams

7.3.1 Cross profile of dams

The cross profile of gravity massive dams is determined on the basis of ensuring the stability and strength of the dam and the foundation when the main loads are applied to the dam, its own weight, hydrostatic pressure and filtration pressure, and that it has a triangular shape, mainly with a vertical upstream face. The optimal profile differs from the triangular profile by the extension at the bottom.

The preliminary dimensions of the profile of the gravity dam can be determined from the condition that all the acting forces R pass through the bottom point of the middle third of the dam bottom (Figure 7.1). This condition guarantees the absence of tensile stresses for the design case of normal operation. For a triangular dam with a vertical upstream face loaded only with hydrostatic water pressure and concrete weight (Figure 7.1a), the dam width b at the bottom is determined by the formula [57]

$$b/h = tg\beta = (\gamma_w/\gamma_c)^{1/2} \tag{7.1}$$

where

γ_w and γ_c are the volumetric weights of water and concrete; other designations are shown in Figure 7.1a.

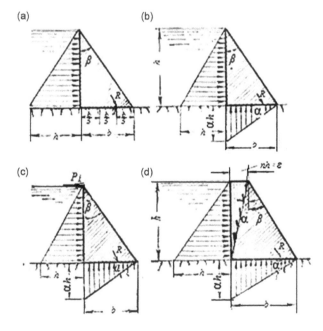

Figure 7.1 Schemes for determining the preliminary dimensions of the gravity dam profile.

Table 7.2 Weighing coefficient a

Type of foundation	Dam height h (m)		
	h < 25	25 < h < 50	h ≥ 50
Uniform waterproof rock	0.33	0.33–0.50	0.50–0.67
Rock with minor flaws	0.33–0.50	0.67–0.75	0.75–1
The rock is highly fractured and heavily cemented	0.75–1	0.75–1	1

If there is uplift at the foundation of the dam (weighing and filtration) taken into account, according to a triangular diagram (Figure 7.1b), the width of the dam at the bottom is determined by the formula

$$b/h = tg\beta = [1/(\gamma_c/\gamma_w - a)]^{1/2} \tag{7.2}$$

where a is the weighting coefficient taken according to Table 7.2.

If in addition to uplift at the foundation of the dam there is also the ice pressure P_i (in t per 1 linear meter of dam length) applied at the elevation of the dam crest (Figure 7.1c), the width of the dam is determined from the formula given below:

$$b/h = tg\beta = (6P_i + h^2)(\gamma_c/\gamma_w - a)h^2]^{1/2} \tag{7.3}$$

The upstream face is assumed to be vertical as in both previous cases.

If the upstream face is inclined, it is dense, and there is uplift taken into account according to the triangular diagram (Figure 7.1d), the width of the dam from the bottom is determined by the formula

$$b/h = tg\alpha + tg\beta = 1/(1-\varepsilon)\gamma_c/\gamma_w + (2-\varepsilon) - a]^{1/2} \tag{7.4}$$

where
$\varepsilon = (h \ tg\alpha)/b$.

In the above formulas, the weighing coefficient a was proposed back in 1950 by M.M. Grishin [57]. In modern SR, [14], the coefficient of effective uplift area of the foundation α_{2f} is used for this purpose, the value of which is recommended to be taken according to the results of special studies. Prior to performing these studies as well as in preliminary calculations, it is recommended to take $\alpha_{2f} = 0.5$ in the compressed zone of the foundation and $\alpha_{2f} = 1$ in the tensile zone of the foundation. In fact as a rule is taken into the stock $\alpha_{2f} = 1$ just as before was taken equal to unity.

Typically, the top of the triangular profile is taken at the level of the NHL (according to the norms of the USA at the level of the SRL or near it [142]).

The first dams (Puentes 1791, Grobois 1838) had heavy cross profiles (Figure 7.2) close to a trapezoid or even a rectangle. But in the future, with the development of calculation theory, more economical profiles of dams of curvilinear or polygonal shape appeared. However, the profile turned out to be the most economical basically representing a triangle with some deviations and corrections caused by the operating conditions of the dam. Currently, in most cases, gravity dams are designed in a triangular profile.

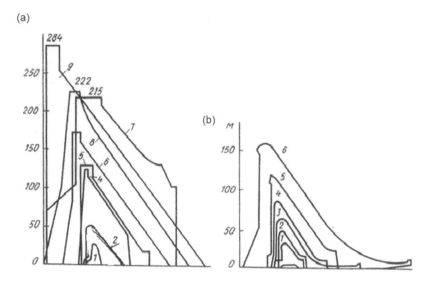

Figure 7.2 Profiles of some gravity dams: (a) deaf: I – Grobois (1838), 2 – Puentes (1791), 4 – Chambon (1936), 5 – Bratsk (1965), 6 – Grand Cooley (1942), 7 – Toktogul (1986), 8 – Hoover (1935), and 9 – Grandr Dixence (1961). (b) spillway: I– Tsimlyansk (1951), 2 – Dnepr (1932), 3 – Noris (1935), 4 – Grand Cooley (1942), 5 – Krasnoyarsk (1970), and 6 – Shasta (1944).

In many cases, the profiles of the dams differ from the triangular and are performed with an inclined upstream face with a broken or curvilinear outline, with an expansion of the profile to the foundation due to the consideration of seismic and temperature effects, pressure of waves, ice, and sediment, the phased construction and loading of the dam, and the real parameters of the rock foundation [75,77].

The broadening of the bottom of the dam contributes to the reduction of the decom-paction zone under the upstream face of the dam both under static loads and seismic effects. In recent decades, during the construction of dams in high seismic sites, the upstream face is usually taken with an inclination toward the upstream along the entire height or in the lower part of the dam (see Table 7.1).

In such dams with a large area of the contact surface with the foundation, under seismic effects, there is an increased emission of vibration energy into the foundation, which determines a high level of dispersion of the vibration energy of the "dam-reservoir-foundation" system and contributes to their seismic resistance [118].

Gravity dams are very earthquake-resistant structures, and most dams suffered earthquakes almost without damage.

To prevent damages to dams, it is necessary to take into account the site seismicity and seismic loads and their variability in time and space when choosing a profile and design.

The vertex device especially a massive creates an additional load on the dam, which has a negative effect on its SSS especially during seismic effects, which can cause an increase in the incline of the faces. To increase the seismic resistance of the dam, it is advisable to design the vertex lightweight, for example, in the form of a frame or buttress structure (Figure 7.3) and provide for the maximum possible reduction in the mass of concrete in the upper part of the dam (at a height of about $0.25h$, where h is the height of the dam).

Figure 7.4 shows the profiles of a number of gravity massive dams built in Japan in sites with high seismicity; dams are characterized by a cross profile with an incline of the upstream face, which starts both from the crest and from an intermediate point, and in addition, the incline increases in the lower part of the dam. The downstream face is mainly made straightforward with a laying in the range of 0.7–0.9.

The downstream face of a spillway dam usually has the same incline as a deaf dam.

The cross profile of the dam has smooth outlines without sharp changes in the contour of the faces. Changing the incline of the downstream face along the height of the dam to reduce its volume is used for high dams. In this case, incoming angles that are stress concentrators are not allowed, especially during seismic effects as was the case with the Koyna dam in India, which suffered serious damage during an earthquake in December 1967 (Figure 7.5).

The feature of this dam with a height of 103 m is a compressed profile with an average ratio of the width of the bottom to the height of 0.667, as well as a massive vertex

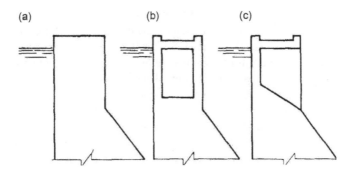

Figure 7.3 **Variants of the design of the dam vertex: (a) monolithic, (b) frame, and (c) buttress.**

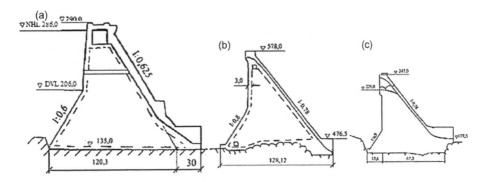

Figure 7.4 **The profiles of dams in Japan in high-seismic sections: (a) Miyegase, $h = 155$ m; (b) Sakaigawa, $h = 115$ m; and (c) Asahi-Ogawa, $h = 84$ m.**

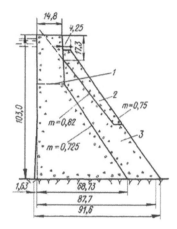

Figure 7.5 The profile of the Koyna dam after being strengthened: I – cracks formed during the earthquake, 2 and 3 – dam reinforcement with buttresses and cast concrete.

and an unsuccessful configuration of the downstream face with a sharp change in the laying in the upper part of the dam, which contributed to the concentration of dynamic stresses during an earthquake [118,175]. Based on the analysis of dam damage after the earthquake and computational studies, it was found that seismic acceleration reached 0.7 g on the dam crest. In the central sections in the upper part at the place of a sharp change in profile, cracks formed, through which enhanced filtration was observed. Also, increased filtration was observed through some intersectional joints as a result of damage to the seals. For repair work, the drawdown reservoir needed to be worked out and about 200 thousand m^3 of concrete was laid. The dam was strengthened by broadening the profile with laying monolithic concrete from the downstream face and the buttress device in the upper part (Figure 7.5).

To reduce the compressive stresses in the foundation under the downstream face of the dam and ensure its slide stability with weak rock foundations, it may be necessary to broaden the dam bottom by increasing the incline of the faces. An increase in the incline of the upstream face allows increasing the slide stability of the dam due to the use of water loading.

To increase the slide stability, the dam bottom can be run not horizontally but with an incline toward the upstream, such as at the Grande Dixence Dam in Switzerland (see Figure 1.19) and Alpe Gera in Italy (see Figure 1.15). However, such a solution leads to an increase in the volume of excavation of the rock and concrete and an increase in the rigidity of the foundation from the side of the upstream face, which can have an adverse effect on its stress state.

The elevation of the dam crest is determined on the basis of the requirements of SR [14], and the size of the reserve, taking into account the parapet arranged on the crest of the dam, is determined depending on the class of dams. The width of the dam crest is determined by the parameters of the road, the location of the crane tracks, the service conditions of the gates, lifting mechanisms, and other operating conditions. In some cases, to reduce the width of the crest, the highway is carried out on consoles, as on the dam of the Boguchan HEP (see Figure 7.26). In the absence of a road and other requirements, the width of the crest should be at least 2 m. Bridge crossings are performed within the spillway sections.

In spillway dams, spillway profiles usually fit well into the cross profiles of gravity dams designed based on the conditions of their static work [26].

Depending on the composition and layout of structures, the purpose of the dams, their position in the site and accordingly changes in their height, the dam profile along the length of the site can change in the HP.

7.3.2 Concrete zoning

In gravity dams, to ensure reliability and durability as well as manufacturability and economy, the zonal distribution of concrete of different classes is of great importance.

Depending on the working conditions of concrete, four zones are usually distinguished in separate parts of the dam (Figure 7.6):

I. external zones that are influenced by atmospheric effects, primarily air temperature fluctuations and not washed by water;
II. external zones within the fluctuation of water levels in the upstream and downstream as well as dam elements periodically exposed to the flow of water (weirs, spillways, etc.);
III. external zones below the minimum operational water levels in the upstream and downstream as well as zones adjacent to the foundation of the zone;
IV. inner zone of the dam.

The concrete of these zones has different requirements for

*   conditions of concrete strength of all zones in accordance with their SSS;
*   frost resistance in severe climatic conditions to concrete of the external zones I and II; the width of the zone should be not less than the depth of freezing of concrete;

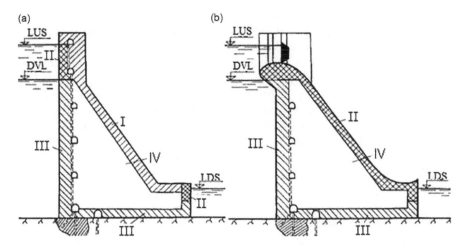

Figure 7.6 Distribution scheme of the concrete of the dam body in the zones: (a) deaf dam and (b) spillway dam; I–IV – dam zones.

- waterproof to concrete of the external zones II and III; the width of the zone is determined by the allowable gradient of filtration;
- resistance to aggressive effects of water to concrete of the external zones II and III;
- resistance to cavitation at a water velocity on the concrete surface of 15 m/s or more in terms, resistance to abrasion by a flow of water in the presence of a significant amount of suspended and entrained sediments to concrete of zone II in areas of weirs, spillways, and outlets;
- deformability to concrete of the zone adjacent to the foundation;
- heat during hardening of concrete to concrete of all zones.

The thickness of the external zones is taken into account in the design features of the dam, the technology of erection, SSS, the head, and the influence of external air temperatures. For dams made of ordinary concrete, the thickness of the external zones is usually taken as at least 2 m.

In accordance with the working conditions of concrete in different zones, the number of classes of concrete in the dam is determined. It is usually thought that the number of concrete classes in a dam does not exceed four, but in high and large dams, the number of concrete classes may be greater. Zone distribution of concrete should be linked to dam concreting technology.

Differences in the zoning of concrete in dams made of RCC are associated with the use of two types of concrete: ordinary vibrated and RCC and the peculiarities of their laying technologies. Depending on the specific conditions and structural and technological features, the volume of RCC can average up to 70%–95% of the total volume. With a rational distribution of concrete, a decrease in the volume of concrete of the external zones is achieved using concrete with a low cement content (80 ... 100 kg/m^3) for the inner zones of the dam. It is also possible in the inner zone to zonal distribution of concrete with different strengths along the dam height in accordance with the stress field [89,162,195]. On the highest gravity dam Gibe-III with a volume of 6.3 million m^3, seven classes of concrete were used.

From vibrated concrete, all external zones and the adjoining zone to the foundation can be performed as is customary in Japan dams; this design ensures reliable operation of the dam especially in harsh climatic conditions and with high seismicity. Sometimes from vibrated concrete, only the upstream face, adjoining to the foundation, vertex and weir sections are made, and from the RCC – the inner zone of the dam. In Japan, dams with an upstream face zone width of ordinary concrete about 3 m wide are being erected, at the Tashkumyr dam in Kyrgyzstan, the width of the zones is 3.7 m, and at the Bureya dam in Russia, it is under 14 m; on many dams, this zone is thinner, 0.3–1.4 m wide, for example, Galesville, Elk Creek in the USA, Salto Caxias in Brazil, and Beni Horoun in Algeria.

Within the zone of ordinary concrete, seals and drainages of expansion joints, galleries, and shafts are arranged, reinforcement is carried out, and CME is installed.

In harsh climatic conditions, RCC dams are usually made with external zones of ordinary concrete. Moreover, due to the fact that the concrete of the external zones has a higher deformation modulus in comparison with the RCC of the inner zone, higher compressive stresses will act under the action of operational loads in the external zones compared to a gravity dam of the same profile made of ordinary concrete. In addition, tensile stresses on the downstream face of the dam in winter decrease (up to 10%) [200].

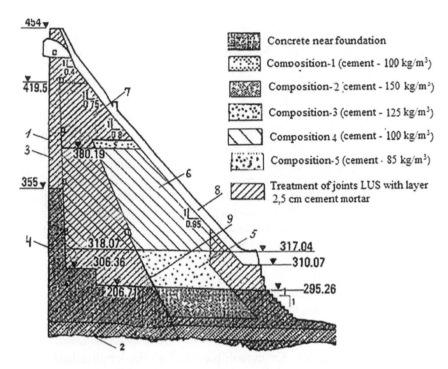

Figure 7.7 Miel-I dam: I – screen from a two-layer film (PVC); 2 – ordinary concrete near foundation; 3–7 composition I–5 accordingly; 8 – ordinary concrete of piers and weir; and 9 – longitudinal seam.

In recent years, dams of RCC in the zone of external faces and in the zone of adjacency to the foundation have begun to widely use

- RCC with a high consumption of binders, which has a high density, strength, and water resistance with a thin layer (up to 2.5 cm) of cement mortar laying between layers of RCC;
- vibrated RCC enriched with cement mortar (GEVR);
- RCC enriched with cement mortar (GE-RCC), which allows refusing from laying in the external zones of ordinary concrete or sharply reduce its volume. For example, on the Miel-1 dam (Figure 7.7), ordinary concrete was laid only in the areas adjacent to the foundation, vertex, and weir.

In this case, climatic conditions and accordingly cooling conditions for RCC with a high binder content and thermal conditions at various stages of construction include [93,198,201]:

- uneven cooling of RCC layers in the inner zone of the dam relative to its faces;
- the effect of galleries on the thermal regime;
- change in thermal conditions during uneven construction of the dam with adjacent sections of different heights.

It should be noted that on some dams in Spain with a high binder content (more than 200 kg/m³), cracks formed, and a high content was not required either to provide strength or to provide density but was caused by the desire to increase the adhesion and waterproofness of the joints between the layers [79].

7.3.3 Deformation joints

Gravity dams are divided along the length into separate sections by constant, transverse, deformation, and vertical joints, which ensures the independence of their movements and avoids dangerous tensile stresses and the "unorganized" cracks caused by them in the dam body under temperature influences and with inhomogeneous foundations. In sections of dams, additionally constant vertical temperature seam-incisions can be made. Transverse deformation joints usually perform normal to the axis of the dam [122].

In gravity dams, during construction with cutting into concrete blocks, temporary (construction) vertical joints are formed, which then by the time the dam is put into temporary or permanent operation are monolithic (cemented or concreted) including:

- longitudinal temperature joints mainly when erecting a dam with columnar blocks and in some cases in high dams from RCC;
- during the construction of a continuous dam to provide spatial work in operating conditions, for example, the lower part of the Toktogul dam.

The terms and the procedure for monolithic (closing) the longitudinal joints are appointed after the temperature of the concrete pillars is equalized to close to many year average and the joints will open. When cementing the joint between the first and second columns, it is possible to improve the SSS of the first column due to compression of concrete and the foundation from the side of the upstream face [22,183].

Studies of the behavior of intern block joints in the construction period were performed for the concrete dam of the Boguchan HEP on the mathematical model "section No. 8 – rock foundation". In the model, the grooves (Figure 7.8) were reproduced in the form of a trapezoid (height 37 cm and bases 75 and 30 cm) as well as the sequence of construction of the section.

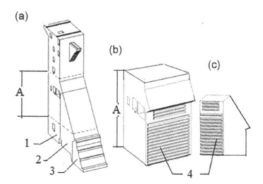

Figure 7.8 Section No. 8 of the Boguchan HEP (a), I, 2, and 3 – pillars No. I, 2, and 3, respectively; (b) and (c) – fragment A of pillars No. I and No. 2; 4 – grooves in the intercolumn joint; No. 2 pillar rotated 90°.

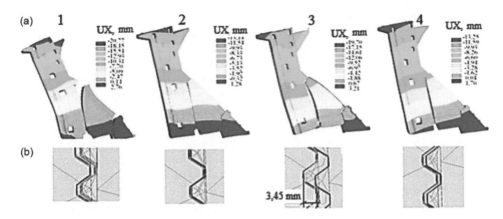

Figure 7.9 Horizontal displacements UX, mm, section No. 8 (a): I – January 05, 2 – April 14, 3 – July 13, 4 – October 21; (b) opening of the joint between the No. I and No. 2 pillars, mm. Movement scale 1000:1.

During 1992–2005, the construction of the Boguchan HEP station was interrupted when the concrete dam was almost completely erected. The temperature calculations performed with average 10-day fluctuations in outdoor temperatures relative to the average annual temperature of –3.2°C showed that a quasi-steady state temperature regime in section No. 8 was formed in the third year. The presence of inter-column joints despite their opening did not affect the temperature distribution in the dam.

An analysis of the temperature distribution and the opening of the joints of the dam was performed for four points in time: January 05 (winter), April 14 (spring), July 13 (summer), and October 21 (autumn). From Figure 7.9, it is seen that opening of joints do not occur simultaneously in height and at different marks at different times of the year (Figure 7.9a), which complicates the choice of time for their cementation. Figure 7.9b shows the openings of the joint between No. 1 and No. 2 pillars at different times of the year, from which it follows that there is almost always has a contact on inclined faces of the grooves (for example, July 11, despite the opening of the joint at 3.45 mm, the grooves has contact on inclined faces).

No. 1 and No. 2 pillars leaned into the upstream due to the eccentric position of the self-weight forces of the No. 1 and No. 2 pillars relative to the bottoms. Despite the slope of the pillars in the upstream face, the contact of the dam with the foundation is compressed.

Permanent transverse seams. The dimensions of the dam sections separated by constant transverse seams depend on the following:

- type and height of the dam, placement of water inlets and turbine conduits in sections of the station dam, and opening in the sections of the spillway dam, including for passing construction discharges during the construction period;
- methods and technologies of dam construction;
- shape of the dam site and engineering–geological conditions of the foundation;
- climatic conditions of the construction area.

When assigning the distance between the permanent joints, on the one hand, there is a desire to increase the size of the sections to improve the conditions of construction, and on the other hand, it is necessary to limit their sizes according to the conditions of cracking under temperature influences during the construction period.

In modern gravity dams made of ordinary concrete, the distance between permanent expansion joints is mainly 9–20 m. With a greater distance between permanent joints in the dam, vertical transverse cracks in the middle of the section and longitudinal cracks can form, as was the case with the Dworshak, Revelstoke, and other dams [65,193]. With a heterogeneous foundation, permanent joints are arranged in places of changes in the deformability and fracture of the foundation contour. Permanent joints are made between spillway, station, and deaf dams; in spillway dams – between the extreme bulls of adjacent sections, or in the middle of spillway spans and in station dams – between the water inlets of one or more pressure head pipelines of HEP units.

The location of the joints and the distance between the permanent transverse joints are determined by the characteristics of the rock foundation, the design features of the station and spillway dams, the technology and pace of concrete laying, climatic conditions, and the results of calculations of the thermal stress state [195].

Despite the measures taken as shown by long-term field observations, cracks form in the gravity massive dams during the construction and operational periods (including poorly cemented joints open), which depending on their parameters (depth, location, and opening) and quantity, can lead to not monolithic dams and have a significant impact on their operational condition. It should be noted that the nonmonolithic dam formed during the construction period usually increases (rarely decreases) under the influence of force and temperature factors when filling the reservoir and operation.

Based on the influence of cracks that occur in dams at different stages on the reliability of their work, they can be distinguished as follows:

- surface temperature-humidity cracks observed practically on all dams that do not affect operating conditions and do not weaken the dam profile;
- cracks that do not weaken the profile and do not lead to enhanced filtration through concrete of the upstream face and can self-heal in the process of filling the reservoir;
- through cracks (including open inter-column joints with poor cementation) weakening the working section and reducing the slide stability of the dam, requiring repair work, and in some cases when the reservoir is empty.

In dams erected from RCC with a low binder content, the number of permanent joints can be reduced with a corresponding increase in the width of the sections to 30–60 m, which improves the concreting conditions and increases the monolithic of the dam.

The distance between the permanent intersection seams and the notch seam-incision is determined based on calculations of the thermal-stressed state of the dam. Experience in the construction of RCC dams at the initial stage, for example, Galesville dam (1985), Upper Stillwater (1987), Elk Creek (1988) in the USA, Townd (1994) in South Africa, showed that an unreasonable increase in the distance between the expansion joints leads to the formation of transverse vertical cracks that can become sources of hazardous filtration, as well as longitudinal cracks that require repair work.

(a)   (b)

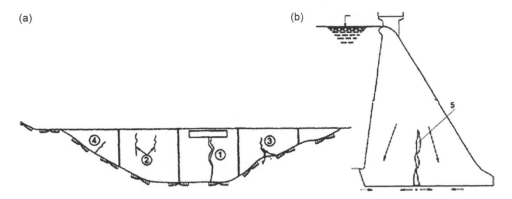

Figure 7.10 Cracking patterns in high dams from RCC: (a) view from the upstream; (b) transverse section: 1–4 – types of transverse cracks; 5 – longitudinal crack.

Schemes for the formation of transverse and longitudinal cracks in high dams of RCC with a large distance between the transverse joints are shown in Figure 7.10. Vertical transverse deep cracks 1 in the middle of the section from the bottom of the dam can cut the sections into half; cracks 2 can form in the upper part of the upstream face, extending mainly to a depth of 2 m (for example, the Rialb dam); cracks 3 and 4 can form at the foundation and abutments at the points of profile fracture and changes in the deformability of the foundation. At the same time, in many dams with a joint distance of up to 45 m, no cracks were observed and in a number of dams with a large distance, for example, Miel-1 dams (distance 60 m) in Colombia, Pangue (distance 50 m) in Chile, and Santa Eulalia, 84 m high (distance 60 ... 90 m), in Spain. To prevent longitudinal vertical cracks in the Miel-1 dam, a longitudinal joint was made [93,205,227].

In the Jinanqiao dam with a height of 160 m and a crest length of 640 m built in 2010 in China in an area with high seismicity (the peak horizontal acceleration during an earthquake repeated every 10,000 years is 0.475 g), to include the entire dam massif in the deformation joint, each layer of RCC is cut into 2/3 of its thickness. For this, an electric cutter with a blade width of 1 cm was used. With a layer thickness of RCC 30 cm, the seam-incision depth was 20 cm. The seams were filled with nonwoven fibers with a seal. Nonlinear calculation of the dam taking into account seismic loads showed that, compared with the variant with completely cut expansion joints, the maximum displacements of the dam along the flow decreased from 8.5 to 7.6 cm and across the flow from 5 to 2 cm [194].

Temperature seam-incisions in sections of the dam with a distance between them of 7–20 m depending on climatic conditions are arranged between permanent intersection joints and are made from the surface of concrete to a depth of usually 5–7 m within the zone of annual temperature fluctuations in concrete.

Permanent joints should have a sufficient width that allows individual sections to move independently, have water resistance, and provide the ability to control their operation and repair seals. The reinforcement must not pass through the joints.

The width of constant temperature joints is determined on the basis of calculations of the temperature regime, taking into account the fact that at the highest temperatures, the joints close, and at low temperatures they are maximally opened:

- 0.5–1 cm at a distance of not more than 5 m from the faces and crest;
- 0.1–0.3 cm inside the body dam, where the temperature varies little and is equal to the annual average.

Permanent joints can be formed from precast concrete elements, can be cut, etc.

Permanent deformation joints are generally flat which ensures independent operation of the dam section. At the same time, the conditions for drainage of water that filtered into it from the joint are also facilitated. Permanent joints are extremely rarely performed with grooves or "hinged" joints, with the help of which it is possible to transfer forces from loaded sections to neighboring ones. The disadvantages of such joints are the complexity of the design, the concentration of stresses in the places of kinks, which can lead to the formation of cracks, and the deterioration of the conditions for removal of filtered water (Figure 7.11).

Permanent joints are performed using removable and more often fixed formwork and precast concrete blocks and in dams made of RCC – also by cutting joints in freshly laid concrete with vibratory knives.

Seals – permanent joints. In permanent seams, the following are arranged:

- seals providing water tightness at all possible deformations of adjacent sections of the dam (temperature, shrinkage, and deformations caused by irregular settlements) as well as protecting the joint from the effects of ice, high-speed flow, and clogging;
- drainage devices for organized drainage of filtered water into the joint through seals or bypassing them;
- viewing shafts and galleries for monitoring the condition of joints and repairing seals.

Joint designs with the arrangement of seals are shown in Figures 7.11 and 7.12.

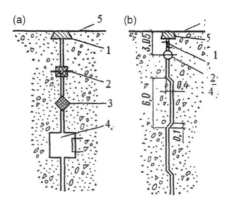

Figure 7.11 Types of deformation joints: (a) flat; (b) grooved; 1 – circuit seal; 2 – copper sheet; 3 – asphalt mastic; 4 – drainage; and 5 – upstream face of the dam.

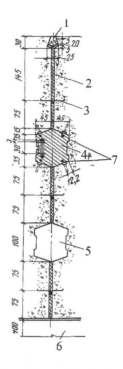

Figure 7.12 The section along the deformation joint of the dam of the Bratsk HEP (dimensions in cm): 1 – reinforced concrete beam, 2 – cold asphalt plaster, 3 – brass diaphragm, 4 – asphalt dowel, 5 – viewing shaft, 6 – viewing longitudinal gallery of the dam, and 7 – electrodes for heating asphalt filling.

Seals in the joint are divided into contour (external and internal) and main vertical.

Contour internal seals are arranged in the joints around the viewing galleries or other internal cavities in the form of metal, rubber, and plastic (polyvinyl chloride, etc.) waterproof diaphragms located at a distance of 0.2–0.5 m from the surface of the cavity walls.

Outline outer seals that protect the joint from the effects of ice, high-speed flow and clogging and reduce water permeability are made in the form of reinforced concrete or concrete beams, slabs, rubber and plastic materials (for example, polyvinyl chloride tapes), and metal strips laid on the preparation of asphalt materials (bitumen mats and asphalt mastic) (Figures 7.11 and 7.13b).

In moderate climates, contour seals are not performed limiting themselves to beveling the outer edges of the joint and coating its surface.

The main vertical seals (Figure 7.13a) ensuring the water tightness of the joint are usually located at a distance of 1.5–2 m from the upstream face and are made in the form of one or more metal, rubber, plastic diaphragms, asphalt keys, and in some cases – injection (using cementation and bitumen). Reliable operation of deformation joint from the main seals primarily depends on the operation [62,65].

To create a volumetric stress state of concrete in the upstream face zone due to hydrostatic pressure from the intersectional seams, Canadian experts propose dams

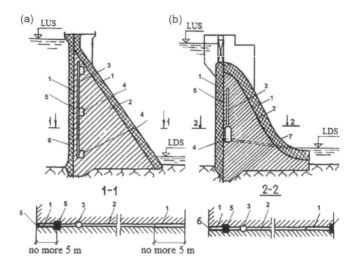

Figure 7.13 The arrangement of seals in permanent deformation joints: (a) deaf section; (b) spillway section; 1 – joint (0.5–1 cm), 2 – joint (0.1–0.3 cm), 3 – drainage, 4 – gallery, 5 – main vertical seal, and 6 and 7 – circuit seals.

with a slight inclination of the upstream face, so that with vertical seals of the joints, the distance in the joint from the upstream face to the seal in the lower part of the dam increases to 7 m. This solution aimed at closing vertical cracks formed during the construction period was applied at Revelstoke and Itaipu dams [193].

Compensators are necessarily arranged in metal diaphragms, which ensure deformation of the seal without breaking when the joint width is changed, or the adjacent sections are moved. The materials of these diaphragms are low alloy steels and alloys with high resistance to corrosion, copper, and brass sheets. Metal diaphragms providing reliability and durability (retain properties for the entire estimated period of operation) however have a higher cost.

All seals should be fixed at the foundation and terminated at the dam crest, and the seams passing in the middle of the span of the spillway sections should be brought to the spillway vertex (Figure 7.14).

When several rows of metal diaphragms (or from other materials) are installed in the joint, the distance between them should be at least twice the depth of the diaphragm in concrete. Usually, drains are placed between the two rows of seals leading to the drainage gallery. When using rubber profiled seals, the rubber must satisfy the conditions of tensile strength (at least 20 MPa) and deformability and requirements for frost resistance and long-term operation (durability).

Plastic diaphragms made of polymeric materials including polyvinyl chloride tapes (PVC), fiberglass, vinyl plastic, etc. must also satisfy the requirements of strength, deformability, frost resistance, and durability.

The widely used asphalt keys are made in the form of vertical wells in a concrete block adjacent to the joint of adjacent sections of the dam, filling the key cavity with asphalt or a bitumen mixture of bitumen and mineral aggregate. Under appropriate climatic conditions, an electric heating is arranged in the key cavity to heat the

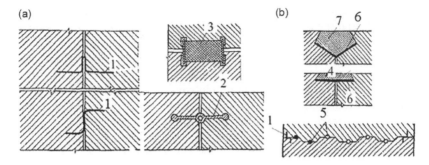

Figure 7.14 Seals of deformation joints: (a) main internal seals in the form of diaphragms (metal, rubber, and plastic masses), bitumen dowels, and injection (cementation and bituminizaion); (b) external seals; I – metal diaphragms, 2 – profiled rubber, 3 – asphalt mastic, 4 – reinforced concrete slab, 5 – hole cementing, 6 – rubber diaphragm, and 7 – key.

aggregate during operation (for example, if it is necessary to replenish the aggregate due to leakage). The cross-sectional dimensions of asphalt dowels vary widely reaching 0.8 × 1 m or more. The dimensions of the keys should be determined based on the operating conditions of the seals: water head, displacement of the dam sections, and geological and climatic conditions. Large section keys are characterized by significantly more favorable conditions and reliable operation, especially in high dams. The key cavity is not allowed to be made of precast concrete slabs, since bypass filtration may take place in the plane of the joint formed when the slabs adjoin the concrete. Reliability of asphalt keys is ensured by fencing them adjacent to the joint with diaphragms or when arranging diaphragms (metal, rubber, etc.) in the joint near the keys.

In many cases, inspection shafts are also arranged in permanent joints, which serve to control the operation of the seals and to drain the joint, including the collection and removal of filtered water and if necessary, to repair the seals or to arrange and seal them.

In thin joints up to 5 mm, the walls of the joint from the upstream face to the inspection shaft are processed by painting with hot bitumen, and with a larger thickness, asphalt plaster or other waterproofing materials are used.

When installing on the upstream face of the groove structures of the gates, the main seals in the joints should be placed behind the groove structures.

For the effective operation of the main seals, it is extremely important to ensure the high density and water tightness of concrete around the seals in order to prevent increased filtration along their contour.

To do this, the following requirements must be observed:

• the material of the seals must be directly adjacent to the concrete mass forming the joint, in connection with which it is not allowed to use nonremovable elements (prefabricated reinforced concrete, metal, etc.);
• compressive stresses on the contact of the asphalt keys with concrete should be higher than the external hydrostatic pressure of water;

- the average gradients of the head of the filtration flow through concrete along the contour of the weld seals are usually equal to 20 (the critical gradient for gravity dams is 25, and the reliability coefficient is 1.25).

When determining the average head gradient, the total filtration path is taken equal to

- when the temperature of the concrete in zone changes within 4°–6° and below, the sum of the filtration paths bypassing the seals (along their contour) and along the length of the cemented and bituminized joints between the seals;
- when the concrete temperature changes above 6° – only filtration paths bypassing the seals.

The number of rows of seals varies in height of the dam and depends on the head, the length of the embedment (penetration into concrete) of the diaphragms, and in general the overall filtration path.

In RCC dams, seals in permanent deformation joints are usually arranged in the form of diaphragms on sections from the upstream made of ordinary concrete and on a number of dams, of enriched concrete and in dams without the use of ordinary concrete, in the form of a PVC film on the upstream face overlapping the joint (for example, a two-layer film on the Miel-1 dam) and drainage from geotextiles.

Temporary joints. Vertical longitudinal joints formed in dams during columnar cutting are monolithic before filling the reservoir at a concrete temperature close to long-term average temperature, in order to achieve which artificial pipe cooling of concrete is used in many cases. Such joints are usually performed by grooving. For cementation of joints, cards with dimensions of, for example, 9×15 m, on the Bratsk dam (Figure 7.15), are allocated in them, where pipe outlets of cement mortar, equipped

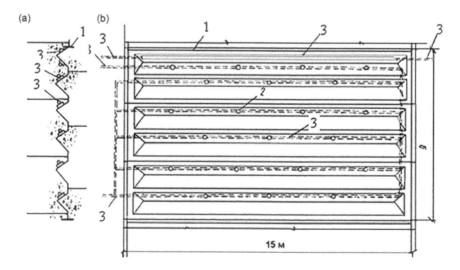

Figure 7.15 Temporary joint construction (Bratsk dam): (a) vertical section along the grooved joint; (b) a plan; I – out joint seal, 2 – outets of cement mortar for monolithic joint, and 3 – tubes, conductive mortar for cementation and venting air key.

with valves and air pipes are laid. Cement mortar is injected into them from below through a discharge manifold and feed pipelines.

After filling with a solution of the card cavity, the excess solution is removed through the return manifold to the cementation unit, and the air displaced by the solution from the card cavity is removed by pipes outside the joint. Each card is protected by seals in the form of diaphragms [22,65,161].

A similar pattern is used for cementing intersection vertical joints in dams in narrow sites operating according to a spatial pattern.

Other expansion joints. In high RCC dams, to prevent longitudinal cracks, a longitudinal deformation joint can be made in the middle in the lowermost massive part of the dam.

In gravity dams, under certain conditions to improve the SSS in the contact zone, one or more horizontal seam-incisions with seals can be arranged from the side of the upstream face [83,226].

At high seismicity, in order to reduce the intensity of seismic effects on the dam, it is possible to arrange horizontal seismic quenching layers in the lower part of the dam made of a material significantly different in its dynamic properties from concrete [82].

7.3.4 Antifiltration devices

Damping devices in the body of dams are provided to prevent concentrated pressure filtration through concrete and leaks of concrete joints and cracks, as well as to minimize filtering pressure in the body of the dam and prevent leaching of concrete [21].

As such antifiltration devices, various types of screens arranged on the upstream face and drainage of the dam body are used.

It should be borne in mind that filtration anisotropy of the dam body associated with horizontal joints between layers of RCC is possible in dams made of RCC.

The drainage of the dam body is mainly carried out in the form of vertical tubular drains 10–30 cm in diameter located in the dam body along the upstream face with a distance between the axes of the drains, usually within 2–3 m, or horizontal drains of rectangular or trapezoidal section with an area of 400–800 cm^2 and the distance along the height of the dam between them within 2–3 m [71,195].

Vertical tubular drains are usually arranged by drilling in concrete wells from drainage galleries with their tiered placement along the dam height, as well as installing prefabricated blocks of ordinary or porous concrete in which tubular cavity drains are made [84].

Water filtered through the outer zone of the concrete enters the drainage wells and then into the longitudinal horizontal galleries and through the transverse galleries is diverted to the downstream.

Horizontal drains are formed from prefabricated blocks of ordinary or porous concrete, in which tubular cavity drains are made, tied to the concreting tiers with the drain exit to the inspection shafts located in the intersection joints of the dam. Filtered water is diverted through them to the downstream.

The parameters of vertical and horizontal drains (diameter, cross-sectional area, and distance between drains and to the upstream face) are determined on the basis of calculations taking into account the technology of dam construction and existing operating experience. The distance from the upstream face of the dam to the axis of the dam drainage (to the longitudinal gallery) is determined depending on the critical average head gradient, which depends on the concrete mark for water resistance. So, with a concrete mark for water resistance W8, the ratio of head to the thickness of the outer zone can be in the range of 15–20.

With such a layout for the drainage of the dam body as shown by field observation, water penetrating the construction joints and cracks saturates the concrete with water in a very limited volume of the upstream face with minimal uplift.

Free unloading of the filtration flow occurs mainly in drains and a small remaining part of the flow through the downstream face of the dam. Moreover, in severe climatic conditions, freezing of concrete in the zone of the downstream face may affect the conditions of unloading the filtration stream [65].

Galleries are designed for collecting diverting drainage water, monitoring the operation of the dam, foundation drainage, and the state of the concrete dam and joint seals, for placing CME laying various communications (mains, air ducts, control cables, etc.), for performing body drainage work of dam, cementation, and drainage curtains at the foundation as well as repair work on cementation and drainage curtains and drainage of the dam. When placing the gates of deep spillways in the body of the dam in large galleries (cavities) arranged above them, mechanical and electrical equipment is installed. Approach galleries should provide transportation of gates and other equipment.

Longitudinal and transverse galleries are located along the height of the dam after 20–40 m and pass along the entire length of the dam. One of the longitudinal galleries should be placed above the maximum level of the downstream for the possibility of gravity diverted of drainage water from the entire upper part of the dam. From the lowest gallery, which also receives drainage water from the foundation of the dam, water is pumped out. In all cases diverted of drainage water to the downstream is provided below the minimum level of the downstream.

Galleries of various tiers are connected by mines in which marching stairs and elevators (passenger and freight) are arranged. Emergency exits to the overlying one are arranged from each lower tier of the galleries, the distance between the emergency exits should be no more than 300 m, and each longitudinal gallery must have at least two emergency exits. Cross galleries are also used as exits from longitudinal galleries. Entrances to the galleries are arranged either from the bank (in the downstream) or from the side of the crest of the dam bulls.

Galleries and mines provide for lighting and ventilation.

The dimensions of the galleries for cementing the foundation and construction joints of the dam creating and restoring drainage in the body of the dam and in the foundation should ensure the transport and operation of drilling, cementing, and other equipment, taking into account the placement of cable and other communications and pipelines when using artificial cooling of concrete. At the same time, their width should be at least 2–2.5 m and the height should be at least 3–3.5 m. The minimum sizes of other galleries should be as follows: width – not less than 1.2 m and height – not less than 2 m.

The floor of the galleries used for collecting and diverting drainage water is taken with a slope of not more than 1:40 toward the spillway.

In high dams, it is necessary to consider the arrangement of separate galleries for performing cementation and drainage curtains at the foundation.

It is advisable to use prefabricated elements, mobile formwork, etc. to form walls and overlap the gallery including in the form of a vault.

Given that the galleries are stress concentrators, reinforcement is usually performed to prevent cracking along their contour, and it is advisable to place the transverse galleries in the plane of intersection joints.

In dams made of RCC, the galleries are mainly performed in the zone of ordinary concrete from the side of the upstream face using prefabricated elements or in direct adjoining to it.

In low RCC dams, in some cases, galleries were not performed, for example, in the Winchester dam, 23 m high, and in the Stacy spillway dam, 31 m high, in the USA [212].

7.3.5 Constructional solutions of faces

Structural solutions of the dam faces are determined by climatic conditions, concrete zoning, construction technology, and operating conditions. Under normal conditions, zoning of concrete with the laying of appropriate concrete on the faces of the dam provides protection for the dam and normal operating conditions. In a number of dam structures, made of RCC and ordinary, special protective coatings of the dam faces were used.

Upper face. In gravity dams, the upstream face is usually formed using formwork.

In dams made of RCC, the formation of the upstream face is determined by the applied structural and technological solutions, such as the protection zone of ordinary concrete or of vibrated RCC enriched with cement mortar (GEVR), screens of various types, and RCC with a high consumption of binders.

When performing the upstream face zone from ordinary concrete, two fundamentally different structural and technological schemes are used:

* with a tiered construction providing for the upstream and downstream zones or only the upstream zone of ordinary concrete and the inner zone of RCC in parallel, for example, Upper Stillwater and Elk Creek in the USA, Tamagawa, Miyegase, and Takizawa in Japan, Guaninge in China, and Olivettes in France;
* with independent execution of the upstream zone of ordinary concrete and the inner zone of RCC.

In the scheme of tiered construction, in most cases, the laying of the tier begins with the execution of the zone of ordinary concrete, the downstream surface of which is performed with a slope, and the height of the tier of ordinary concrete can equal the height of two to four layers of RCC, such as in dams in Japan. On a number of dams in the USA (Upper Stillwater and others), the tiers of the zone of ordinary concrete with the creation of the upstream and downstream faces were carried out in the form of border elements using mobile formwork [159]. Further layers of RCC of the inner zone were rolled onto the inner surface of the tier from ordinary concrete.

On a number of dams, for example, Olivettes in France, laying began with the execution of one or more layers of RCC in the inner zone followed by concreting the tier from ordinary concrete.

In a number of dams, prefabricated reinforced concrete panels were used to form the upstream face.

The disadvantages of the tiered scheme include the following:

- the need for technologically tight connection of concrete laying of the outer and inner zones using different concreting methods with the volume of RCC averaging from 60% to 90%, and as the dam grows the ratio of volumes of rolled and vibrated concrete is constantly changing, which complicates the technology of concreting;
- increase in the number of horizontal construction joints in the area of ordinary concrete from the side of the upstream face.

When building large dams from RCC in areas with harsh climates and long winters where construction usually stops in winter (Upper Stillwater, Tamagawa, Guaninge dams, etc.) technology with independent execution of the outer upstream zone from ordinary concrete and the inner zone of RCC is used. This technology allows to overcome the shortcomings of the tiered scheme and conduct construction all year round, including the winter period. Using this technology, the inner zone of RCC and the protection zone from the downstream face of vibrated concrete are erected in the warm season, and the protection zone of the upstream face of the dam is made of vibrated concrete, which provides water resistance mainly in the cold season with concrete laying in multilayer blocks. In this case, the temperature regime in such blocks is controlled mainly by pipe cooling and concreting can be carried out under the protection of a self-raising tent [159]. This technology was implemented during the construction in harsh conditions of the dam of the Bureya HEP from RCC [93,161,219].

In a scheme with independent execution of the outer zone from the upstream face of ordinary concrete and the inner zone of RCC, a complicating condition is to ensure reliable interfacing between the zones, and therefore, special measures are required to ensure their joint work, including grooving of the resulting longitudinal joint and cementing it.

Improving the construction technology with independent execution of the outer and inner zones especially in severe climatic conditions can be achieved by applying the scheme [89,233] according to which prefabricated concrete elements of a trapezoidal section are installed between the upper and inner zones with a device along their faces (or in the body of the elements) of the drainage cavities forming the dam drainage (Figure 7.16). These elements are similar in parameters to the prefabricated elements used in a number of dams to form a downstream face.

The prefabricated elements installed on top of each other form a surface with their upstream faces, with which the upstream zone of the dam (or screen) mates, and with their downstream faces at an angle of inclination of 45°. stepped sawtooth surface, which provides a reliable interface with the internal area of rolled concrete. To ensure reliable interfacing with the upper zone from ordinary concrete, the upstream face of the precast elements is made with the device of vertical and horizontal ledges-ribs or with an inclination toward the upstream, and the overlying prefabricated elements are shifted toward the upstream with the formation of a stepped surface with their upper faces.

With this scheme, advanced concreting of the inner zone is performed by tiered installation of prefabricated elements 1 m high and layer-by-layer laying of RCC. Laying of ordinary concrete in the upstream zone is lagging behind in the most favorable period of the year with high blocks with a sharp decrease in the number of

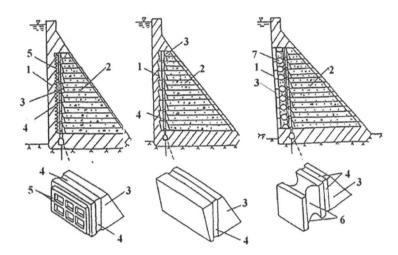

Figure 7.16 Dam with prefabricated elements between upstream and inner zone: I –
upstream zone, 2 – inner zone of RCC, 3 – prefabricated elements, 4 –
cavity drains, 5 – ledgesribs, 6 – groove cavity, 7 – concrete monolithic.

horizontal joints. Two technological streams using different concreting methods are
not interconnected, which improves the conditions of construction and creates favorable conditions for the use of continuous technology and high intensity concrete work
all year round.

Additional advantages are obtained during construction in harsh climatic conditions with the execution of the inner zone of RCC only in the warm season and the
upper zone of ordinary concrete with a lag and continued work in the winter (using
warm formwork, self-lifting tent) [83,233].

The reliability of such a combination of monolithic vibrated concrete with the
upstream face of prefabricated elements is confirmed by extensive experience in the
construction of many HEPs, the normal operation of these structures under long-term
operation, and field observations [159]. The performed experimental studies including
those on reinforced concrete fragments showed that reliable contact is ensured by the
contact of the precast element with RCC. In this case, the design resistance to slide
along the contact surface reaches 2 MPa, with the action of compressive stresses along
the contact surface slide resistance increasing. Prefabricated elements connect the upstream zone with the inner one and ensure their joint work. The width of the precast
element is about 1 m, i.e., less than 0.25–0.35 width of the upstream zone on average.
A computational study of the stress state of a dam 88 m high under the action of the
main usual loads in the horizontal joint zone showed that the combined operation of
the upstream and inner zones is ensured.

It should be noted that the dam in which the upstream and inner zones are connected by prefabricated elements is less susceptible to the influence of temperature
cracking in the construction and operational periods [93]. This is due to more favorable conditions for cooling the concrete in the absence of direct contact of the upstream
and inner zones (with the possibility of opening the joint by contact). In addition,

this contributes to a decrease in the number of horizontal joints in the upstream zone with its independent concreting. The possibility of applying new solutions in projects should be justified by appropriate calculations and research [176].

In recent years, on many dams built in favorable climatic conditions (for example Jiangya and Longtan dam in China and Son La dam in Vietnam), the upstream face was formed from rolled vibrated concrete enriched with cement mortar (GEVR), which increased manufacturability and ensured favorable conditions for the construction of dams from RCC [43,198,246].

Protective screens on the upstream face of the dam providing waterproofing of the dam are used in

- dams of ordinary concrete and RCC;
- dams in highly seismic areas due to the possibility of tensile stresses and cracks on the upstream face during an earthquake;
- dams of a compressed profile with the assumption of cracks in concrete on the upstream face of the dam;
- dams in severe climatic conditions in the zone of variable water levels;
- during the repair and strengthening of old dams.

The screens of concrete dams can be made of different materials: polymer, reinforced concrete, asphalt concrete, metal, etc. The responsible element of the dams with the screen is the interface unit of the screen with the foundation.

A 3 mm thick stainless steel metal screen with drainage behind it was made in 1964 at the Alpe Gera dam (Italy) (see Figure 1.15) on which a new technology was applied with the laying of hard concrete with 0.8 m layers with cutting vibratory knives of deformation joints without sealing devices [159].

A metal screen made of stainless steel with a thickness of 2 mm with horizontal compensators in the form of a corrugation was proposed as a dam of a compressed profile (see Figure 4.29) [27]. Behind the screen was an elastic gasket on a bitumen basis providing minimal friction between the screen and the upstream face of the dam.

In the construction of dams from the RCC in moderate and warm climatic conditions, asphalt screens and screens in the form of membranes made of synthetic film materials were used [162].

A 6 cm thick asphalt screen is made on the upstream face of the dam from RCC Kengkou (China, 1986). For the screen arrangement, prefabricated reinforced concrete panels were installed on the upstream face and the space between them and concrete was filled with liquid asphalt. Such an asphalt screen was used on a number of RCC dams in China, including Mianhuaton (2002), Baise (2007), and Guangzhao (2008).

In the new dam design, it is proposed to use prefabricated elements of trapezoidal cross section for screen formation in which groove-cavities and drain-cavities are made along the lateral faces (see Figure 7.16) [104,226]. After the installation of prefabricated elements and concreting of the inner zone of the dam, a screen forms when filling the groove-cavities with cast asphalt concrete, and the drain-cavities serve as the drainage of the dam.

The use of synthetic materials from PVC and others for screens of concrete dams was based on the positive experience of their use as seals for deformation joints of concrete dams and antifiltration geo-membranes in dams made from local materials.

Such screens in the form of open and closed geo-membrane (with a protective element) type used in the construction of concrete dams, their reconstruction, and repairs are characterized by high water resistance for a long time, are able to withstand significant movements, and provide protection when cracks are formed on the upstream face caused by seismic and temperature effects [78,162]. Screens in the form of open-type geo-membranes on the upstream face of dams are also designed for wave, temperature, and ice effects, and they should provide movement in places of deformation joints.

Under certain conditions, the use of screens can improve the conditions for the construction of dams from RCC.

A screen in the form of a closed-type geo-membrane of prefabricated reinforced concrete slabs, 8–15 cm thick, with a polyvinyl chloride coating, 6.5–8 mm thick, on their rear face was erected on Winchester RCC dam, 23 m high, in the USA in 1985 and then on Siegrist, 40 m high, in 1993 and a number of other dams. Behind the screen, a layer of ordinary concrete about 0.5 m thick was laid in the upstream zone. The technology turned out to be somewhat busy, but it provided high water resistance to the upstream face.

The same screen made of prefabricated reinforced concrete slabs with polyvinyl chloride coating on their rear face was applied on the Capanda RCC dam (Angola, Hydroproject project, Figure 7.17), Urugua (Argentina), Cindere (Turkey), and several others.

An open screen on the upstream of the dam made of a two-layer PVC film [37] and geotextile drainage was used on one of the highest RCC dams on the Miel-1 dam in tropical climates (see Figure 7.7), as well as on a number of other RCC dams, including Xibin (China), Balumbano (Indonesia), Mujib (Jordan, Figure 7.18), Olivenhein (Figure 7.19), and Porce-2 (Colombia).

The arrangement of screens makes it possible to sharply reduce filtration through the upstream face of the dam. So, on the Miel-1 dam, in 2002, it amounted to 2.5 L/s and on the Balumbano dam to about 1 L/s.

In total, by 2007, screens of various types using geo-membranes were used on 30 operated and under construction gravity dams made of RCC [161,162], but in

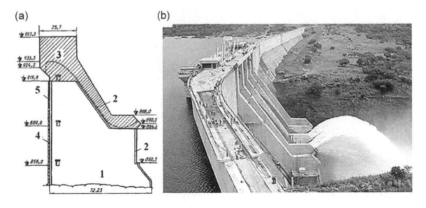

Figure 7.17 Capanda Dam: (a) spillway dam profile: 1 – RCC, 2 – ordinary concrete, 3 – ordinary concrete of piers and spillway, and 4 and 5 – reinforced concrete panels with PVC film; (b) view from downstream and right bank.

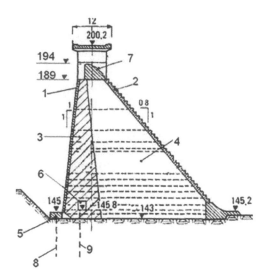

Figure 7.18 Mujib dam profile: I – upstream face with a geo-membrane and a layer of ordinary concrete, 2 – downstream face from ordinary concrete, 3 – RCC with mortar laying between layers, 4 – RCC, 5 – concrete apron, 6 – gallery, 7 – vertex of a spillway from ordinary concrete, 8 and 9 – cementation and drainage curtains.

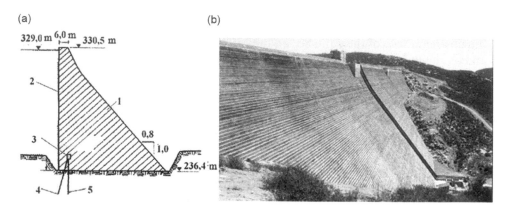

Figure 7.19 Olivenhain Dam (USA): (a) profile: I – RCC, 2 – screen in the form of an external geo-membrane, 3 – gallery, and 4 and 5 – cement and drainage curtains; (b) view from the downstream.

subsequent years, they were used extremely rarely. Open-type screens are widely used in the repair of old concrete dams, for example, Silvretta (Austria) and Toules (Switzerland).

In severe climatic conditions, sections of the concrete surface in the zone of variable water level are subjected to the most intense destruction, which is associated with complex temperature-humidity processes under conditions of alternating freezing and thawing of concrete. To protect the concrete in these areas, the feasibility of using thermal insulation coatings from asphalt materials, asphalt-polymer concrete, etc. should be considered.

In addition to antifiltration screens, special screens can be considered that localize and reduce seismic effects in the form of waterproof containers filled with air and mounted on the crest of the dam and its upstream face or suspended on floating pontoons (Figure 7.20). A variant of a seismic insulating screen in the form of air-holding containers suspended on pontoons (Figure 7.20c) [88,182] was developed in relation to the arch dam of the Chirkey HEP.

In another seismic insulating system, a device is provided in the reservoir from the side of the upstream face of the dam of the air or air-bubble curtain (Figure 7.20b). A full-scale experiment to test such a curtain was carried out in 1991 at the gravity dam of the Krivoporozhsk HEP in Russia [182].

In accordance with the proposal of [88,233], continuous screens covering the entire upstream face of the dam can combine the functions of protective antifiltration and seismic insulating screens. The waterproof screen connects the dam with flexible connections, and a cavity is formed, filled with removable closed elastic elements divided into compartments in height (Figure 7.20g). Under operating conditions, the height of air pressure in the compartments increases to the dam foundation corresponding to the hydrostatic external pressure, due to which the screen forces are reduced when the external water pressure is transferred to the elements. For the possibility of repair, the screen is designed for one-sided external water pressure within the replaced element. During repairs after lowering the element, each compartment is filled with compressed air using a compressor. Vertical elements can be opened from above and filled with water of the upstream. In this case, the screen is practically not loaded, which allows us to simplify its design (Figure 7.20e).

Due to the damping effect, screens with cavities filled with compressed air or water can not only reduce the wave and hydrodynamic effects on the dam, but also ensure its waterproofness.

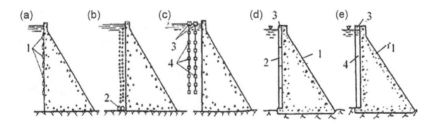

Figure 7.20 Variants for reducing hydrodynamic pressure on the upstream face of the dam: (a) a screen in the form of containers with air; (b) air bubble veil; (c) containers with air suspension worn on pontoons; I – containers with air mounted on the upstream face, 2 – perforated pipes for compressed air spirit, 3 – pontoons, 4 – tanks with air on pontoons, (d) screens divided into compartments and filled with air; (e) screens open on top and filled with water: I – dam, 2 – screen, and 3 and 4 – removable elements divided into compartments and open on top.

The use of such solutions requires special justification.

Down face. In gravity dams, the downstream face is usually formed using formwork. In dams made of RCC, its implementation may have certain features.

On the Olivettes dam (France), a mechanical vibro-compactor was used to form the downstream face, which made it possible to form and compact slopes of RCC with a laying of 0.75 without the use of formwork.

In dams made of RCC, it is widespread that the downstream face is stepped of rolled or ordinary concrete using inventory climbing formwork, both within deaf and spillway dams with the formation of step weirs [104,162]. At the same time, its formation is simplified, manufacturability is increased, and a favorable aesthetic perception, accessibility for examinations, the possibility of creating step-type spillways, and safe flood overflows through dams during the construction period are achieved. On a number of dams, the downstream face is formed from prefabricated concrete blocks, for example, in China, Mianhuatan, Shibansshui, and Tongjiezi, in Greece, Ano Mera and Steno, and in the USA, Nort Fork.

In dams erected in severe climatic conditions (when the amplitude of the monthly average fluctuations in the outdoor temperature is more than 17°C and the average long-term temperature is below 0°C), the temperature effects lead to the disclosure of inter-block construction joints on the downstream face of the dam in winter and as a result to the disclosure of joints from the side of the upstream face. In particularly harsh climatic conditions (at an average annual air temperature below −3°C with an amplitude of average monthly fluctuations of more than 18°C), the depth of opening of joints (cracking) on the downstream face can reach 5 m, and therefore the calculated section of the dam profile is significantly reduced and the concrete of this zone in practice plays the role of a heat-protective layer and gravitational loading. This can lead to an increase in destructive processes in concrete in the area of the upstream face due to increased filtration along the opened joints, as well as in concrete of the downstream face due to the effect of freezing and thawing, which in turn reduces the durability of concrete in the outer zones, as shown by field observations of dam conditions [68].

The calculations of such dams on the loads and the effects of the operational period should be carried out taking into account the possible disclosure of construction joints at the downstream and upstream faces and accordingly a decrease in the actual width of the calculated horizontal sections in the body of the dam and its bottom. To correct the situation, it will be necessary to expand the dam profile with an increase in the slope of the downstream face.

At the Boguchan HEP, within the sections of the concrete dam No. 20–22, where a step spillway is made, seam incisions are arranged at the marks of the steps, the opening of which prevents uncontrolled formation of cracks on the downstream face [44].

To reduce the negative impact of seasonal fluctuations in air temperatures in gravity dams erected in harsh and especially harsh climatic conditions, a device is used for constant thermal insulation of the downstream face in the form of

- a special heat-shielding wall;
- cavities with heated air;
- heat protection zone on the downstream face.

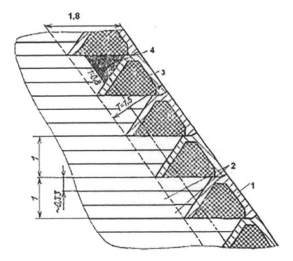

Figure 7.21 Dam with prefabricated elements from the downstream face: I – prefabricated elements, 2 – inner zone made of RCC, 3 – heat-insulating material, and 4 – additional warming.

Heat shielding devices on the downstream face of dams were rarely used; heat shielding walls are known on several small dams (arch and buttress) in Norway.

For thermal protection of the downstream face of gravity dams and regulation of the thermal stress state porous concrete, foams, foam epoxy compounds, and soil materials can be used.

To create a heat-shielding zone, precast reinforced concrete blocks of a triangular profile with an internal cavity filled with heat-insulating materials, for example, cellular concrete, polystyrene, etc. can be used (Figure 7.21).

In this case, the amplitude of seasonal fluctuations in the temperature of concrete in the inner zone can be significantly reduced to 10%–15% of the amplitude of fluctuations in the outside air [39,75]. The heat-shielding zone can also be formed by arranging parallel to the downstream face of the dam of the inclined wall from prefabricated elements with filling the formed cavity with soil.

To simplify the design, the cavity is made of prefabricated cellular blocks in which flat longitudinal plates made with an inclination in opposite directions on the one hand are formwork for RCC and on the other hand form the downstream face of the dam. Such blocks can be assembled from flat slabs and have a minimum volume of concrete (Figure 7.22).

During the construction of the dam, the prefabricated blocks of each tier are mounted on top of each other after which the internal cavities are filled with soil (rock mass), and then the layers of RCC of the inner zone are laid, and the work on laying the rock mass and RCC is similar; it is allowed to use the same mechanisms in a single process stream. The rock mass having high thermal insulation properties (with a thermal conductivity coefficient 5–10 times lower than that of concrete) with a zone width of 3 m provides reliable protection of concrete in the inner zone. The performed computational studies of the temperature regime and SSS of a conventional design dam and

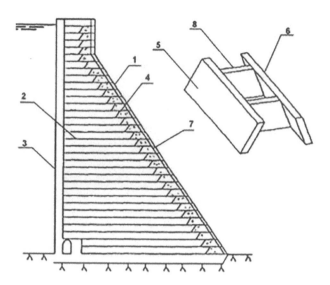

Figure 7.22 Dam with prefabricated elements with soil filling from the downstream face: 1 – downstream zone, 2 – inner zone of RCC, 3 – upstream zone of vibrated concrete, 4 – soil backfill, 5 and 6 – longitudinal plates of the precast element 7, 8 – piece assembly 7.

dam in Figure 7.23 from force and temperature effects in severe climatic conditions showed that the following are achieved [88,233]:

*   thermal protection of the inner zone with a sharp decrease in the amplitude of seasonal fluctuations in the temperature of concrete (the minimum temperature in concrete from the downstream face was 0°C);
*   compressive stresses of 0.5–2 MPa on the downstream face in the winter period in the absence of tensile stresses (in the dam of traditional design, it reached +3 MPa), as well as a decrease in the opening of inter-block joints on the upstream face and improvement of the contact area SSS.

With this design, it is possible to reduce the volume of concrete by compressing the dam profile.

It is interesting to note that the naturally occurring ice cover on the spill faces of the Ust-Ilim and Bratsk dams due to water leaks from the gates, acting as thermal insulation, has made it possible to significantly reduce the temperature component of the stress and strain of the spill sections. Such ice cover under certain conditions could be created artificially [68].

7.3.6 Reinforcement of dams

The approach to reinforcing gravity dams differs in different countries and depends on the experience of design and operation, existing traditions, climatic conditions, seismicity, and design features [75,82,175,195].

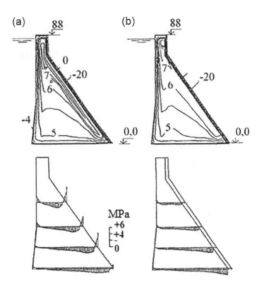

Figure 7.23 Distribution of temperature (degrees) and vertical stresses σ_Y (MPa) in the dam during the winter operating period (January 15): (a) in the dam of usual design; (b) in a dam with soil backfill; "-" compression.

Structural reinforcement (without calculation) is usually installed around galleries and mines, in areas of overflows or within toe-spring board, around deep spillways, gutters, and in severe climatic conditions on the upstream face in the zone of variable water levels as well as on the downstream face of the dam for cracking limitations.

Constructive reinforcement of the upstream face in the zone of variable water levels mainly made in the form of grids from reinforcement of class AII with a diameter of 16–25 mm (up to 4 rods per 1 m). Structural reinforcement can slightly reduce surface cracking in concrete caused by temperature fluctuations and shrinkage, improving the conditions for repair work. Reinforcement is carried out separately within each section, separated by a deformation joint.

Design reinforcements are installed on the basis of SSS calculations of dam sections around large openings, galleries, cavities, deep spillways, and pressure pipelines, as well as in water inlets, overflow goby, etc.

Under conditions of high seismicity, reinforcement is performed in areas of high tensile stresses and changes in the dam profile.

On the Jinanqiao dam, on the faces where tensile stresses during seismic impacts reached +3 MPa, two rows of reinforcement with a diameter of 28 mm with a pitch of 20 cm were installed [207].

The consumption of reinforcement in gravity massive dams can be from 0 to 14 kg/m^3. At the Kurpsay dam, the valve consumption amounted to 4.8 kg/m^3 and in the severe climate of Siberia, (excluding turbine pipelines) at the Krasnoyarsk dam, it amounted to 13.2 kg/m^3 and at the Mamakan dam to 11.1 kg/m^3 [80].

The amount of reinforcement in concrete dams should be minimal so as not to complicate the construction conditions. The installation of reinforcement should be provided in the form of reinforcing nets, reinforcing frames, or including it in prefabricated reinforced concrete elements, for example, when installing galleries.

7.3.7 Spillway and station dams

The installation of spillway water inlets and water pipes of HEP in gravity dams is their most important advantage over dams made of soil materials, but it complicates the design and construction conditions. Therefore, when choosing structural-technological solutions for spillway and station gravity dams, and especially dams made of RCC, an important task is to ensure favorable construction conditions while minimizing their negative impact on the technology.

Spillways. Operational spillways are intended for flood discharge, partial preflood drawdown of the reservoir for cutting the flood peak, deep emptying of the reservoir during repair work, quick drawdown of the reservoir in case of deviations from normal operation of the dam, and foundation and the avoidance of emergency situations. They are intended for water supply, irrigation, shipping, environmental releases, ice and sediment, and regulation of the filling of the reservoir.

Spillways should provide a flood discharge equal to the maximum estimated discharge without significant damage to the culvert. The erosion of bedrock and the reformation and deposition of erosion products in the downstream should not affect the reliability of the HP and the conditions for their normal operation and should not lead to a decrease in the head of the HEP. When passing test discharge, erosions are allowed in the downstream not threatening the reliability of the main structures.

Spillways in the body of gravity dams are of two types: surface weirs and deep spillways [63,65,134,180].

Surface weirs have been the most widespread and constructed on 90% of HP with high gravity dams [130] due to the simplicity of design and operating conditions. They are used mainly as nonvacuum, practical profile with the installation of gates on the crest, and without gates, automatic actions, unregulated with the passage through the weir discharges when the water level in the reservoir rises above the mark of the weir crest. Vacuum spillways are extremely rare. Automatic spillways require an increase in the spillway front, an increase in the level above the NHL and accordingly additional flooding in the upstream when the flow passes through the weir. They are used mainly in HP at relatively low flood discharges (up to 4 thousand m^3/s). On large HPs, weirs are usually carried out with the installation of flat or segment gates, which allows reducing the spillway front, to ensure that the flood discharges are passed, mainly at NHL, except for the passage of extreme discharges at SRL.

The formation of the vertex is carried out on the basis of the triangular profile of the dam, taking into account the shape of the vertex and the placement of gates, piers, bridge crossings. In this case, the vertex can be partially carried out to the upstream in the form of a console beyond the limits of the calculated dam profile.

The weir face depending on the conjugation scheme of the upstream and downstream ends with a toe-springboard to discard the stream or mates with a stilling pool or water well. The weir on both sides must be surrounded by side walls, which should be of sufficient height to ensure that the designed flow discharge can pass without water spreading along the downstream face. When designing them negative pressures from the water, flow should be taken into account [63].

The dimensions of the bulls of overflow dams depend on the type and design of the gates and their lifting mechanisms, the dimensions of the overflow openings, the placement of deformation joints, operational and emergency exits from the galleries, and the parameters of bridge crossings. The thickness of the bulls is mainly 2–3.5 m with a thickness of the groove isthmus in all cases not less than 0.8 m.

The outlines of the bulls in terms of the upstream and downstream should provide favorable hydraulic conditions with a smooth supply of water, minimum flow compression, so that the flow rate of the spillway is as large as possible, and in severe climatic conditions, also provide favorable conditions for the passage of ice.

In dams of RCC, the spillway face is made with a smooth surface of ordinary concrete, for example, Beni Haroun dam (Figure 7.24a), Tamagawa (Figure 7.24b), Jiangya (Figure 7.24c); and stepped from ordinary or RCC Capanda dam (see Figure 7.17). In such dams, with the installation of gates, the dam vertex and spillway face are made of ordinary concrete.

Figure 7.25 shows a longitudinal profile of a spillway dam Nam Theun-2 made of RCC with a height of 48.2 m under construction in Laos.

Previously on many, mainly low and medium-high, dams of RCC at low specific discharges (usually up to 25 m²/s) and recently on high dams with large specific discharges, spillway face is performed stepwise at a step height of the order of 0.8–1.5 m which allows [44,195,246] the following:

- avoid cavitation, exclude cavitation-resistant concrete, and simplify the construction technology;

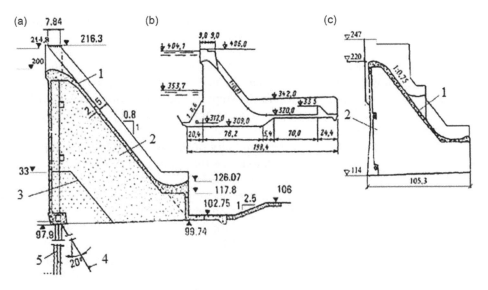

Figure 7.24 Weir dams: (a) Beni Haroun, (b) Tamagawa, and (c) Jiangya; 1 – ordinary concrete spillway surface, 2 – RCC, 3 – cofferdam circuit, 4 – drainage curtain, and 5 – cement curtain.

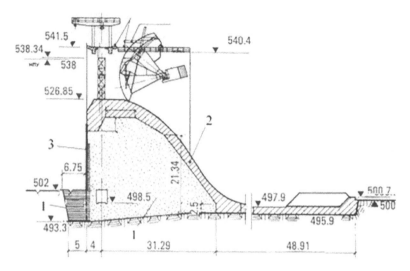

Figure 7.25 Section of the Nam Theun-2 dam: 1 – RCC, 2 – ordinary concrete, and 3 – geomembrane.

- dissipate up to 70% of the kinetic energy of the flow on the spillway face, thereby improving the conditions of energy quenching and reducing the size of fastenings and erosion in the downstream;
- ensure saturation of the discharged stream with oxygen, thereby increasing the self-cleaning ability of water and improving environmental conditions in the downstream;
- safe flood overflow over the dam during the construction period (for example, the 155 m high Ralko dam in Chile).

The process of quenching energy on a step spillway occurs during the interaction of the mainstream with the whirlpool areas formed in the sinuses of the steps.

The device of step spillways where the steps are flow aerators is successfully combined with RCC technology, which is characterized by a step configuration of the downstream face of the dam.

In dams made of RCC, given the simplicity of the device and the reliability of the stepped spillway, it is advisable to maximize the spillway front, based on the topographic conditions of the site and layout of the HP.

At a flow velocity of more than 12 m/s, the stepped spillway face is usually made of ordinary concrete to prevent cavitation erosion, for example, dams: Upper Stillwater with $h = 91$ m and Olivenhein with $h = 97$ m (see Figure 7.19) in the USA, Joumoua with $h = 57$ m and Sidi Said with $h = 120$ m in Morocco, and Jucazinho with $h = 63$ m in Brazil, and a number of dams from RCC with a high binder content, for example, Kinta with $h = 90$ m (with a binder content of 200 kg/m^3) in Malaysia, Ghatghar with $h = 86$ m in India, La Brena II with $h = 120$ m in Spain, and the Gina dam under construction, 135 m high, in Turkey – using precast concrete panels in cells of 1.2 m.

On a Toker dam in Eritrea from RCC (1999), 73 m high, a spillway dam without gates erected with a stepped spillway face from ordinary concrete is designed to pass a maximum flood discharge of 1.61 thousand m^3/s at a specific flow rate of 47 m^2/s.

At the Boguchan HEP, in severe climatic conditions (at an average air temperature from −2.6°C to −4.3°C), an additional spillway No. 2 was made with 5 spans on 10 m width and with a stepped spill face, which helped to improve the working conditions of the water well, which made it possible to reduce its size. A stepped spillway consists of a spillway vertex, a transitional section with steps of 0.5 m, and a spillway with steps of 1.5 m (Figure 7.26) and is designed to allow a flood discharge of 2,740 m^3/s at a NHL of 208 m and 3,540 m^3/s at SRL 209.5 m. The highest bottom velocities on the step spillway tract do not exceed 12 m/s [44,46].

Deep spillways are widely used in gravity dams along with surface spillways. On many spillway dams, deep spillways are combined with weirs, for example, Libby dam (Figure 7.27) in the USA.

Deep spillways during operation in addition to flood discharge are also intended for preflood drawdown and partial or complete emptying of reservoirs in the event of an emergency or during repair work. Thanks to the presence of a deep spillway, it was possible to carry out the necessary work to strengthen the high arch dam Kelnbrein in Austria.

Deep spillways are made permanent and temporary (construction) of a rectangular or circular section; their width is not more than 0.6 ℓ, where ℓ is the width of the dam section. At high heads (more than 50 m) and speeds of more than 20–25 m/s, a metal cladding is arranged. Deep spillways can be head, headless, and partially headless, passing on the downstream face to the weir. In some cases, taking into account the stages of construction, they can be placed in 2–3 tiers along the height of the dam.

In deep spillways, gates can be located on the upstream face at the beginning of the spillway, in the body of the dam, or on the side of the downstream face at the end of the spillway. The vertex deep spillways can be equipped with trash racks. In the dams, special aeration pipes are provided that supply air to the cavity behind the gate to prevent the formation of vacuum.

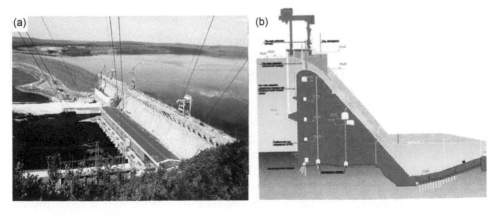

Figure 7.26 **Boguchan HEP: (a) panorama of the HP and (b) section with stepped spillway dam.**

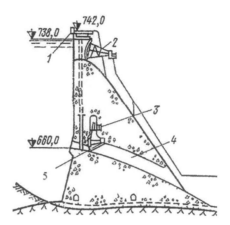

Figure 7.27 Cross section of the spillway Libby dam: 1 – bridge, 2 – segment surface gate of weir, 3 – gate camera, 4 – deep spillway, and 5 – segment gate of deep spillway.

In spillway dams made of RCC, in order to minimize the negative impact on concrete conditions, deep and bottom spillways are located outside the dam contour from the upstream side adjoining the upstream face, and operational gates can also be placed on the toe-spring board from the downstream face. Their water conduits were erected in ordinary concrete in the foundation zone of the dam and considered combining them in a single block of ordinary concrete.

The operating conditions of the HP are significantly influenced by the schemes for coupling the upstream and downstream and quenching of the kinetic energy of the stream passed through spillways.

Coupling of downstream flows when passing through spillways depending on the width of the site, the size of the spillway dam and its height, designed flow discharge, discharged flow energy, engineering-geological and topographic conditions, and level regime in the downstream can be performed using the following basic schemes [61,63,65,180,185]:

- with a bottom hydraulic jump (bottom regime);
- with a surface hydraulic jump (surface regime);
- with the rejection of the jet toe-springboard (usually at an angle of 30°);
- according to the combined scheme.

The selection of the scheme for coupling the downstream flows and the specific costs of the discharged flow is based on a technical-economic comparison of the variants. At the same time, the selected scheme and devices in the downstream should ensure reliable operation of the HP while minimizing costs, protecting structures and their adjacent banks from dangerous washing out, and creating favorable conditions for the operation of the HEP (if it is located in close proximity to spillways). The justification of the hydraulic conditions, the parameters of the fixtures and the washout funnel, the maneuvering patterns of the gates, and the opening of spillway spans is carried out on the basis calculated and experimental studies on hydraulic models.

The bottom regime with a stable location of the jet at the bottom and damping of energy in the well directly adjacent to the dam is usually used in relatively wide sites with spillways of low-, medium-, and high-head HP.

In the bottom regime, the coupling between the spillway of the dam and the water well (water apron) should be provided smooth or with a very small ledge. A water well should provide effective energy quenching, minimal wave formation, and relatively even flow in the downstream of the well.

When quenching energy with a bottom regime, intensive quenching of excess energy is ensured at the site of the water apron and favorable operating conditions for the dam are created with minor erosion of the bedrock behind the concrete fastening.

The disadvantages of this regime include significant bottom velocities, slowly damping along the length of the quenching section, the risk of damage and destruction of the elements of the fastenings of the water well with ice and stones, large volumes of concrete, and accordingly the cost of fixing devices.

The quenching of energy depending on the depth of water in the downstream, specific discharges, and dam height can be carried out in a water well with concrete fastening. Large HPs with spillways located in the bed part of the sites and conjugation of downstream waters through the bottom regime include the following:

- Bhakra in India (see Figures 1.20 and 2.17) with a maximum flow discharge, $Q = 11.2$ thousand m^3/s, and a specific flow discharge in a water well, $q = 141\,m^2/s$;
- Sakuma in Japan, $h = 155\,m$, $Q = 10$ thousand m^3/s, and $q = 132\,m^2/s$;
- Warragamba in Australia, $h = 137\,m$, $Q = 12.7$ thousand m^3/s, and $q = 139\,m^2/s$;
- Sayano-Shushensk in Russia (see Figures 1.28 and 2.6).

With a good quality of the rock in the downstream and relatively small heads (up to 40 m), concrete fasteners may be short or not provided. Behind a spillway dam, 441 m long, of the Djerdap – Iron Gate dam (Romania-Serbia) HP designed for a flow pass of 15.4 thousand m^3/s with a 30-meter drop difference with relatively low rock quality, a small length was fastened (Figure 7.28). An ice passage is also provided through the spillway. Analysis of the spillway after 7 years of operation with a flow rate of 12 thousand m^3/s showed that the erosion has stabilized and is not dangerous for the structure [130].

Under the bottom regime, a number of HPs experienced serious damage and destruction of water wells, separate walls, and piers, which were caused by hydrodynamic, cavitation, and abrasive effects, which required the implementation of large amounts of repair work, for example, at the Sayano-Shushensk HEP in Russia. The spillway of the arch-gravity dam with a height of 242 m consists of deep culverts and open trays on the downstream face with the quenching of energy in a water well. The designed flood discharge through the dam was 13 thousand m^3/s with a maximum specific discharge of $180\,m^2/s$. Serious damage to the fastening of a water well when passing relatively small flood discharges of about 4.5 thousand m^3/s endangered the possibility of its further operation [132,140] and led to the need to build an additional

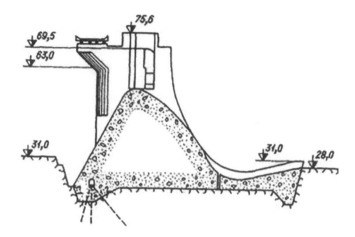

Figure 7.28 Spillway section Djerdap – Iron Gate dam.

bank spillway designed for a flow discharge of 4 thousand m³/s, which allowed relatively small floods to pass only through the dam and HEP [86].

At a number of HPs in the USA, a combined scheme for coupling the downstream is adopted, which provides for often repeated low flow discharges that the water well provides a bottom hydraulic regime, and at high flow discharges, the flow is discarded by a ledge at the end of the spillway with the formation of a surface regime. Such a scheme was used, for example, at Dworshak HEP with a gravity dam, $h = 203$ m, with deep spillways, $Q = 5.4$ thousand m³/s and $q = 155$ thousand m²/s, as well as Libby dam with $h = 136$ m and $Q = 5.9$ m³/s in the USA [180]. With such a scheme erosion, and destruction in the downstream area is possible when the maximum discharge is passed, and it may be necessary to carry out repair work. At the Libby HP (see Figure 7.27) with a flow pass of 1.3 thousand m³/s, which is much smaller than the designed one, concrete fastening slabs and a water-resistant wall, 3.7 m high, were destroyed. Similar damage occurred at the Dworshak HP [132].

For the Koteshvar HP in India with a 97.5 m high gravity dam and dam HEP and a surface spillway in the bed part, a variant with a combined circuit with a ledge at the end of the spillway and a water well was recommended as a result of experimental hydraulic studies. Verification discharge flow at SRL equal to 13.2 thousand m³/s was passed through four spans 18 m wide each [140].

The circuit of the coupling downstream with the surface regime for medium and high-head HP is rarely used.

In the surface regime, a sock-ledge with a horizontal or inclined surface should be provided at the end of the spillway. The surface regime with the location of the jet on the surface of the stream or in close proximity to it makes it easier to mount the coupling and create favorable conditions for the passing ice; however, it has significant disadvantages:

- requires the creation of great depths in the downstream;
- unstable coupling regime and significant fluctuations in water level;

- the formation of high waves in the downstream, leading to intensive processing of the banks;
- weak attenuation of the pulsations of speed and pressure as well as the energy of the waste stream.

Such a scheme was used at the Grand Coulee HP, $Q=35.8$ thousand m^3/s and $q=90$ m^2/s, with a device at the end of the spillway of the ledge, with the help of which a surface jump was formed, as well as the Hells Canyon in the USA with a gravity dam $h=100.6$ m, $Q=8.5$ thousand m^3/s, and $q=174$ m^2/s [195].

The scheme of coupling the downstream with the throw-off spring of the jet, the energy of which is extinguished in the downstream water cushion or in the washout funnel, is most widely used on medium- and high-head HP with fairly strong rocks in the bed and on the banks, and there are no fears that erosions in the bed under the influence of a discarded stream will affect the stability of the dam and bank slopes in the downstream or worsen the conditions of the HEP [132,139,195]. Such a scheme provides a stable regime for coupling the downstream, and therefore, it was used by more than 50% of HP with high dams [180].

The scheme with the rejection of the jet into an unsecured or reinforced channel is especially effective in relatively narrow sites with a significant output flow. With this scheme, in many cases, the device of fasteners in the downstream is not required, which allows reducing the amount of work and cost under appropriate conditions.

The disadvantages of this scheme include the following:

- the disordered nature of the flow in the quenching section during the initial period of operation before the formation of the erosion pit (funnel);
- washing of the banks is possible;
- behind the erosion pit, products of destruction of bedrock in the form of a "bar" can be deposited, causing backwater, which negatively affects the operation of the HEP;
- uneven distribution of specific discharges across the width of the channel;
- the occurrence of water dust, which complicates the operation of HEP equipment and leads to watering of slopes in the downstream and their creeping.

To remove the washout pit from the structures, the mark of the toe-springboard is lowered to increase the speed and range of the jet throw. In this case, the edge of the toe-springboard should be performed above the water level in the downstream. To improve the conditions of energy quenching and reduce erosion in the downstream at the end section of the spillway, it is advisable to provide splitting teeth, multitiered springboards, etc., which increase the aeration of the stream, effectively distribute it over the area and length of the bed, and drop it into the deepest part. The distance from the dam to the jet fall in the downstream can be about 0.6 of the head and in some cases, even more. The need for concrete fastening and its parameters is determined depending on the hydraulic conditions, the quality of the rock foundation, the parameters of the erosion funnel, and the stability conditions of the dam.

Depending on the conditions of energy quenching in case of strong hard-eroded rocks, the jet is usually thrown out onto the natural surface of the bed, and for less durable medium-eroded rocks, it is injected into a previously prepared artificial excavation of a rock in the zone of the jet fall, where in some cases a water well is arranged.

To ensure normal conditions for the operation of a HEP, a separate wall is constructed between the outlet bed of the spillway and the outlet bed of the HEP, or a rock mass separating the bed is left; often between a spillway and a station dam, a deaf dam is being constructed (for example Krasnoyarsk HEP, Son La HEP).

The jet rejection scheme was used at many HPs with gravity dams:

*   Tucuri in Brazil, $h = 108$ m, $Q = 100$ thousand m^3/s, and $q = 200$ m^2/s (see Figure 2.1);
*   Pine Flet in the USA, $h = 34$ m, $Q = 11.2$ thousand m^3/s, and $q = 125$ m^2/s;
*   Srisalam in India, $h = 145$ m, $Q = 33$ thousand m^3/s, and $q = 132$ m^2/s;
*   Xinanjiang in China, $h = 105$ m, $Q = 14$ thousand m^3/s, and $q = 78$ m^2/s [180];
*   Krasnoyarsk (see Figures 1.21 and 2.4);
*   Bratsk (see Figures 1.22 and 7.5) in Russia;

and also on most recently built and under-construction high-head HPs:

*   Khlong Ta Dan in Thailand, Beni Haroun in Algeria (Figure 7.24a);
*   Miel-1 in Colombia (see Figures 2.20 and 7.7);
*   Capanda in Angola (see Figure 7.17);
*   Miyegase in Japan (see Figure 1.24);
*   Jinanqiao in China;
*   Bureya in Russia (see Figures 2.7 and 7.29);
*   Son La in Vietnam (see Figure 2.10);
*   Longtan (see Figure 2.21);
*   Three Gorges in China (see Figure 1.45);
*   Gibe-III in Ethiopia.

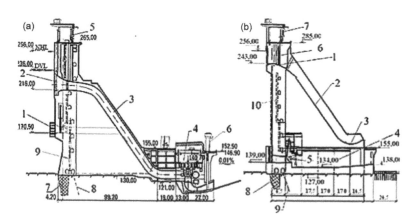

Figure 7.29 Bureya HP: (a) station dam: 1 – time water intake, 2 – permanent EP inlet, 3 – head water with a diameter of 8.5 m, 4 – building of HEP, 5 – crane water intake, 6 – portal crane, 7 and 8 – cementing and drainage curtains, 9 – drainage dam bodies; (b) spillway dam: 1 – vertex of the spillway, 2 – spillway surface, 3 – toe springboard, 4 – bottom spillway, 5 – segment gate, 6 – flat gate, 7 – crane, 8 and 9 – cement and drainage curtains, 10 – drainage of the dam body.

At the Hyong Dieng HP in Vietnam with a dam, $h = 82.5\,\text{m}$, $Q = 7.68$ thousand m^3/s, and $q = 110\,\text{m}^2/\text{s}$ (see Figure 2.19), the bed in the downstream of alluvial deposits was cleared for the downstream dam to prevent their removal into the outlet bed of the HEP. According to hydraulic studies, the maximum depth of the erosion funnel in rock soils can be 15 m.

The dam of Three Gorges HP (China) with a height of 175 m was built in 2010 as part of a HEP with a capacity of 22.5 million kW (see Figure 1.45). The dam is divided into spillway sections 21 m long with a spring-toe for rejection of the jet, station section 24.5 m long, and 13.2 m deaf section (Figure 7.30) [74,184].

Model studies of the hydraulics of the Bureya HP during the work of the operational spillway with the toe-springboard flow rejection with the gates fully open showed [40,219] the following:

- with the recommended rules for the passage of discharges with the opening of three to four spans and a flow discharge of up to 6.5 thousand m^3/s, the minimum impact of the flow on the bed and the banks is achieved due to its intensive expansion on the springboard and beyond;

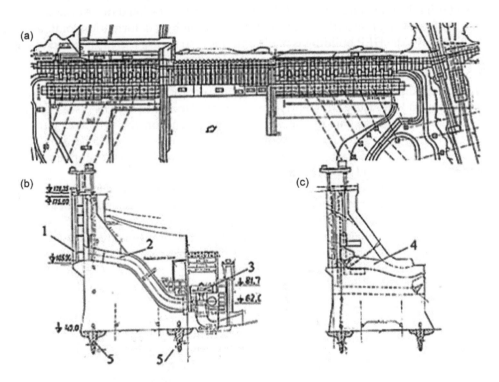

Figure 7.30 Three gorges dam: (a) plan; (b) the station dam section; and (c) the spillway dam section: 1 – water intake, 2 – turbine water conduit, 3 – building HEP, 4 – deep spillway, and 5 – cement curtain.

• if a flow discharge of 1% probability is equal to 11.7 thousand m³/s through all eight spans, the depth of the erosion pit can reach 45 m at a distance of 180–200 m from the dam with the formation of a bar behind it with a crest mark of 144 m, which may lead to backwater HEP; at a flow velocity near the right bank of 10 m/s and a wave height of up to 7 m, a combined fastening with a concrete facing and an oversized stone is required.

Serious damage took place on the Keban gravity dam, 210 m high, in Turkey with a jet-to-springboard rejection at flow discharges of 0.5–1.3 thousand m³/s: the mounting plates behind the spring-toe were damaged, the depth of the erosion funnel reached 18 m [132], and the maximum estimated flow discharge was 17 thousand m³/s.

Cases of damage to the downstream mounts behind the springboard socks at low discharges when the jet is not rejected and the flow immediately after the springboard falls on the concrete mount should be noted.

In the scheme of coupling the downstream with the waste stream, it is necessary to determine the shape and size of the erosion funnel and its effect on the stability of the dam and bank slopes, as well as the location and size of the deposits of the destruction of the rock foundation (bar). The location of the HEP should be chosen so that the products of the destruction do not impede the outlet of water from the HEP and do not lead to an increase in the water level in the outlet channel.

Water intakes and turbine conduits of *HEP*. When assembling a HP with a dam HEP, a water intake and pipelines of HEP are located in the station dam. At the same time, the HEP building is usually separated from the dam by a deformation joint, although in some HPs the HEP building is made as a single structure with a station dam.

The water intakes of the dam HEP are buried under the DVL and are carried out in the body of the gravity dam or outside it from the upstream side, which provides more favorable conditions for the construction of dams from RCC. HEPs of this type were made at the Yeywa (see Figure 2.18) in Myanmar and the Hoover in the USA (see Figure 1.11).

The overall dimensions of the water intake depend on the type and parameters of the trash racks, gates (emergency-repair, repair), and lifting mechanisms and the dimensions of its flow part, which provides favorable conditions for supplying water to the HEP conduits and minimal head losses.

Behind the gates, aeration pipes are arranged that provide air supply when emptying the water pipes or emergency closure of the gates so that vacuum is not allowed, as well as air removal when filling the water pipes.

Turbine head water conduits are made with metal cladding and are placed in the body of the gravity dam or outside it mainly on the downstream face in the form of open steel-reinforced concrete or metal water conduits.

Turbine water conduits both in the body of the dam and on the downstream face have a significant impact on its SSS. Large cross sections of water in the dam body weaken the dam profile especially at the downstream face where maximum compressive stresses act and seasonal temperature stresses manifest themselves to the maximum extent in severe climatic conditions. Water conduits brought to the downstream face increase the stiffness of the sections and the dam profile as a whole and reduce the seasonal temperature stresses on the downstream face, acting as thermal insulation [43,75].

With remote water conduits, favorable erection conditions for dams made of RCC are provided.

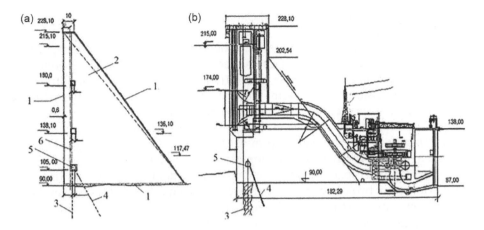

Figure 7.31 Son La dam. Cross section of the deaf (a) and station (b) dams: 1 – enriched concrete (GEVR), 2 – RCC, 3 and 4 – cement and drainage curtains, 5 – gallery, and 6 – dam body drainage.

Figure 7.32 The Lai Chau HP. View from the downstream.

Open steel-reinforced concrete conduits on the downstream face of the dams were made at many dams, for example, at the Krasnoyarsk (see Figures 1.21 and 2.4) and Sayano-Shushensk HEP (see Figures 1.28 and 2.6), Son La HEP (Figure 7.31, see Figure 2.10), and Lai Chau in Vietnam (Figure 7.32).

7.4 Dams in narrow sites

In narrow sites, to increase the stability of the gravity dam, it is advisable to use the spatial nature of the work [89], which under certain conditions can be achieved by

• making transverse joints with grooved, articulated, or monolithic with the formation of a jointless dam;
• special cutting with transverse joints fan-shaped in plan, for example, at the Toktogul dam (see Figure 2.23);

- supplying the water of the upstream into a tight transverse joint in the central part of the dam with compression of its lateral parts by hydrostatic pressure.

The grooved joints allow for the combined operation of all sections of the dam under static and seismic effects. Given that the sections of the dam have different heights, during an earthquake, they vibrate with different frequencies and amplitudes, which eliminates the possibility of resonant vibrations, achieves a favorable redistribution of seismic load between sections, and increases the seismic resistance of the dam.

Solutions with grooved, punched, or monolithic joints are implemented on a number of dams. In the dam of the Kurpsay HEP (Kyrgyzstan) with a height of 113 m and a crest length of 364 m, the construction of intersectional joints allows the sections to work together under load [95,111]. The loads on the dam in this case as in arch dams are distributed in two directions – cantilever and beam. In the gravity dam of Lucyasia in China, with 147 m high and 240 m long divided by joints into sections 16 m long, after filling the reservoir, joints with grooves were cemented (except for the upper zone 24 m high). According to the designers, the static loads – dead weight and hydrostatic pressure – should be perceived as separate sections and seismic loads – as a whole dam, working on a spatial pattern. The calculation results showed that due to the spatial effect, both the strength and stiffness of the dam increased, which made it possible to reduce the concrete volume by 18% under conditions of high seismicity [91].

Computational studies performed for the Tabelout gravity dam (Algeria) from RCC, 121 m high and 360 m long along the crest, showed that due to the curvature of the dam axis in plan and cementation of intersection joints after filling the reservoir, the seismic resistance increased due to the arch effect of the dam, which allowed reducing the volume of the concrete [261].

Fulfillment of the dam axis in the plan of curvilinear (arched outline) leads to compression of the dam body due to longitudinal compressive forces, both under statistical and dynamic effects, to increase the seismic resistance of the dam [118].

The Kowsar dam is 144 m high and was erected in Iran in a narrow gauge 10 m wide along the bed and about 50 m at a height of 75 m; above the bank slopes expand symmetrically at an angle of 45–50. Within the narrow lower part of the canyon, weathered and unloaded rock was practically absent; the dam was erected without vertical joints in it (Figure 7.33).

The upper part of the dam is cut by intersectional joints into separate sections. The construction of the dam began with a metal bridge, while construction discharges were passed along the riverbed; after completion of dam construction in the dry season, a concrete plug was arranged in the bed part, which is separated from the main dam by a joint [104,154]. The cementation of the joint was performed only in the area of 5 m from the side of the upstream face. Due to the spatial work of the dam, its profile was decreased with downstream face and made vertical within the lower narrow part of the canyon, where the dam was only 62 m wide.*

---

* The Kowsar dam project was completed at the International Institute of Geomechanic and Hydrostructures (IIGH).

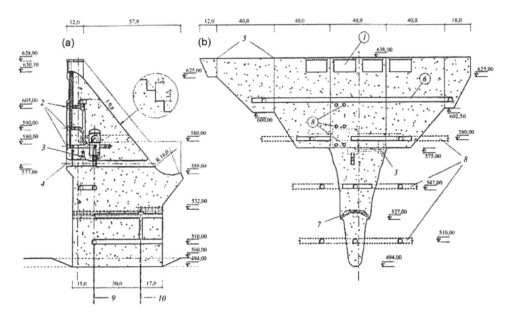

Figure 7.33 Kowsar dam: transverse (a) and longitudinal (b) sections; I – surface spillway, 2 and 3 – irrigation water fence and water outlet, 4 – deep spillway, 5 – intersectional joints, 6 – look gallery, 7 – bridge, 8 – cement galleries, 9 – axis cementing curtain, and 10 – axis drainage.

7.5 Spread dams

In 1992, P. Lond and M. Lino proposed a new dam design using particularly lean RCC (at a cement rate of $50\,\text{kg/m}^3$) with a symmetrical triangular profile and faces of 0.7 [229], calling it a "hard embankment" ("hardgile dam") with an increase in volume of almost double compared to a conventional dam (Figure 7.34). The water tightness of the dam is ensured by the device on the upstream face of the reinforced concrete screen (as in stone-filled dams), under which drainage is performed.

With a symmetrical profile, the dam resulting from hydrostatic loading and dead weight is located in the center of the dam bottom, which ensures uniform distribution of compressive stresses in contact with the foundation, and at a dam height of 100 m, the maximum compressive stresses will be about −5 MPa. In such dams, an improvement in the SSS is achieved and stresses in the body and foundation of the dam are reduced in comparison with conventional gravity dams, which allows them to be raised to a relatively weaker rock foundation and in areas with high seismicity. Under high seismic effects in the dam and foundation, no extension zones are formed.

Due to a significant reduction in stresses, RCC has more stringent requirements compared to conventional RCC, which reduces cement consumption, simplifies erection conditions, and increases the intensity of concrete laying.

The first dam, Marathia, with a symmetrical profile with laying faces of 0.5, $h = 28\,\text{m}$, $L = 265\,\text{m}$, and concrete volume of 48 thousand m^3 (including RCC 31 thousand m^3), was built in Greece in 1993. The dam is located on a granite foundation, its faces are formed by concrete blocks in a sliding formwork, a reinforced concrete screen 0.3 m

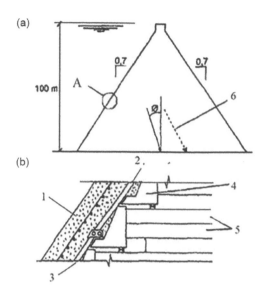

Figure 7.34 Dam of symmetric profile: (a) dam profile; (b) node A; I – reinforced concrete screen, 2 – drainage in the form of perforated PVC tubes, 3 – drainage gallery, 4 – prefabricated concrete formwork blocks, 5 – layers of RCC,
and 6 – resultant forces taking into account seismic acceleration 0.2 g.

thick is arranged on the upstream face, and a drainage in the form of 10 cm PVC pipes
leads into a drainage gallery. The screen is separated from the gallery by a contour
plate (Figure 7.35a). The binder consumption is $70 \, kg/m^3$, including cement $55 \, kg/m^3$.
A dam of a similar design, Ano Mera, with a height of 32 m and with a 0.5 laying face
was also built in Greece in 1997. A Moncion dam with a laying face of 0.7 was built in
the Dominican Republic (Figure 7.35b).

The Can Asujan dam with a symmetrical profile with a face of 0.6, $h = 42 \, m$, $L = 136 \, m$,
and concrete volume 85 thousand m^3 (including rolled 75 thousand m^3) with a cement
content of $100 \, kg/m^3$ in RCC was built in the Philippines in 2005 [161,264]. In Japan,
dams with a symmetric profile were constructed with a face of 0.8: Okukubi, $h = 39 \, m$
and $L = 461 \, m$; and the Choukai dam, $h = 81 \, m$, $L = 365 \, m$, and volume 1,690 thousand
m^3 [203].

In the design of the Papadiana dam (Greece), a symmetrical profile 108 m high, at
the foundation of which there are very weak quartzites and phyllites, the laying of
faces is accepted to be 0.9 [237,238]. The upstream face of the dam was formed using
precast concrete blocks with horizontal and vertical voids-drains, forming a drainage
system and diverting filtered water to the drainage gallery. To prevent concrete from
draining void-drains when concreting a reinforced concrete screen, prefabricated elements were protected by a film that also reduced friction under the screen, thereby
minimizing tensile forces in the screen and the risk of cracking. Seals are installed in
the perimeter joint between the screen and the contour block with the gallery.

The Oyuk dam (Turkey) with a symmetrical profile $h = 100 \, m$, $L = 212 \, m$ was built in
an area with a seismicity of nine balls in a canyon whose sides are composed of gneisses

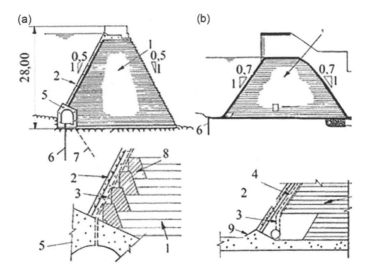

Figure 7.35 Dams with a symmetrical profile: (a) Marathia; (b) Moncion: 1 – RCC, 2 – reinforced concrete screen, 3 – drainage, 4 – porous concrete, 5 – drainage and cementation galleries, 6 and 7 – cement and drainage curtains, 8 – concrete blocks in sliding formwork, and 9 – circuit plate.

and in the upper part are weathered mica schists. Calculations of the SSS of the dam showed that under the action of seismic acceleration of 0.40 g, no tension occurs in the dam and the maximum vertical compressive stresses in the foundation from the down-stream side are −3 MPa with only hydrostatic pressure −1.6 MPa [93,189]. The dam was designed with the laying of faces of 0.7 from RCC (cement consumption of 50 kg/m³ and fly ash 100 kg/m³) with the exception of the lower 7 m adjacent to the foundation and the upper block in the bed from ordinary concrete as well as the lower block from RCC with a high binder content (Figure 7.36).

A reinforced concrete screen, made after the dam was erected in a sliding form-work with joints across 12 m, is interfaced with a contour antifiltration plate and an upstream block in the bed using a contour joint. Under the contour slab, matching grouting and cementing curtain are provided with a depth of 40 m in the bed to 70 m in the abutments. Screen thickness was determined by the formula

$$t = 0.3 + 0.00235h \qquad (7.5)$$

where h is the height (m) from the crest to the corresponding section.

The drainage behind the screen is made of perforated tubes with a diameter of 10 cm, which exit into the drainage gallery. The formation of both faces is carried out using precast concrete blocks. In blocks of the upstream face, drainage tubes are installed to divert rainwater from the surface of the layers of RCC into the subscreen drainage. On the downstream face of the dam, a spillway has been designed for a flow discharge of 530 m³/s.

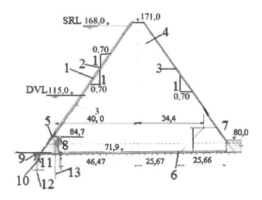

Figure 7.36 Oyuk dam profile: 1 – reinforced concrete screen, 2 – perforated drainage pipes, 3 – prefabricated blocks, 4 and 7 – RCC, 5 – upstream block, 6 – rock concrete, 8 – drainage gallery, 9 – contour plate, 10 – anchors, 11 – area cementation, 12 – cementation curtain, and 13 – drainage.

Figure 7.37 General view of the construction of the Cindere dam.

The Cindere dam of the same design, $h=107$ m, $L=280$ m, and $V=1.68$ (1.5 million m³), of concrete (cement consumption 50 kg/m³, pozzolan 20 kg/m³) of a symmetrical profile with face laying of 0.7 was built in Turkey in 2005 (Figure 7.37) [264].

The Koudiat Acerdoune high dam with a symmetrical profile with 0.65 face laying, $h=121$ m, $L=500$ m, $V=1.85$ (1.65) million m³, and a binder content of 164 kg/m³, including cement number 77 kg/m³, was built in Algeria [264].

In spread dams, the upper and middle zones are nonloaded. A more uniform stress distribution can be achieved by performing faces with a variable slope increasing at the foundation of the dam. This reduces the volume of the dam.

In spread dams, compared with gravity dams, the volume of concrete increases significantly and the binder content in RCC in high dams can be 70–160 kg/m^3. Such dams allowing to improve the SSS of the "dam-foundation" system can be used in specific conditions including a relatively weak rock foundation and high seismicity.

7.6 Gravity dams constructed by stages

Due to the relatively long construction time of large HPs, the erection of dams by stages with phased filling of the reservoir has become widespread. By reducing the "necrosis" of capital investments and accelerating their return on investment by commissioning structures during the construction period with minimal workloads and capital investments – their economic efficiency is increased. When dams are commissioned in stages, then the generation of electricity at hydroelectric power plants, water supply and irrigation are carried out at unfinished HP. The technical solutions in this case are also used to build up previously constructed and operated dams.

The existing experience fully confirmed the effectiveness of phased commissioning of large HP with concrete dams of various types and HEP including, for example, gravity ones: Grande Dixence in Switzerland, Bratsk and Ust-Ilim in Russia, Tukuri in Brazil, Guri in Venezuela, and Three Gorges and Longtan in China, with arch and arch-gravity dams: Sayano-Shushensk and Chirkey in Russia, Inguri in Georgia, and others [65,98,233].

The phased commissioning of high-pressure HP in some cases makes it possible to recoup the costs of their construction during the construction phase and makes the additional costs economically advantageous associated with the installation of intermediate HEP intake and installation of temporary impellers of hydro turbines operating at lower heads, as was done at the Sayano-Shushensk HEP. During the phased construction it is necessary to ensure the safety of structures at all stages and not to allow the deterioration of SSS and the weighting of the dam structure.

When constructing dams with stages or building them up, a set of issues should be resolved related to the conditions and technology for erecting subsequent stages, their combined work, the phased application of loads to the dam, and the formation of the SSS of the "dam-foundation" system, as well as the operation of HEP and spillways during construction.

To achieve the required SSS and regulate it during the phased construction process, strict compliance with all design measures is required to ensure the monolithic of the dam profile at all stages of construction and temporary operation.

Dam building without profile expansion with prestressed anchoring was extremely rare. Extension of dams in relatively narrow sections can also be performed without expanding the profile of the first stage by cementing intersectional joints, so that the dam begins to work according to the spatial pattern.

The main solutions used for the phased construction and extension of dams [35,233] are shown in Figure 7.38.

In the simplest first scheme with performing only the upper part of the dam in the second stage, it is necessary at the first stage to build a dam with a full profile to the upper part and lay a large volume of concrete, which is not required by the strength and stability conditions for the first-stage dam (Figure 7.38a).

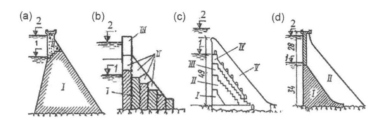

Figure 7.38 Schemes for the phased construction of gravity dams: 1 – intermediate NHL; 2 – NHL after completion of construction of dams; I – initial profile; II – V-profiles follow stages of dam construction.

Such a scheme was used, for example, during the construction in two stages of the Longtan dam from RCC with an increase in height from 192 to 216.5 m (see Figure 2.21), the volume of concrete in the second stage was 0.9 million m^3 (17% of the total volume of concrete).

The second and third schemes are carried out by adjoining to the downstream face and crest of the dam of the first (previous) stage of concrete of subsequent stages with the formation of a single profile. So, for example, the second scheme was used in the phased construction of the Bratsk and Ust-Ilim HEP dam in Russia (Figure 7.38b), and the third scheme was used in the construction of dams such as Grande Dixence in Switzerland, Bolark in Spain, Odomari in Japan, and Guri in Venezuela (Figure 7.38c).

In the fourth scheme, which is used extremely rarely, buttresses are carried out from the downstream face of the dam of the first stage on which reinforced concrete slabs are supported, for example, the Lajis dam in Brazil (Figure 7.38d)

The main advantage of the second and especially the third schemes is the reduction to the required minimum volume of concrete of the first and subsequent stages under the conditions of strength and stability of the dam. The disadvantage of this scheme is a significant complication of the conditions for the construction of the second and subsequent stages; to ensure the combined work of different stages of the dam, special events are necessary. These measures are aimed at creating favorable conditions for the mating parts of the dam under external loads, taking into account temperature-shrinkage deformations and differences in the properties of old and new concrete. In a number of cases, the position of the downstream face and the increase in concrete volume were required.

A typical example of the use of the third scheme is the construction in 1962 of the highest Grande Dixence gravity dam with a columnar section in four stages (see Figure 1.19): at the first stage with a dam height of 182 m (0.64 of full height), the volume of concrete was minimal – 1.85 million m^3 (0.3 of the total volume). To ensure favorable conditions for the combined work of the stages, the downstream face of the dams of the first, second, and third stages was carried out with protrusions.

Considering the possibility of the formation of tensile stresses on the upstream face and at the foundation of the dam with such an erection scheme, the concrete blocks of the second (subsequent) stage were separated by wide longitudinal joints, the concreting of which was carried out during the drawdown of water into the reservoir.

The Guri dam 162 m high with a 10 million kW HEP in Venezuela, the construction of which was completed in 1986, was constructed in two stages (Figure 7.39). The first stage with a height of 110 m had a concrete volume of 1.05 million m³ and the volume of concrete of the second stage was about 5 million m³. According to tests when building up the second stage after 15 years, the elastic modulus of old concrete was 1.5 times higher than the new one.

To ensure reliable contact between old and new concrete, a notch was made on 50% of the area of the downstream face of the first-stage dam, and the remaining area was treated with a sandblasting machine. Also provided was the drainage of the contact of old and new concrete [243].

A 67-meter-high San Vicente gravity dam in the USA designed for water supply and located in an area with high seismicity was built using RCC; the height of the dam was increased to 102.7 m with a crest length of 436 m. In the existing dam, the upstream face was made with a slope of 0.05–0.1 and the downstream face was 0.76; in the extended dam, the upstream face above the old dam was made vertical, and the downstream face is stepped with a slope of 0.7 (Figure 7.40).

Contact joint drainage is provided. In the central part of the dam, an automatic spillway with a length of 83.8 m was built designed for a flow pass of 1.3 thousand m³/s [242,271].

In some gravity dams erected in two stages, part of the contact oblique joint was not cemented, for example, when the Loscop dam was built up on 43 m in South Africa [128].

The performed experimental and computational studies of the SSS under static and temperature effects in severe climatic conditions for the variant of the gravity dam of the Katun HEP when erecting in two stages with a nonmonolithic step joint showed

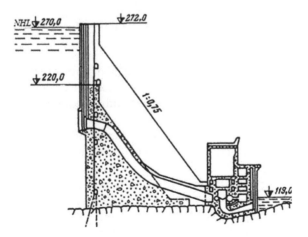

Figure 7.39 **Guri dam profile.**

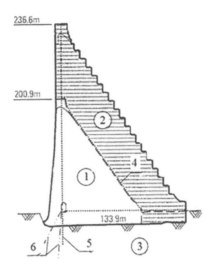

Figure 7.40 San Vicente dam profile after building: I – preexisting dam; 2 – new part of the RCC dam; 3 – rock foundation; 4 – drained joint; and 5 and 6 – drainage and cementation curtains.

that such a design generally meets regulatory requirements. However, this leads to a redistribution of compressive stresses with an increase in their values in the region of the nonmonolithic joint and a decrease in the area of the upstream face including the contact zone which is a negative factor (Figure 7.41) [174].

When designing dams made of RCC according to the staged construction scheme, it is necessary to provide for

- favorable conditions for the construction of the second stage;
- reduction in the volume of work of the first stage;
- reliable combined operation of the mating parts of the dam by creating the appropriate SSS dam and foundation.

The construction scheme of the dam from RCC in two stages proposed [89,104] uses constructive solutions of the dam with concrete apron and arch ceilings (see Section 4.4.3) and provides the possibility of independent construction of the dam of the second stage in conditions of filled reservoirs of the first stage without its drawdown. Behind the dam of the first stage, a dam of the second stage of the required height is erected from RCC, which is connected to the first by means of an arch ceiling (Figure 7.42).

Arch overlapping can be performed using prefabricated elements or mobile formwork. After increasing the reservoir level to the NHL, the flooded dam of the first stage works like a concrete apron flexibly connected by an arch overlap with the body of the dam of the second stage (Figure 7.42a). To ensure the independent operation of the dams of the first and second stage, a joint is arranged between them (Figure 7.42b). In both cases, a generally favorable SSS and a fairly uniform distribution of compressive

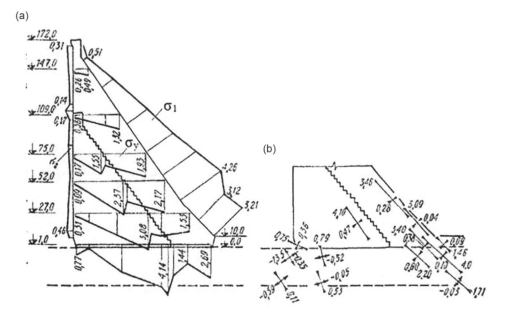

Figure 7.41 Stresses, MPa, in the dam built in two stages (from the main loads without temperature effects): (a) stresses in the dam (σ_1 – main and σ_y – vertical); (b) main σ_1 and σ_2 in the foundation and dam. "-" Tension.

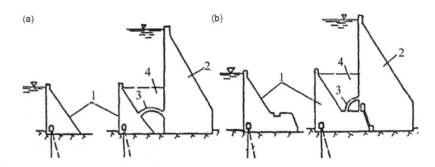

Figure 7.42 The dam erected in two stages: with a joint between (b) the dams of the first and second stages and without (a); 1 and 2 – dams of the first and second stage, 3 – arch ceiling, and 4 – soil loading.

stresses at the contact are provided at the foundation of the concrete apron (first stage dam). Such a scheme was considered when designing the expansion of a HP with a massive buttress dam 40 m high, including a spillway and a station part with a dam building of HEP. Behind the existing dam, a second dam was built 115 m high from RCC, which connected the existing dam with an arch ceiling (Figure 7.43).

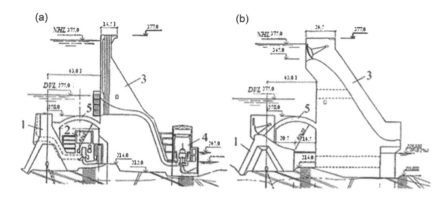

Figure 7.43 Dam building scheme: (a) section station the dam; (b) section spillway dam; I and 2 – existing dam and HEP building, 3 – dam of the second stage, 4 – new HEP building, and 5 – arch ceiling.

Using such a scheme, generally favorable construction conditions can be achieved: reduction in the volume of work and cost, reduction in construction time, and additional generation of electricity from the existing HEP operating during the construction period; however, there is no need for drawdown in the first stage reservoir during the construction period [35,88].

7.7 Adjustment to foundation, banks and other dams

7.7.1 Adjustment to foundation and banks

To ensure reliable work of the "dam-foundation" system in which the foundation is a more complex element, a set of measures is usually carried out related to the large volumes of earth-rocking, cementing, concrete, and other special works (see Chapter 4), which, along with concrete works on the construction of the dam, determine the cost, time construction, and its effectiveness.

In difficult engineering-geological conditions in addition to measures to improve the foundations in some cases, it is necessary to increase the width of the dam profile adjacent to the foundation and accordingly increase the volume of concrete in the dam.

Considering that the removal of a large volume of strong fractured rock leads to an increase in the volume of excavation and concrete of the dam, it is advisable to consider the variant of partial removal of a strongly fractured rock and strengthen the foundation by performing cementation of the remaining part to achieve the necessary physico-mechanical properties. It is possible to remove weaker rocks not under the entire bottom, but under its part in the zone of the downstream face. The choice of variant is based on a technical and economic comparison.

In cases of heterogeneous rock foundations including layered ones with weak interlayers, horizontal or weakly inclined large fissures with low strength, and deformation indicators that determine the slide stability of the dam, it is advisable to consider the variant of reinforcing these interlayers and fissures with concrete dowels, traffic jams, etc. (see Section 4.4.3).

During the construction period, after opening the pit on the basis of an engineering–geological survey with a qualitative characteristic of the open rock zones, tectonic disturbances, and fissuring, it may be necessary to clarify the configuration of the dam bottom and reinforcing measures.

A significant effect on the stress state of the contact zone is exerted by an inhomogeneous foundation with different deformability of the rocks along the width of the bottom of the dam. The dynamic characteristics of gravity dams depend on the deformability of the foundation; with an optimal E_d/E_f ratio, an improvement in static work and an increase in the seismic resistance of dams are achieved.

In bank abutments of dams, which are usually characterized by a higher fissuring (including cracks of on-board rebuff) and heterogeneity of the rocks in height, it is advisable to consider engineering measures to reduce the excavation of the pit, especially at high and steep slopes. At the same time, the selected steepness of the slopes of the pit within the bank abutments should ensure the stability of both the abutments and the dam during construction and operation, including the stability of the rock slope above the crest of the dam.

In the areas of steep landfalls, horizontal berms are arranged (usually in the zone of deformation joints of the dam sections). An important factor that must be taken into account when designing gravity dams in such areas is the decrease in the slide stability of the dam sections along the slope located on inclined parts in bank abutments. In such cases, it is advisable to consider the design of the dam, working on a spatial pattern.

Analysis of the volume of the rock excavation at the foundation and abutments of high (over 100 m) gravity dams showed that the ratio of the volume of the rock excavation to the volume of concrete of the dam can vary significantly, primarily depending on the quality of the rock foundation, as well as on decisions made regarding the depth of removal rock [51,80,176]; below are the ratios of the volume of the rock excavation to the volume of concrete of the dam for some dams:

- 0.14 for the Ust-Ilim and 0.16 for the Bratsk dam, at the foundation of which diabases lie;
- 0.30 for the Krasnoyarsk dam, at the foundation of which granites lie;
- 0.27 for the Dworshak dam in the USA, at the foundation of which granites lie;
- 0.21 for the Bicaz dam in Romania, at the foundation of which sandstones and shales lie;
- 0.38 for the Kurpsai dam in Kyrgyzstan, at the foundation of which sandstones and aleroliths lie;
- 0.44 for the Sarrance dam in France, at the foundation of which gneisses lie;
- 0.71 for the Naglu dam in Afghanistan, at the foundation of which gneisses lie;
- from 1 to 1.36 for the San Myn Xia dam in China, at the foundation of which diorites lie;
- 1.16 for the Libby dam in the USA and 1.36 for the Keban dam in Turkey, at the foundation of which limestone lies;
- 0.08 for the Toktogul dam in Kyrgyzstan, at the foundation of which limestones lie.

The antifiltration measures at the foundation and bank abutments of the dam usually consist of cementation and drainage curtains (see Section 4.4.1). Moreover, it is drainage that is the most effective and reliable measure to reduce filtration pressure.

In conditions of poorly permeable substrates with $K_f < 0.1$ m/day, the need for a cementation curtain with drainage requires special justification.

Cement curtain and drainage at the foundation and bank abutments of the dam provide a decrease in filtration pressure on the bottom of the dam and on the bank abutments, a decrease in the speed of the filtration flow and protection of the foundation and bank abutments from mechanical and chemical suffusion, and reduction of water losses from the reservoir. The main parameters of the cementation and drainage curtains including the depth at the foundation of the dam, deepening into bank abutments, are determined on the basis of engineering–geological surveys, experimental work, and filtration calculations and studies (see Chapter 4). When installing a concrete apron in front of the dam, a cement curtain and drainage are usually arranged at the foundation of the apron. Cement and drainage curtains are made from special galleries (rubbed) in the dam and on bank abutments, which allows them to work on their device regardless of the construction of the dam, to control their work in operating conditions and, if necessary, to carry out repair work. The distance from the upstream face of the dam to the axis of the cementation curtain is usually (0.05–0.1) of the width of the bottom of the dam. Cement curtain is usually vertical, less often with an inclination toward the upstream.

The drainage curtain is performed mainly from one or, less often, two rows of wells. The diameter of the wells is taken as 20–25 cm, and the distance between them is within 2–5 m depending on the head and engineering-geological conditions. The distance from the axis of the drainage to the axis of the cementation curtain can be 2–3 steps of the cementation curtain, but not less than 4 m.

Drainage wells are completed after completion of cementation work.

To increase efficiency and reduce filtration uplift in addition to the drainage curtain, drainage is performed in the form of horizontal transverse and longitudinal galleries (cavities) located on the surface of the dam bottom. The arrangement of drainage galleries complicates the production of concrete work somewhat. Such combined drainage was performed, for example, at the Boguchan and Ust-Ilim dams in Russia (Figure 7.44), and drainage cavities were also made at the foundation of the Krasnoyarsk dam.

In some cases, drainage and cementation curtains are additionally carried out from the downstream face of the dam (for example, at Longtan dam in China). To ensure the stability of bank slopes against collapse, within the adjoining dam and behind it, if necessary, reinforcing measures are carried out, as well as drainage of the rock slope in the downstream of the dam.

7.7.2 Interaction of gravity dams between themselves and with soil dams

The structure of HPs with concrete gravity dams, including deaf, spillway, and station, in many cases includes dams from soil materials.

When connecting concrete dams performing various functions in HPs, it is necessary to ensure the following:

• independence of movements with the device between them deformation joints with antifiltration seals;
• continuity of the cementation and drainage curtains at the foundation of dams;

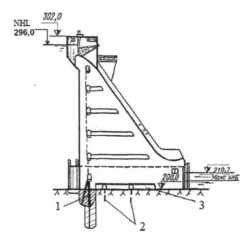

Figure 7.44 **Profile of the Ust-Ilim dam: I – cement curtain; 2 – drainage wells; and 3 – drainage cavity.**

• continuity of galleries in the body of the dams and various communications, including bridge crossings on the crest of the dam.

The most rational is the connection of the upstream faces of concrete dams with a single laying without protrusions. If it is necessary to arrange significant protrusions on the upstream face of the dam, in its calculations, additional hydrostatic pressure acting on the lateral face of its protruding part should be taken into account.

It is important to ensure high-quality connection of concrete and soil dams with each other as part of the HP head front. It is here as practice shows that there can be violations of the stability and strength of structures, including the filtering strength of soils by contact, associated with uneven precipitation of soil and concrete dams in the interface, with the occurrence of tensile stresses and accordingly the formation of cracks in the soil and concentrated path filtration.

When connecting concrete dams with dams from soil materials of various types, it is necessary to ensure the following:

• prevention of the formation of cracks, weakened zones at the contact of the dam from soil materials with concrete, and in its waterproof elements as a result of uneven settlements;
• continuity of the antifiltration devices at the foundation of the dams with the prevention of an increase in filtration pressure under the zone of contact of the dam from soil materials with the section of the concrete dam.

The connection of a concrete dam with a dam of soil materials is performed with

• the device of concrete piers in the upstream and downstream, limiting the soil dam;
• the location of the concrete dam in the body of the dam from soil materials;
• according to the combined version.

Concrete pier practically repeats the profile of the dam from soil materials, and its maximum height is equal to its height. The face of the extreme section of the concrete dam in order to ensure a tighter fit to the antifilter element of the dam from soil materials to it is made inclined with a slope of 0.1; to lengthen the filtration paths, a concrete diaphragm is installed, which is included in the dam's filter element (for example, in the core). The antifiltration element at the abutment site is expanded in plan, and drainage is performed from the downstream side. In the contact zone, usually more plastic soil with high humidity is laid. These measures make it possible to exclude the formation of a continuous crack at the contact of concrete and soil and to organize the filtered flow in an organized manner.

The piers in the upstream of the concrete spillway dam provide favorable hydraulic conditions for supplying water to the spillway and in the downstream take hydrodynamic effects when the flood passes and protect the dam from soil materials from washing out.

When a concrete dam is interacted with a soil dam, the pier device is excluded. So, at the San Simão HEP in Brazil (1978), the connection of a concrete dam 95 m high with a stone-earthen dam with a core is made in the form of a section of a concrete dam (Figure 7.45a), which enters the stone-earthen. To ensure a denser contact of the soil core, the upstream face of the concrete dam is made with a slope with 0.15 m depth [130].

With the combined variant, the length and height of the pier and the length of the section of the concrete dam entering the dam body from soil materials are reduced. Such an abutment can also be used during the construction period as a longitudinal cofferdam, enclosing at the first stage the foundation pit of the main concrete structures, and at the second stage the foundation pit of the concrete dam section. So, at the Kang Don HP in Vietnam (2004), connection of the spillway dam and a homogeneous loam dam were mated according to the combined variant (Figure 7.45b).

The most complex and critical element of the head front of the Boguchan HEP is the connection of the concrete dam and the asphalt concrete diaphragm of the rock-fill dam, which ensures the water resistance of the head front and the independence of the movement of dams during construction and operation with significant seasonal temperature fluctuations in the outside air.

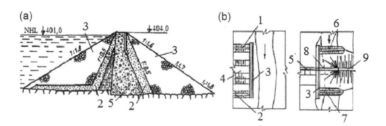

Figure 7.45 Connection of a concrete dam with a dam from soil materials: (a) connection of stone-earth and concrete dams of Sãn Simão (cross section): 1 – clay core, 2 – gravel transition zone, 3 – stone sketch, 4 – sand filter, 5 – section concrete dam; (b) the connection of Kang Don dams from soil materials and concrete (plan): 1 and 2 – cofferdams of the first stage, 3 – concrete longitudinal cofferdam pier, 4 – pit of a spillway dam, 5 – weir dam with bottom construction openings, 6 and 7 – cofferdams of the second stage, 8 – concrete dam, and 9 – soil dam.

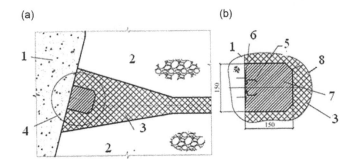

Figure 7.46 The design of the interface section No. 34 of the concrete dam with asphalt concrete diaphragm of rock-fill dam: (a) plan, (b) key 4; I – section No. 34, 2 – stone-filling dam, 3 – asphalt concrete, 5 – metal key No. I, 6 – anchor Hilti, 7 – asphalt mastic, and 8 – link keyway.

A special interface was developed (Figure 7.46) consisting of a keyhole 8 located in the expanding part of the asphalt concrete diaphragm 3 adjacent to section No. 34 of the concrete dam 1 and filled with asphalt mastic 7.

Monitoring the operation of the interface is carried out using displacement sensors on the contact of concrete and dowels. The assessment of the state of the connection is carried out by comparing the measured movements with the safety criteria for the movements calculated in the design process.

Chapter 8

Arch dams

"All is embraced by the world
All is like a parachute
All is an uplifting vision."

Boris Pasternak (1890–1960),
Russian poet

8.1 Evolution of arch dams

The high bearing capacity of the arch was known even to ancient builders who widely used this principle in the Roman Empire for the construction of bridges and aqueducts. *Chris* of Alexandria, the dam builder (circa 550 CE), wrote as follows: "... a crescent-shaped arch directed against the flow of water best resists the force created by the water."

Until the XIX century, builders of arch dams (for example, the Kurit dam in Iran in the XIII century, Elhe in Spain in the XVII century, and Pontalto in Italy in the XVII century) used the rules adopted for the construction of arch bridges; since then, there were no methods for calculating dams. In this regard, the results of studies of the Elhe arch dam of modern methods are of interest: it turned out that the maximum stresses in it are close to the designed compressive strength of the masonry.

The beginning to the static work analysis of arch dams was laid by *M. Zola* – the father of the famous writer E. Zola. The 47-m-high Zola arch dam was named after its designer and built despite numerous protests in France in Provence in 1847–1854 and became a kind of monument to its creator (Figure 8.1). Its active operation continued until 1877, and to date, it has been maintained in working condition. Static analysis of the dam was performed by the average (cross-section) stresses calculated in horizontal arches using the boiler formula:

$$\sigma = \gamma_w H R / d \tag{8.1}$$

where
  γ_w – specific gravity of water;
  H – the head;
  R – the axial radius of the arch;
  d – the thickness of the arch.

Figure 8.1 **Zola arch dam (France).**

As far back as 1885, the French engineer *M. Levy* on the basis of the causes of the accident at the Busy dam in France from masonry 25 m high suggested that tensile stresses in the dam body must not be when taking into account uplift in the masonry; in other words, compressive stresses σ in the dam from its own weight and water pressure on the upstream face should be greater than $\gamma_w H$. Subsequently, in the 1960s of the last century in the USSR, in the code for the design of concrete dams, this condition was substantially facilitated, and the condition $\sigma > 0.25\,\gamma_w H$ should be fulfilled; as they said then: "a quarter of the rule of M. Levy." In the 1970s in BR 2.06.06-85 [15], the *M. Levy* rule was completely excluded: zero compressive stresses were allowed without taking into account uplift in horizontal sections of the dam, however, if there is drainage in the dam beyond the upstream face.

An arch dam on the river Psyrtsha in Abkhazia was built in 1882. With the help of a dam 8.6 m high and the crest long 21 m (Figure 8.2), river flow regulation was carried out for the needs of the New Athos Monastery. The constructed HEP was one of the very first in the territory of tsarist Russia. In Soviet times, the building of the HEP was destroyed and restored only in 2012 for the needs of the New Athos Monastery. The dam suffered 5–6 balls earthquakes in 1915, 1922, and 1966 without noticeable damage.

The Beer Valley arch dam erected in the USA in 1884 turned out to be heavily loaded, which became clear later when its static calculation was performed: the maximum compressive stresses reached −4.27 MPa. For that time, it was a very brave structure (Figure 8.3).

By 1899, a record ratio of ℓ/h was achieved (where ℓ is the length of the crest and h is the height of the dam) equal to 7 and 10 for arch dams Wellington and Madgle in Australia, respectively; the height of both dams is 15 m and the bottom thickness is only 3 m. These were the first arch dams built entirely of concrete.

Figure 8.2 **Arch dam on the river Psyrtsha in Abkhazia.**

Figure 8.3 **Beer Valley arch dam (USA).**

The beginning of the XX century began to look for the shape of arch dams based on the analysis of stresses and deformations. When designing the Buffalo Bill dam in the USA (built in 1910), horizontal (arches) and vertical (console) elements were considered between which the hydrostatic load on the dam was distributed, based on the equality of the deflections of the center console and a number of arches in the key. The dam was 100 m high, a record for that time. Thus, the method of "arch-center console" was developed, which was widely used in the world.

In 1879, the French engineer *Peltro* put forward the idea that the radii of arches should decrease from the crest to the bottom. This type of dam called the "constant angle dam" was used by the Norwegian engineer *L.R. Ergenson* to design the Salmon

Figure 8.4 Salmon Creek arch dam (USA).

Creek Dam in the USA in Alaska, which was built in 1914 and became the prototype of future arch dams of this type. In order to reduce the inclination of the dam in the upstream, the thicknesses to the heels of the arches were increased due to a decrease in the radius of curvature of the downstream face of the arches. The key console at the lower elevations had a thickening on the downstream side, which made it possible to reduce vertical tensile stresses in it (Figure 8.4).

The optimal value of the central angle of the dam according to the calculations of *L.R. Ergenson* was 120°. In a relatively narrow and high canyon, an arch dam with a constant angle made it possible to save 33% of the concrete volume of the gravity dam, while the safety factor of the arch dam is two times higher compared to the gravity variant. Salmon Creek was the first arch dam built in the USA entirely from concrete. To avoid cracks from shrinkage of concrete, the dam was first divided by two vertical joints into three sections.

Around the same time, Lake Spoulding arch dam designers in the USA in California for the first time provided for monolithic dam joints using well cementing.

In 1929, an arch dam Pakoima in Greece was erected with a constant angle, a record at that time, and 116 m height at $\ell/h = 1.58$ with a change in the central angle from 111.5° in the upper part to 70° in the lower part of the dam.

One of the high dams designed with a constant radius, the Tignes dam, 180 m high, was erected in France in 1952. An advantage of dams of this type is that the downstream face is inclined to the upstream, and it is convenient to place spillways on it. Famous built dams are La Aigle, 95 meters high (built in 1945), Orzhu Les Board – 121 m (1951), and Shasta – 25 m (1952) in France; Dam Bao – 107 m (1960) and Aldeadavila (Figure 8.5) – 140 m (1963) in Spain

Swiss engineer *H.E. Gruner* transferred the American experience of erecting arch dams to Europe and on the Monsalvens arch dam built in 1920, used noncircular arches delineated along the pressure curve with bulges on his heels to reduce stresses on contact with the foundation. His compatriot *A. Stukki* introduced into the analysis in addition to the key, side consoles, which made it possible to improve the shape of arch dams in vertical sections.

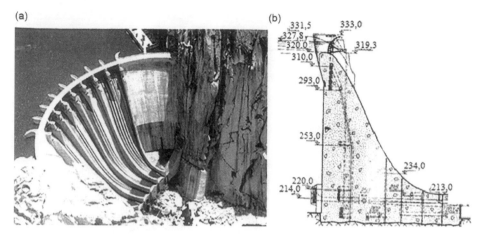

Figure 8.5 **Aldeadavila arch-gravity dam (Spain): (a) view from the downstream and (b) cross section.**

In the 20s of the XX century, studies of the bearing capacity of arch dams on models began, and experimental centers appeared in Portugal (laboratory of civil engineers in Lisbon), Italy (ISMES Institute in Bergamo), England (Royal College), and Spain (Central Laboratory in Madrid). American researchers after testing a number of models drew attention to the fact that dams as a rule are destroyed in the upper third of the height; below, the dam remains monolithic. After 20 years, French engineers took this circumstance into account when designing and constructing the Roseland 150 m dam, the central part of which was erected in the form of a shell with a "diving" crest (lowering from the key to the banks) on which buttresses rest.

The remarkable ability of thin arch dams to adapt to different working conditions was demonstrated by two arch dams in the USA: Moye built in 1924 in Idaho and Lake Lanier built in 1925 in North Carolina, which lost support on the bank due to erosion of weak rock mass but did not collapse.

A year after the start-up, the filtration stream washed the highly fractured rock in the left-bank abutment of the Moye arch dam; as a result, the 16-m dam turned out to be unsupported in height of almost 14 m (Figure 8.6). Through the resulting gap, the reservoir underwent drawdown, but the dam itself resisted. The dam had a crest length of 47 m, a bottom thickness of 1.6 m, and a crest of 0.6 m; the upstream face of the dam is vertical and cylindrical, outlined with a radius of 20 m. The dam was reinforced with railway rails – horizontal belts at two elevations and vertical belts at two sites (consoles). The dam has survived to the present day.

In the abutments of the Lake Lanier arch dam, there were weak sandstones, which were removed to a depth of 1.2–2 m, after which the piers from the cyclopean masonry were erected and faced with concrete on the upstream side. The dam had a height of 19 m, a crest length of 72 m, a bottom thickness of 3.68 m, and a crest of 0.60 m. In 1926, as a result of leaching of rock at a height of 8.5 m, the pier from the masonry was not having support to the bank and was overturned into a downstream; the reservoir underwent drawdown, but the dam reinforced in the area of the crest did not collapse (Figure 8.7). The dam was subsequently restored.

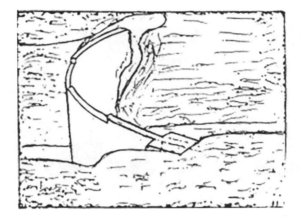

Figure 8.6 Moye arch dam (USA).

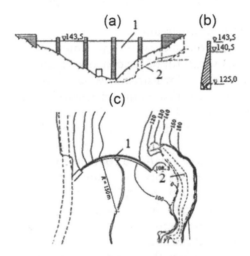

Figure 8.7 Lake Lanyo arch dam (USA): (a) view from the downstream, (b) section along the center console, and (c) plan; 1–dam and 2–washed part of the bank.

The Sant Francis arch-gravity dam, 62 m high along the crest of 210 m and a thickness at the base of 52 m, was erected in 1927 to supply water to the city of Los Angeles (Figure 8.8a).

Geologists warned the designers that the dam was located along the geological fault, and the base consisted mainly of clastic rocks subject to erosion and dissolution.

In 1926, as the reservoir was filled, cracks appeared in the dam, and leaks were discovered at the foundation, which by the beginning of 1928 had become rampant.

Three minutes before midnight on March 12, 1928, the St. Francis dam collapsed, killing more than 600 people. The keeper of the dam and his family were the first victims of the 38 m wave that surged through the canyon of San Francisco.

The reconstruction of the events showed that the eastern sections first collapsed and the gushing water began to unfold the dam, as a result of which the western sections

Figure 8.8 **Saint Francis arch-gravity dam (USA): (a) the beginning of the dam destruction and (b) after the disaster.**

collapsed when the reservoir was already half empty. Only the central section remained (Figure 8.8b), and the eastern and western sections, broken into large pieces, were carried away 800 m downstream.

The first arch dam of double curvature was built in Italy in 1925 (Gurtsia dam). In order to reduce the overall tilt of the consoles in the upstream, the top of the key console was thrown into the downstream. When filling the reservoir at the base of the center, console cracks formed on the upstream face due to high tensile stresses. On Ozil dome dam, 77 m high with a perimeter joint built in 1939, this circumstance was taken into account, and the lower part of the cantilever was tilted into the upstream.

For the first time, the idea of creating a concrete foundation (saddle or the so-called "pillow") at an arch dam was put forward by the famous Italian engineer *Guido Oberti* in 1935–1938; the first dam with a saddle, Osilietta, 76.8 m high, was built on porphyry gneisses in 1937–1939. Then was built the Lumia arch dam (Figure 8.9), the highest in Europe at that time, 136 m high, separated from the saddle by a classic perimeter seam.

Using a concrete saddle at the foundation, many high arch dams were built in Italy, Spain, and Portugal. A saddle with a perimeter joint was used in connection in the Inguri arch dam with the foundation in Georgia (see Figure 8.39).

A saddle cut off by a perimeter joint from the arch part of the dam allows

- reducing the degree of influence of violations (faults, cracks, and weak zones) in the foundation on the SSS of the dam and rock foundation;
- reducing the static indeterminacy of the dam and consequently reducing tensile stresses in the body of the dam especially on its upstream face and at the abutments;
- ensuring freedom of deformation of the dam under the influence of temperature changes and seismic effects and reducing the possibility of cracking in concrete;
- creating a symmetrical dam even in a highly asymmetric site, for example, the almost symmetrical Piave di Cadore dam, 55 m high, erected in a highly asymmetric site with a maximum depth of 112 m.

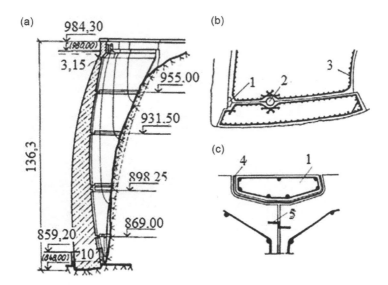

Figure 8.9 Lumia arch dam (Italy): (a) section along the spillway, (b) perimeter joint, and (c) a seal of the joint; I – reinforced concrete block with a seal, 2 – drainage, 3 – reinforcement, 4 – antifiltration composition, and 5 – copper sheet 10 mm thick.

The disadvantage of the perimeter joint in the arch dam according to French engineers is the lowering of the static indeterminacy of the dam, which does not allow the full use of the strength of the dam concrete and the ability of the arch dams to adapt to a certain extent to the geological conditions of the site. Therefore, as a rule, dams with a perimeter joint have a greater thickness than statically indeterminate arch dams. French engineers did not build a single arch dam with a perimeter joint. According to Russian engineers, the installation of a perimeter is advisable in the construction of arch dams in areas of high seismicity and especially when they are placed in wide sites.

In France, *A. Coin* first used the double curvature and inclination of the central console in its lower part to the upstream on the Mareges arch dam, 89.5 m high and 198 m long (ℓ/h = 2.21), built in 1935 (Figure 8.10). This is the first dam in the world with a spillway springboard.

When designing domed arch dams, much attention was paid to the selection of the rational outline of the axes and thicknesses of the arches. If previously circular arches mainly of constant thickness were mainly used, then arches with a more complex outline of axes are spread: three-centered, outlined by a parabola, hyperbole, and other curves, often with an increase in thickness to heels. At the same time, it is sought to obtain both a more favorable stress state of the dam and an improvement in the working conditions of the bank abutments.

In the 60s of the XX century, thanks to the studies of the French engineer *M. Leroy,* ii is began to use arches outlined in a logarithmic spiral. The first dam with such arches is the Vouglans dam, 130 m high, built in France (Figure 8.11). Using a logarithmic

Figure 8.10 Mareges arch dam (France).

Figure 8.11 Vouglans arch dam (France).

spiral, it is possible to smoothly mate the central thin part of the dam with thicker parts connecting to the foundation and banks.

The equation of a logarithmic spiral depending on two parameters R and α in polar coordinates has the form

$$\rho = R^{\alpha v} \tag{8.2}$$

where
$\theta = 0$ and $\rho = R$ in the arch key.

The parameter α is $ctg\beta$; at $\alpha = 0$, the angle $\beta = 90°$ and the spiral turns into an arc of a circle; as $\alpha \rightarrow \infty$, the angle β and the curvature of the spiral tend to zero. For deep narrow gorges (at $\ell/h \leq 1$), it is recommended to take $\alpha = 0$. For relatively wide gorges, $\alpha \neq 0$

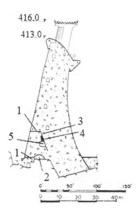

Figure 8.12 Henrik Verwoerd arch dam (South Africa): 1 – joint incisions, 2 – cavity, 3 – joint slip, 4 – key, and 5 – gallery.

and the wider the gorge, the greater the value of α. The angle of contact of the arch to the foundation depends on the value of R: the larger the R, the greater the angle of contact (which is better for SSS dam and bank stability), and on the other hand, the greater the R, the greater the bending stresses in the arches. Under these conditions, a compromise solution is needed, which is sought using iteration.

The Henrik Verwoerd arch dam, 88 m high (Figure 8.12), designed with arches outlined in a logarithmic spiral has a record ℓ/h ratio of 7.45.

In Portugal, three-center and elliptical arches were used in the design (for details on the shape of arches and arch dams, see Sections 8.3.4 and 8.3.5).

The best form for an arch dam from the point of view of reducing bending deformations and torsion (i.e. obtaining uniform compression and limited tension in the dam), taking into account all the existing loads and effects, is a surface with a double curvature. In this case, the radii of curvature of the arches increase from key to heel, and the side consoles are thrown into the upstream so as to compensate for the tensile stresses at the base of the consoles due to their own weight.

Often under the dam construction, it is necessary to limit the console inclination to both the upstream and downstream. When the reservoir is empty, there is a problem associated with tensile stresses on the downstream face of the consoles thrown back into the upstream; the problem is compounded in a seismically active area. From this point of view, it is desirable to tilt the consoles into the downstream.

For circular arches, the optimal central angle is 120°, while the axis of the arch forms a 30° angle with the horizontal at the corresponding elevation; for dams of double curvature, the optimal central angle decreases to 80°–90°, which creates much better conditions for the dam to adjoin the base.

In 1953, a professor at the University of Grenoble (France) *A. Bourget* performed an analysis of the static work and concrete volumes for four types of arch dams designed in one canyon (dam height – 50 m and canyon width at the level of the dam crest – 70 m):

- type A – a dam with a constant height radius of arches and variable angles;
- type B – a dam with a constant angle and variable radii of arches;

- type C – a dam with variable radii and angles of arches;
- type D – a dam with a pronounced slope in the downstream.

Permissible compression stress was taken for all types of the same and equal to 3 MPa. *A. Bourget* received the following results:

- the volume of the type A dam is the largest; the thrust in the lower arches is negligible, and to increase it, it is advisable to increase the angles of the arches;
- the type B dam has a very significant drawback: the side consoles are tilted into the upstream so that during reservoir drawdown, there is a danger of them overturning;
- a dam of type C is devoid of the disadvantages of dams of type A and B;
- the type D dam turned out to be the best in both volume and stress state; it was concluded that this design allows you to solve the main issues that arise in the design of arch dams; however, subsequent design practice has shown that in many cases, this conclusion is unacceptable.

Due to certain successes in arch dam construction, in the 50s of the XX century, in different countries, thin arch dams are built in narrow sites.

The Lajanuri arch dam, 69 m high and 127 m long along the crest ($\ell/h = 1.84$), was erected on the Lajanuri river in Georgia; the thickness of the dam varies from 7.6 m down to 2.5 m on the crest (Figure 8.13). On the crest of the dam are three spillways with a span of 7 m each designed to pass up to 120 m³/s of water. In addition, water can be discharged through a construction bypass tunnel with a diameter of 6 m.

A number of arch dams had central consoles tilted strongly into the downstream in order to avoid tipping the side consoles into the upstream.

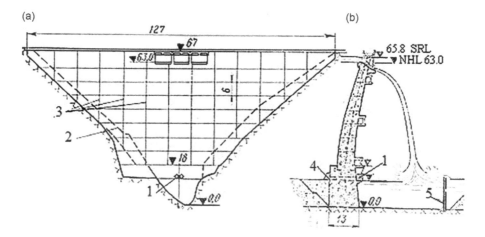

Figure 8.13 Lajanuri arch dam (Georgia): (a) view from the downstream and (b) section along the spillway; 1 – gutter, 2 – natural surface, 3 – intersectional joints, 4 – grating, and 5 – three-hinged experimental arch dam with a height of 14 m.

Observations of the dams showed that the inclination of the center console into the downstream led to the appearance of cracks on the upstream face at the bottom of the Ehone and Rio Fredo dams. The reason was the insufficient taken of the dead weight of the dam in the formation of the stress state. To compensate for the vertical tensile stresses in the lower part of the dam, the consoles began to bend into the upstream. To avoid overturning of pillars in the upstream during the construction period, special supporting structures have been developed. This solution was first used in Spain on the 98-m-high Valdecanas dam.

In connection with the progress in the selection of concrete composition, there is a tendency to increase the acting stresses in arch dams and therefore the concrete classes. In modern arch dams, compressive stresses reach 10–14 MPa, in the Inguri arch dam – 14 MPa, in the Sayano-Shushensk MPa, and Gage (France) – 10 MPa, etc.

Along with the tendency to reduce the concrete volume of arch dams due to the search for a rational shape of the dam and increase the level of current stresses, RCC is being introduced into the construction of arch dams. The world's first arch-gravity dam made of RCC is the Knelport dam (South Africa), 50 m high, built in 1990. The dam has a vertical upstream face, and the downstream face is made stepped with a slope of 1: 0.6. Another arch-gravity dam Wolverdans (South Africa) with a height of 70 m has a similar design. Later on, arch dams from RCC began to be built in China. Of these, the highest dams are as follows: Shapai built in 2002 with a height of 132 m, a crest length of 250 m, and concrete volume of 392 thousand m^3 and Junloghe III (2008) with a height of 135 m, a crest length of 119 m, and a concrete volume of 183 thousand m^3 [194].

Arch dams are highly reliable structures. There are cases when arch dams experienced significant overload and remained almost intact. Thus, the Italian Karfino dam (height – 40 m and thickness along the crest – 1.5 m and the bottom – 7 m) remained intact after a strong earthquake during which all structures in the vicinity collapsed.

Many arch dams (in Japan, Iran, Greece, USA, Russia, Georgia, and China) were built in areas with high seismic activity (8–10 balls). The great reliability of the arch dams is evidenced by the well-known disaster on the Vaiont reservoir.

Achievements in the construction of arch dams, improvement of structures and technology of their construction and calculation methods using computer technology, increasing the general level of knowledge about the combined work of dams with a rock foundation, and their high reliability and competitiveness ensured the further widespread construction of arch dams in the second half of the XX century and at the beginning of the XXI century [31,76,121].

The height of the arch dams increases significantly reaching 271.5 m on the Inguri dam in Georgia (see Figure 1.28), on the Xiluodu dam 285 m (Figure 8.14), and 294.5 m on the Xiowan dam (Figure 8.15) erected in China in 2015 and 2010, respectively. The construction of the Jinping-1 dam, 305 m high, was completed (Figure 8.16).

Xiluodu arch dam was built on the Yellow River and is designed to generate electricity (13.8 million kW HEP), irrigation, shipping, flood prevention, and flow regulation. In addition, as conceived by the designers, the 12.9 km^3 reservoir should accumulate a solid river flow in order to reduce the siltation of the Three Gorges reservoir, which was built downstream. In Figure 8.14, it can be sees that a huge volume of rock excavation had to be filled to create a water well and equip the banks of the downstream.

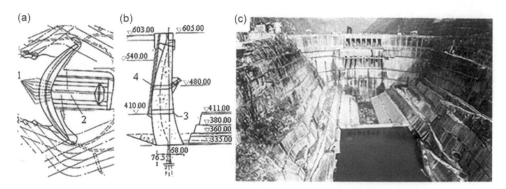

Figure 8.14 Xiluodu arch dam (China): (a) plan; (b) section along the spillway; 1 – axis of seven spillways, 2 – watering well, 3 – time bottom spillway, and 4 – deep spillway; (c) view from the downstream.

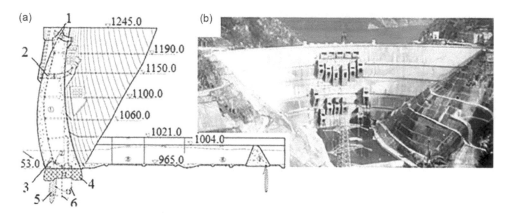

Figure 8.15 Xiowan arch dam (China): (a) section along the center console and (b) view from the downstream: 1 and 2 – surface and deep spillways; 3 – cementing gallery; 4 – strengthening cementation, and 5 and 6 – cementation and drainage curtains.

One of the highest arch dams in the world Xiowan erected on the river Lantsanjiang (a tributary of the Mekong in the Himalayas) was built in 2010 and is designed to generate electricity (HEP with a capacity of 4.2 million kW) and regulate flow (Figure 8.15). The highest arch dam in the world Jinping-1 with a height of 305 m is designed for generating electricity (HEP with a capacity of 3.6 million kW), preventing floods, and regulating flow (Figure 8.16).

The length by the crest is 568 m, the width of the dam along the crest is 16 m, the base is 63 m (the ratio of the width at the base to the height is $b/h = 0.207$), and the concrete volume is 7.4 million m^3.

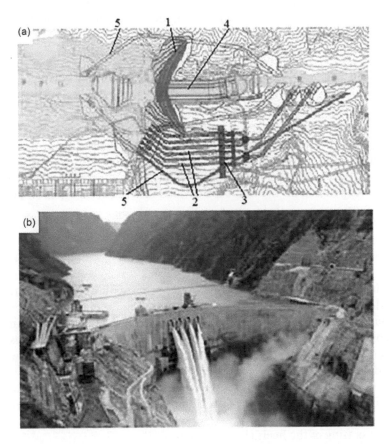

Figure 8.16 Jinping-1 arch dam (China): (a) plan, 1 – dam, 2 – penstock HEP, 3 – HEP, 4 – plunge pool, 5 – diversional tunnels; (b) view from the downstream.

It should be noted that China took the first place in the world in the construction of arch dams of different heights and various designs:

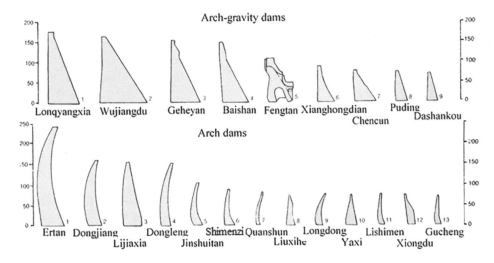

Among the high dams with a height of more than 200 m constructed and under construction in the world in 2000, concrete dams accounted for 78% including arch dams – 60% (see Sections 1.2 and 1.3).

8.2 Features of arch dams

8.2.1 Basic types and structures

The arch dam is a spatial structure in the form of a vault transmitting the loads acting on it from the pressure of water, sediment, seismic, temperature, and other effects mainly on the banks of the gorge. At the same time, due to the action of thrust forces on the banks of the gorge, effective use of high concrete compressive strength is achieved. Arch dams depending on the shape are divided into the main types (Figure 8.17) with

- constant central angle of the arches;
- constant radius of arches or cylindrical;
- with double curvature or domed.

Dams with a constant central angle are used in sites with a triangular cross-sectional shape. Moreover, the central angle $2\alpha_0$ usually does not remain constant in height, but somewhat decreases downward (especially for the lowest arches). Cylindrical dams with a constant radius characterized by the simplest form are used in sites with a trapezoidal cross-sectional shape. Dome dams which have become the most widespread in the world are used for various cross-sectional sites [179].

Depending on the design features and the conditions for connection with the foundation, the following main types of arch dams are distinguished (Figure 8.18) with

- elastic cutting of the heels (Figure 8.18a);
- bank gravity piers (Figure 8.18a) and without piers;

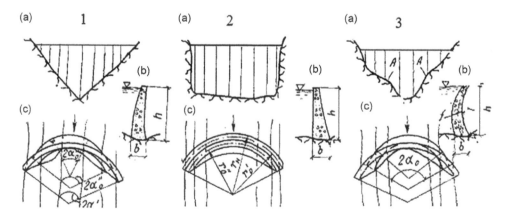

Figure 8.17 Arch dams with: 1 – constant angle, 2 – constant radius (cylindrical), and 3 – double curvature (dome); (a) view from the downstream, (b) key console, and (c) plan.

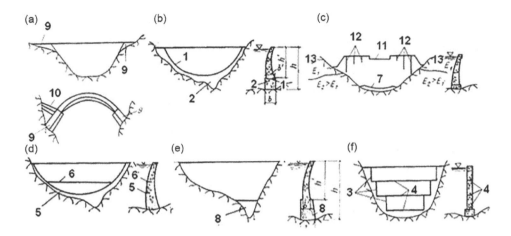

Figure 8.18 Types of arch dams with: (a) elastic of the heels, (b) contour or perimeter joint, (c) "diving" upper arches, (d) horizontal joint notches, (e) plug, (f) divided by joints into separate arches; 1 – contour joint; 2 – "saddle", 3 – hinges, 4 – interarch joints, 5 – contour joint notch, 6 – joint notch, 7 –partial contour joint notch, 8 –plug, 9 –gravity pier, 10 –gravity wing, 11 – spillway, and 12 – vertical joint notches.

- perimeter (contour) joint (Figure 8.18b);
- "diving" upper arches, with vertical joint notches on the crest (Figure 8.18c);
- horizontal joint notches (Figure 8.15d);
- a plug (Figure 8.18e) and without a plug;
- two or three-hinged arches (Figure 8.18f);
- the dam divided by joints into separate arches (Figure 8.18f).

Arch dams are self-regulating systems that adapt to changing working conditions due to their hyperstatic property (multiple static indeterminate) as a result of which local attenuation and displacement of banks in limited sizes are not dangerous for the strength of the dam and cause only a redistribution of forces. In this case, the load perceived by the dam is transmitted to more rigid sections of the foundation unloading its weakened sections. With an increase in horizontal load in the arch dam, thrust forces increase, which as a rule improve the stability conditions of bank abutments.

When designing the Tolla dam to the island Corsica, the bearing capacity of the arch dam was exaggerated: the dam at a height of 88 m and a thickness on the crest of 1.5 and 4.2 m at the bottom was extremely thin. Even during the construction period, temperature cracks began to form, and when force cracks formed during the initial filling of the reservoir, a decision was made to significantly strengthen the dam (Figure 8.19).

The point of view is based on the hyperstatic property according to which the device in the arch dam of various kinds of joints (including perimeter) is considered

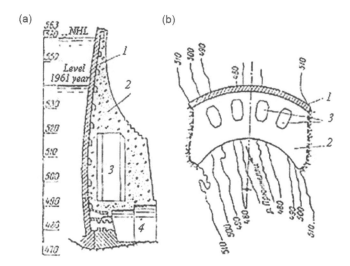

Figure 8.19 Tolla arch dam (Italy): (a) section along the center console, (b) plan section at mark 510: 1 – initial dam, 2 – dam reinforcement, 3 – cavity in the reinforced dam body, and 4 – spillway.

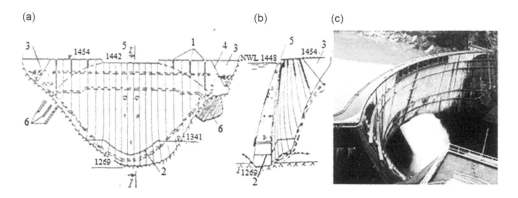

Figure 8.20 Kurobe-4 arch dam with a "diving crest" (Japan): (a) top view, (b) section I-I: 1 – vertical joint notches, 2 – partial perimeter joint, 3 – gravity wings, 4 – deformation joint, 5 – spillway, 6 – filling by concrete weakened rocks of the foundation; (c) view from the downstream.

impractical since they reduce static uncertainty to one degree or another (French and Austrian schools) (Figures 8.20, 8.21, 8.23–8.25).

There is another point of view according to which the arrangement of joints (organized cracks), including perimeter joints equipped with antifiltration devices, makes it possible to relieve tensile stresses in an arch dam (Italian school). Such dams are widespread in Italy. In Georgia, one of the highest arch dams in the world Inguri was built with a perimeter joint (Figure 8.23).

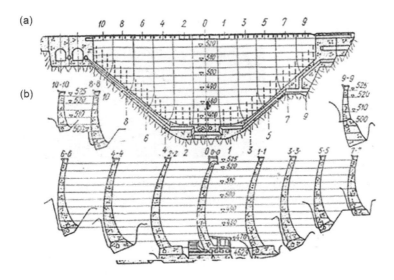

Figure 8.21 Santa Eulalia arch dam (Spain): (a) view from downstream and (b) cross sections.

Perimeter seams

- allow preventing the occurrence of tensile stresses in the area connecting to the foundation, which may be dangerous; to improve the conditions for the contact of the dam with the antifilter curtain;
- give greater symmetry to the arch part of the dam and adjust the pressure transmitted from the dam to the rock foundation by changing the width of the saddle; with proper reinforcing of the saddle on a weak rock section, you can transfer the load in this zone to neighboring stronger rock sections (for example, Ambiesta dam, see Figure 8.33).

On some relatively thin dams, for example, Lumia and Bearegard (Figure 8.24), bitumen lubrication is arranged in the perimeter joint to allow horizontal displacement of the dam along the saddle (Figure 8.22).

Perimeter seams can be made along the entire contour of the dam connection to the foundation or on part of the contour. On a number of arch dams, partial joints reduce stiffness of consoles and thereby reduce stresses in them and in the foundation under them, creating a more favorable stress distribution without tensile stresses on the upstream face and in the foundation near the antifiltration curtain (Figures 8.23 and 8.24).

Such joints are made on dams in wide sites: Schifferen (Figure 8.25), Lauza (see Figure 8.48), and Les Toules (see Figure 8.56) as well as on arch dams by Lumia (see Figure 8.9), Kurobe-4 (Figure 8.20) and others.

At Kurobe-4 dam, the joint was initially planned around the entire perimeter of the support; however, it was replaced by a partial joint located in the bed part of

Figure 8.22 Two-arch dam Hongrin (Switzerland): (a) and (b) southern and northern dams, (b) view from the upstream.

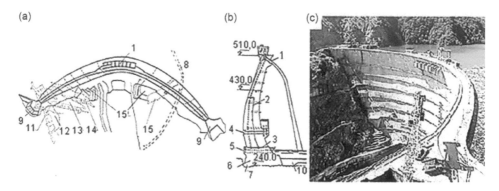

Figure 8.23 Inguri arch dam (Georgia): (a) plan, (b) section along center console; 1 – spillway, 2 – circuit of the first stage (project), 3 – perimeter joint, 4 – deep spillway, 5 – construction opening, 6 – concrete plug, 7 – cement curtain, 8 – diversional tunnel, 9 – bank piers, 10 – plunge pool, 11–15 – fissures in the foundation filling by concrete; (c) view from the left bank.

the dam, as model studies showed that the joint above does not improve the stress state of the dam. In addition, the designers feared that with a full perimeter joint, the dam might move along it under high seismic intensity (designed seismicity of 10 balls), given that the dam at the top marks does not rest on a weak rock foundation within 40 m in height. Perimeter joints (including partial ones) can be both non-cemented (Inguri dam, Val-Gallina, and Vaiont) and cemented (dam Les Toules, Kurobe-4).

Cemented perimeter joints leave the area uncemented from the upstream face to the internal seal, turning it to a certain extent into a joint notch. This ensures the removal of tensile stresses on the upstream face above the uncemented area; in the case of insufficient depth, a joint may open at the beginning of the cemented area with tensile stresses.

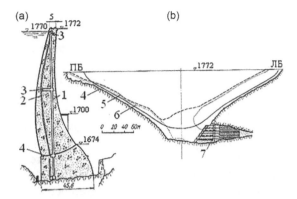

Figure 8.24 Bearegard arch dam (Italy): (a) cross section, (b) sweep along the down-stream face; I – shaft, 2 – drainage, 3 – gallery, 4 – perimeter joint, 5 – natural surface of the canyon, 6 – contour of rock excavation, and 7 – concrete plug.

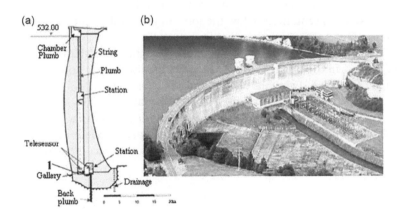

Figure 8.25 Schifferen arch dam (Switzerland): (a) section along the center console: I – partial perimeter joint; (b) view from the downstream.

Typically, cementation of perimeter joints is performed when the reservoir is filled to an intermediate level.

In some cases, the use of longitudinal joint notches with seals on the upstream face of the dam allows you to improve the stress state of the arch dam; for example, Henrik Verwoerd dam (see Figure 8.12) and Lauza dam (see Figure 8.46).

As part of a contract with a design institute in Kunming (China), Hydroproject specialists at the end of the 1990s investigated three design variants for connecting the Xiowan arch dam with a rock foundation [234]:

• elastic sealing;
• perimeter joint in the dam;

- joint notch at the contact of the upstream face of the dam with the foundation (partial perimeter joint);
- a subvertical joint at the foundation under the upstream face of the dam at a depth of 20 m in the bed and 7 m in the abutments at the upper elevations.

Computational studies were performed by the FEM for the "dam foundation" system for loads: dead weight of the dam, hydrostatic pressure and sediment pressure on the upstream face, uplift on the bottom of the dam, and seismic effect. The tensile strength of the perimeter joint, joint notch, and subvertical joint at the foundation is taken equal to zero.

In the variant with elastic sealing, the main tensile stresses reached + 6.37 MPa on the upstream face of the dam near the contact with the foundation (Figure 8.26, point 3); the contact of the upstream face turned out to be stretched along the entire perimeter of the foundation (points 11 and 12), and in the foundation under the upstream face, a triaxial tension was obtained (points 21–23).

In the variant with a perimeter joint, the latter is almost squeezed around the entire perimeter; under seismic effect, the perimeter joint was opened in the bed and bank parts at about the lower 2/3 of the dam height. Due to the opening of the perimeter joint, there are no tensile stresses in the arch part of the dam (i.e. above the joint); however, tensile stresses remained below the joint in the saddle and the foundation under the upstream face.

In the variant with a joint notch, the contact opened to a depth of 40% in the bed part and by 15% in the bank parts. Due to the opening of the contact, tensile stresses on the upstream face disappeared and tensile stresses at the foundation decreased from +4.25 MPa to +3.09 MPa (point 23). The calculation results showed the feasibility of constructive design of the joint notch to the depth of opening.

The subvertical joint in the foundation under the upstream face of the dam had on the whole a favorable effect: tensile stresses on the upstream face practically disappeared

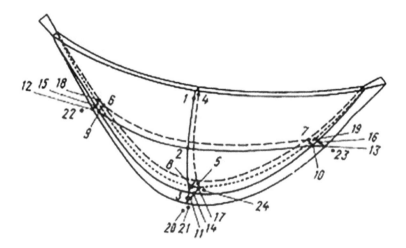

Figure 8.26 **Xiowan arch dam (China): 1–24 – points at which stresses were analyzed.**

(points 1–3); tensile stresses on the contacts decreased significantly (points 11 to 12); and the area of triaxial extension at the foundation under the dam disappeared (points 21–23).

As a result of the discussion of the variants considered with Chinese engineers, a subvertical joint in the foundation under the upstream face of the dam was recommended for further design. However, under a government agreement between France and China, the French bureau Coin and Bellier was involved in the design. French engineers reject any joints in arch dams arguing that the joints reduce the degree of hyperstatic property and thereby weaken its bearing capacity.

Dam Xiowan was erected without any constructive measures to connection with the foundation.

In order to enhance the arch effect, in some cases, the arch dam is divided into belts by subhorizontal joints, which are especially effective in narrow canyons (Vaiont arch dam, see Figure 1.29).

The dam in which the arch belts have three hinges, and the belts are separated by sliding joints is statically determinable, and there are no tensile stresses at the main loads, and compressive ones are evenly distributed over sections. Changes in ambient temperature and bank displacements slightly affect the stress state of such a dam. However, under seismic effects, especially across the gorge, its working conditions sharply worsen.

During the construction of the Lajanur arch dam in 1959, an experimental dam was constructed from three-hinge belts in the foundation pit as a downstream coffer dam (Figure 8.27). The dam, 13 m high, was cut with horizontal joints into three arches with a height of 5, 3.5, and 3.5 m, respectively. The purpose of the horizontal joints was to ensure the independent operation of the three arches (i.e., to exclude the dam from working in the vertical direction) for which the sliding joints were filled with bitumen mastic (30% bitumen and 70% cement).

Three hinges (in the key and heels) were formed by narrowing the cross section of the arches by 2.5 times and constructively reinforcing these zones. The studies showed

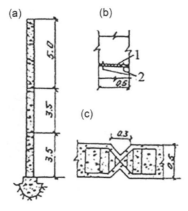

Figure 8.27 Experienced arch coffer dam in the downstream of the Lajanur arch dam: (a) transverse profile, (b) joint, and (c) hinge in the heel and key; 1 – bitumen mastic and 2 – key.

good agreement between the calculated and experimental values of stresses, displacements of arch belts, and hinge rotation angles [65].

However, excessive dividing of the dam with a sharp decrease in its rigidity requires serious justification given the variety of loads acting on the structure including seismic effects as well as the fact that the hinges and joints are to some extent "weak points", require waterproofing, complicate the work, and others.

With weak rock abutments, a joint notch can be arranged at the upper marks of the dam forming a "diving" crest of the gravity profile as on the Kurobe-4 dam (see Figure 8.20).

To reduce asymmetry and give the arch part of the dam a smooth contour and some reduction of its span on many dams bank gravity piers were made and plugs in the lower narrowing part of the gorge.

In arch dams, preliminary compression is sometimes used to improve its stress state especially when waiting for uneven deformations of the foundation. Thus, preliminary compression in the horizontal direction was carried out using flat jacks of thin arch dams: Nambe Falls (Figure 8.28), Caccia (Costa Rica), Belesar (Spain), and Clark (Australia).

By the *coefficient of shapeliness* $\beta = b/h$, where b is the thickness of the dam down and h the largest dam height, arch dams are divided as follows:

- thin ($\beta < 0.2$);
- thick ($\beta = 0.2–0.35$);
- arch-gravity ($\beta = 0.35–0.65$).

If there is a plug in the base of the dam (Figure 8.18e) or a saddle (Figure 8.18b), the shapeliness coefficient is often determined for the arch part of the dam.

By *height,* the arch dams are divided as follows:

- low at $h < 25$ m;
- average at 25 m $\leq h < 100$ m;

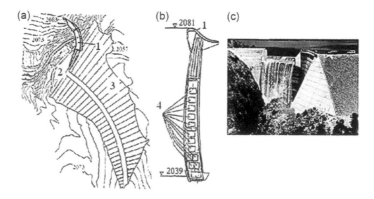

Figure 8.28 Nambe Falls arch dam (USA): (a) plan, (b) cross section, and (c) view from the downstream; 1 – spillway, 2 – abutment, 3 – dam from local materials, and 4 – jacks.

- high at $100\,\text{m} \leq h < 150\,\text{m}$;
- ultrahigh at $h \geq 150\,\text{m}$.

8.2.2 Application area of arch dams

For arch dams, the most favorable are the sites with a relatively small width of the gorge and with a V-shaped symmetrical cross-sectional shape composed of rocks that are sufficiently strong and uniform in deformability.

With increasing relative width of the gorge, the efficiency of arch dams decreases.

However, the large width and asymmetry of the site are not an obstacle to the construction of arch dams [76]. The ℓ/h ratio often reaches 5–6 and even 10: the arch dam of the Sayano-Shushensk HEP with a height of $h = 242\,\text{m}$ and $\ell/h = 4.56$ (see Figures 1.26 and 8.29), Valley di Ley ($h = 138\,\text{m}$ and $\ell/h = 4.6$), Punt dal Gall (Figure 8.30, $h = 130\,\text{m}$, and $\ell/h = 4.15$), Pangola (Figure 8.31, $h = 89\,\text{m}$, and $\ell/h = 4.25$), Kariba (see Figure 1.33, $h = 128\,\text{m}$ and $\ell/h = 4.88$), Porgolapurt ($h = 89$ and $\ell/h = 5.04$), Tul ($h = 86\,\text{m}$ and $\ell/h = 5.35$), Schlegeis (see Figure 8.4, $h = 131\,\text{m}$ and $\ell/h = 5.53$), Lauza (see Figure 8.48, $h = 51\,\text{m}$ and $\ell/h = 5.69$), Henrik Verwoerd ($h = 86\,\text{m}$ and $\ell/h = 8.83$, see Figure 8.12), Piave di Cadore ($h = 57\,\text{m}$ and $\ell/h = 7.20$), Schifferen; ($h = 47\,\text{m}$ and $\ell/h = 8.87$, see Figure 8.25), Moulin Ribou ($h = 16.2\,\text{m}$ and $\ell/h = 10$), and other dams in relatively wide sections have been built and have been successfully exploited for many years.

In the SR [14], it is recommended to design arch and arch-gravity dams at $\ell/h \leq 5$. Under favorable natural conditions, the volume of concrete in arch dams can be reduced by 50%–70% compared with the volume of concrete in gravity dams.

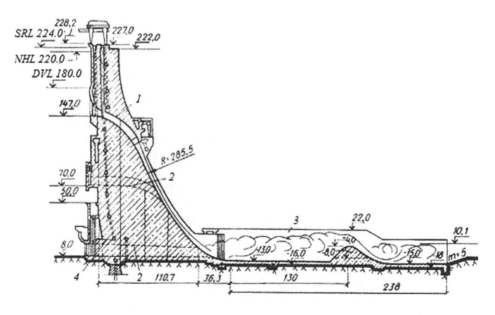

Figure 8.29 Arch-gravity dam of the Sayano-Shushensk HEP. Cross section; 1 – operational spillway openings, 2 – construction (temporary) spillways, 3 – water well, and 4 – joint notch (in the bed part).

(a)   (b)

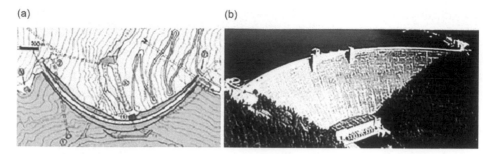

Figure 8.30 Punt dal Gall arch dam (Switzerland): (a) plan, (b) view from the downstream.

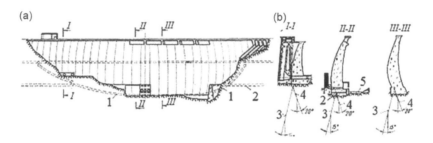

Figure 8.31 Pangola arch dam (South Africa). (a) view from the downstream and (b) cross sections; 1 and 2 – cementing gallery and tunnel, 3 and 4 – cementing and drainage curtains, and 5 – water well.

Arch dams are built as both deaf and spillway with surface and deep spillways including the following:

- with the rejection of a freely falling stream, for example, spillways – Ambiesta (Italy, Figure 8.33), 56 m high and Kurobe-4 (see Figure 8.20) and with deep openings – Kariba (Zimbabwe, see Figure 1.33) and Boundery (USA, see Figure 2.28), 116 m high;
- with a spillway downstream face of arch-gravity dams, for example, Aldead-avila (Spain, see Figures 8.5 and 8.35) and Sayano-Shushensk or arch with special spillway slabs attached from the downstream, for example, Picoti (Portugal, Figure 8.32), 95 m high, and Ovan Spin (Switzerland, see Figure 2.27), 73 m high.

Spillways are also carried out at the piers of dams, for example, Santa Eulalia (Spain, see Figure 8.21), 74 m high, or on the bank parts of the arch dam of Hitoshuse (Japan, see Figure 8.42).

At many HPs behind dams, HEP buildings are located, for example, Glen Canyon (h = 216 m, USA), Nagowado (156 m, Japan), and Chirkei (h = 232.5 m, Russia), while

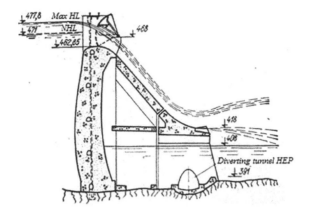

Figure 8.32 Picoti arch dam (Portugal) 95 m high.

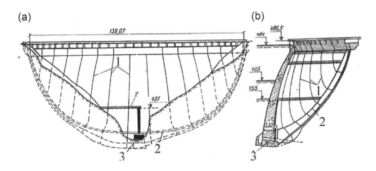

Figure 8.33 Ambiesta arch dam (Italy): (a) view from the upstream and (b) section along the center console; I – temporary construction joints, 2 – perimeter joint, and 3 – chute.

water inlets and head pipelines of the HEP are located in the body or on the downstream face of the dam (Figure 8.33).

In sites with gentle banks, in some cases, an arch dam is erected in the central bed part of the gauge and buttress or gravity dams, for example, Rozeland dam, 150 m high, in France is constructed on the banks (see Figure 1.38).

Arch dams are erected under various engineering-geological conditions, including complex ones, as well as in conditions of high seismicity and in severe climatic conditions.

Significant defects and heterogeneity of the rock foundations cannot serve as a reason for refusing to erect an arch dam, thanks to the great practical experience in engineering reinforcement measures in the foundation, including cementation, anchoring, installation of concrete walls, dowels, gratings, etc. (see Section 4.4) and the development of calculation methods to justify their design decisions. At the same time, technical and economic efficiency is a decisive factor.

Experience shows that in difficult geological and topographic conditions, the quality of the project is crucial. And the point is not only to choose the optimal shape of the dam and design measures but also to accurately determine the deformation and strength properties of the concrete of the dam and the rock mass in order to correctly take into account in the calculations the operating conditions of the "dam-base-reservoir" system based on the mode of drawdown and filling of the reservoir.

The incorrect ratio of the deformation of the rock foundation and concrete as well as the neglect of the annual cycles of drawdown and filling the reservoir laid in the design led to the formation of significant cracks in the rock foundation of the arch dams of Kelnbrein (Austria, Figure 8.34), Zeitzir (Switzerland), and Rud-Elsberg (South Africa). As a rule, rock foundations poorly resist cyclic and especially sign-alternating loads.

The Kelnbrein arch dam, 198 m high, crest 626 m long, and $\ell/h = 3.10$ (dam thickness along the crest 8 m and the bottom 42 m), was erected in 1978 to power generation. The dam was built in massive granite gneisses at an altitude of 1,705 m above the sea level. The dam site is composed of rocks with weak fissuring without serious geological disturbances.

Filling the 1977 reservoir to an intermediate mark did not cause any problems. However, when the level increased by 20 m below NHL, the filtration discharge reached 35 L/s, and a further increase of 10 m caused an unexpected increase in filtration discharges to 130 L/s [79,101].

During 1978–1984, the cementation curtain and drainage at the dam base were reinforced, the foundation was frozen under the upstream face of the dam, concrete apron was installed at the upstream face of the dam, cracks that appeared in the dam were isolated, repair work was performed on the apron, and cracks were found at the upstream face and in the inner zone of the dam body (Figure 8.34b).

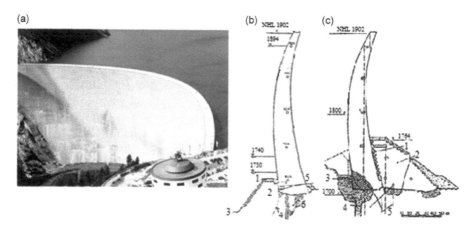

Figure 8.34 Kelnbrein arch dam: (a) view from the downstream; cross sections: (b) before repair; 1 – apron, 2 and 3 – first and second cementation curtains, 4 – drainage, 5 – cracks in the dam, and 6 – cementation; (b) after repair: 1 – neoprene gaskets, 2 – concrete thrust block, 3 – synthetic resins, 4 – cement grout, and 5 – drainage.

Computational studies have shown that the arch dams have a high bearing capacity (safety factor of 4.5) even under the assumption that the dam carries a hydrostatic load only due to thrust in the bank.

Super deep annual drawdown of the reservoir, higher than provided for in the design, and high modulus of deformation of the rock foundation can be considered as the reasons for this work of the dam.

A thrust concrete block, 65 m high, was constructed on the downstream of the dam in the bed part (Figure 8.34c) providing a perception of a slide force of 1.2 million tons, which is 22% of the total hydrostatic load acting on the dam.

8.3 Design of arch dams

The choice of type and design of arch dams depends on the following:

*   natural conditions (climatic, hydrological, topographic, and engineering-geological features of the site);
*   layout of the HP (type and location of spillways and HEP);
*   construction conditions (including passing construction discharges and erection conditions);
*   operating conditions;
*   traditional approaches to the design of arch dams in a given country;
*   technical-economical comparison of variants.

Moreover, all considered variants should satisfy the reliability conditions of the "dam foundation" system.

Based on these conditions, a choice is made in which on the basis of detailed studies and calculations taking into account the data of detailed geological surveys and studies, the design and landing of the dam and strengthening and antifiltration measures at the foundation are specified [107].

The layout of HPs with arch dams with different locations of spillways and HEP buildings, which can have a significant impact on the type and design of the dam are discussed in Chapter 2.

8.3.1 Accounting of topographic and engineering-geological conditions

According to the SR operating in Russia [14], it is recommended to design arch dams with the following:

* $\ell/h \leq 2$ in a triangular shape of the gorge with arches of circular shape of constant thickness or local thickening at the heels;
* $2 \geq \ell/h \leq 3$ in the trapezoidal or close to it form of a gorge of double curvature with arches of variable thickness and curvature;
* $\ell/h > 3$ – arch and arch-gravity dams in which the curvature in the vertical direction is determined from the conditions for obtaining the optimal stress state of the dam;
*   asymmetric cross sections and on inhomogeneous foundations with arches of non-circular shape of variable thickness.

When choosing the location of the dam, it is necessary to ensure reliable support of the dam on the banks of the gorge and the stability of bank abutments.

On this basis, dams should not be placed upstream:

- a sharp expansion of the canyon;
- a ravine that cuts the side of the canyon;
- a sharp turn of the canyon.

When placing the dam in a symmetric site, more favorable working conditions of the arch dam are provided.

In the case of an asymmetric alignment or its irregular shape, it should not achieve symmetry of the support line of the dam due to a significant increase in cuts and volume of the rock excavation.

The adverse effect of the asymmetry of the gauge on the stress state of the dam should be neutralized in the first place by selecting the appropriate law for changing the curvature and thickness of the dam, which allows a relatively symmetrical stress distribution to be obtained. In the case of sections with sharp fractures in the site to give a smooth outline to the contour of the dam in addition to the variant with removing the rock the device the following should be considered:

- gravity piers at the upper elevations;
- plug in the lower narrowed part of the site;
- expanded pillow (saddle) separated from the actual arch part of the dam by a perimeter (contour) joint and which is as it were an extension of the foundation.

A number of dams were built in highly asymmetric cross sections, for example, Piave di Cadore, Guzana (Figure 8.35), and Schlegeis (see Figure 8.49).

When arranging the perimeter joint of the arch part of the dam, full symmetry is not always given, for example, the Frera dam (Figure 8.36).

The shape of the cross section of the gauge has a significant effect on the work of the dam and its volume. With a rectangular, trapezoidal, and triangular cross-sectional

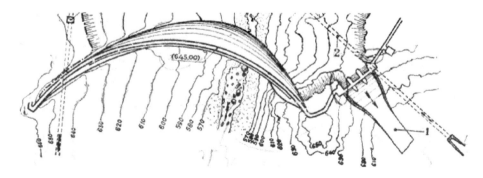

Figure 8.35 Guzana arch dam (Italy).

(a)

(b)

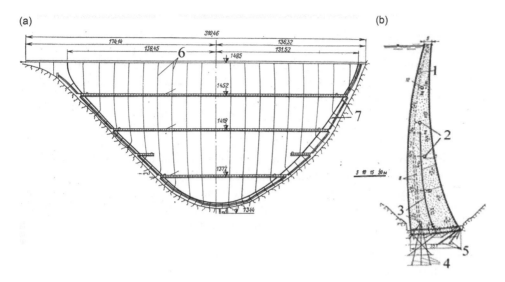

Figure 8.36 Frera arch dam (Italy): (a) view from the downstream and (b) cross section; 1 – drainage of the dam body, 2 – viewing galleries, 3 – longitudinal joint, 4 – cementation curtain, 5 – drainage, 6 – sectional joints, and 7 – perimeter joint.

shape of the gauge with the same ℓ/h ratio, the smallest concrete volume is achieved with a triangular shape, in which the spans of the arches and thickness significantly decrease downward, and in general, the total hydrostatic pressure on the dam is less.

Arch dams in many cases are built in difficult engineering-geological conditions, which requires a set of engineering measures to improve the rock foundation. The composition of reinforcing and antifiltration measures depends on the engineering-geological conditions, including deformation, strength, slide, filtration characteristics of the rock mass of the foundation and abutments, their heterogeneity, fissuring, the presence of large tectonic disturbances, fissures filler, and natural stress state.

Inhomogeneous deformability and the ratio of the deformation modulus of the rock foundation and concrete of the dam have a significant impact on the SSS of arch dams. The most favorable from the point of view of the impact of dam and foundation on the SSS are the E_f/E_c relations of the order of 0.25–0.5. However, arch dams are built both in conditions of very rigid foundations and very malleable, which leads to a certain deterioration in their SSS.

The Schifferen dam, 47 m high, in Switzerland (see Figure 8.25) was built on weak sandstones with the ratio $E_f/E_c = 0.1$. The analysis of the stress state of the Val-Kurner dam with a height of 152 m in Switzerland with a change in E_f/E_c in the range from 0.5 to 0.05 showed that the largest compressive stresses were 5.7 MPa and increased to 8.2 MPa when $E_f/E_c = 0.05$.

An increase in the rigidity of the foundation of the Kariba dam with a height of 128 m led to tensile stresses in the arch of +0.6 MPa at $E_f/E_c = 0.25$, and at $E_f/E_c = 1$, it increased to +4.3 MPa [65].

The increased deformability of the rock massif uniform in area of the bottom of the dam is not the basis for the removal of rock. And in these cases, one should consider variants with the implementation of engineering measures and changes in the design solutions of the arch dam (see Sections 4.3 and 4.4).

8.3.2 Interaction with the foundation

The connection of the arch dams with the foundation is carried out by cutting it into the rock foundation and bank abutments and is characterized by certain features. The rational insertion depth is determined on the basis of the conditions for ensuring the stability of the bank abutments and the favorable stress state of the dam and rock foundation, based on the technical-economic comparison of the variants [59,75,101]. After excavation of the pit, the surface of the foundation pit is treated and cemented; the strength cementation and the antifiltered curtains are carried out to interface the dam with the foundation.

When cutting an arch dam into the site, the desire to maximize the advantage of the arch effect leads to the inevitability of embedding an arch dam into the banks of canyons. In this regard, the volumes of rock insets in arch dams as a rule are larger in comparison with gravity dams. When choosing the type of dam of the Toktogul HEP, it was the large volume of the rock embedding that was decisive when abandoning the arch dam: steep banks at an angle of 65°–75° extending 400–600 m above the crest of the dam in combination with unfavorably oriented cracks of on-board rebuff would necessitate insertion 500 m high above the dam. At the Chirkei HEP where the crest of the dam almost coincides with the plateau, successful embedding under the dam was possible, despite the presence of cracks in the bank unloading.

A problem similar to the Toktogul one arose during the construction of the Kukuan (Taiwan) arch dam, 86 m high, where even a small embedding led to enormous excavations due to the loss of stability of steeply falling quartzite shale formations. To prevent the collapse of these layers, the embedding of the dam into the bank abutments was carried out by the mining method.

On many arch dams, landing was carried out with the help of a saddle, plug, and piers, which allowed to reduce the rock excavation.

Experience shows that arch dams respond quite calmly to possible movements and displacements of their abutments. Studies performed on models at the Massachusetts Institute of Technology (USA) showed that the stress state of the arch dam remains virtually unchanged when the displacements of the abutments are within 15% of the deflection of the dam crest.

Since there was a possibility of movement along the fault in the right-bank abutment of the Inguri arch dam, to ensure the preservation of the arch part of the dam, the saddle was cut by a system of joints (Figure 8.37) the purpose of which was to neutralize the movements.

The presence of a tectonic zone of small thickness at the foundation of the arch dam has practically no effect on the stress distribution in it with the exception of purely

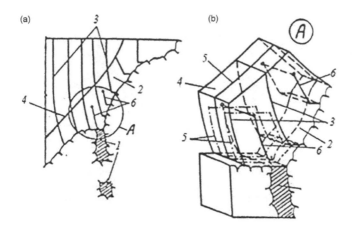

Figure 8.37 The design of the saddle on the Inguri arch dam (a) node A and (b): I − embedding fault, 2 − saddle, 3 − intersectional joint, 4 − perimeter joint and 5 − longitudinal joint parallel to the middle surface of the dam; and 6 − joint system parallel to the fault.

local zones where special measures need to be taken. The problem is compounded when the width of the weakened zones increases to sizes commensurate with the thickness of the dam. In this case, a significant effect on the alignment of the deformation modulus is given by reinforcing cementation.

Strengthening cementation is an effective means of changing the properties of the foundation of arch dams. Studies of limestones at the foundation of the Inguri dam showed that after cementation, the deformation modulus of rock increased 1.5 times and in some cases 2.5 times [145].

Resistance to slide of bank abutments and the foundation of the arch dam is the main factor determining the depth of the embedding. The local strength of the rock foundation under the downstream face of the dam should also be considered when designing the embedding. So, when designing the arch dam Oimapinar (France), the desire to create a favorable stress state in the bank abutments and the necessary direction of the resultant arches on the heels led to a rock excavation with a volume of 650 thousand m^3 with a concrete volume of the dam of 550 thousand m^3.

High water permeability cannot justify additional removal of rock; a decrease in water permeability and a decrease in the force of the filtration flow can be achieved due to the rational design of cementation and drainage curtains.

The support of the dam on the rock foundation should be carried out on a surface normal to the axis of the arches of the dam. At the same time, the conjugation of the arches with the banks by means of radial heels to the maximum extent ensures the involvement of the rock foundation in the work and, however, leads to a large volume of rock excavation. To reduce the embedding and accordingly the volume of the recess of the heel of the arches, a curved or broken outline (Figure 8.38), spoon-shaped,

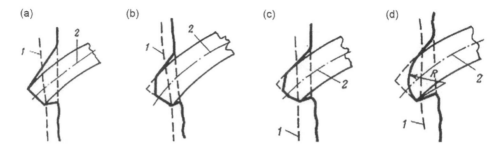

Figure 8.38 The contours of the heels of the arches when embedding into the bank of the site: I – undisturbed rock, 2 – arch axis: (a) radial; (b) semi-radial with cutting of the upper corner; (c) polygonal; and (d) curved (spoon-shaped and circular).

polygonal, and semi-radial, is used. A stepped outline is not recommended due to stress concentrations in the foundation under the heel.

Under the condition of ensuring stability of bank abutments, it is extremely undesirable to remove rocks behind a dam even if unstable massifs are identified on them, since the removal of the outer parts of a rock mass involves as a rule unloading and violation of the rock mass. Such mass is recommended to be fixed using engineering measures (anchors, etc.).

When connecting a dam with a rock foundation, the necessary provisions should be made for (see Sections 4.3 and 4.4) the following:

- reinforcing cementation;
- filling faults, large cracks, and caverns by installing concrete and reinforced concrete gratings, dowels, plugs, or solid concrete masses;
- arrangement of underground reinforced concrete walls and buttresses for transferring forces from the dam into the interior of the rock mass with enhanced strength characteristics;
- use of prestressed and nonstressed anchors, retaining walls and their combination.

To reduce stresses at the contact of the dam with the foundation, one should consider the device of local thickening of the dam along the support contour.

Thus, the size of the embedding of the arch dam into the rock foundation should be determined on the basis of a feasibility study with consideration of alternative solutions for strengthening the rock foundation, changing the structure of the dam and connection structures, changing the site, and other activities.

In the case of elastic cutting, it is usually strived to give the support contour of the dam a smooth convex shape toward the bank since sharp fractures of the support contour provoke the appearance of local stress concentrations, and the concavity of the contour toward the dam can lead to a noticeable deterioration in its overall stress state. To improve the interface between the dam and the foundation, piers, plug, "diving" crest, and other structural solutions can be used.

The arrangement of bank piers in the upper part of the gorge is advisable in case of insufficient strength or increased deformability of the foundation rocks at the upper elevations.

Dams with gravity piers on both banks are known: Lonqyangxia (China); Mansour Eddahby (Morocco, Figure 8.39); Almendra (Spain, Figure 8.40); with a stand on one bank: Dongjiang (China); Hitoshuse (Japan, Figure 8.42), etc.

Sometimes, special wings are included in the structure of piers, with the aim of relieving the pier from the action of hydrostatic loading and filtration uplift: the Inguri dam (see Figure8.23); Almendra (Figure 8.40), La Palis (France), etc.

Figure 8.39 **Mansur Eddahbi arch dam (Morocco).**

Figure 8.40 **Almendra arch dam (Switzerland).**

In gorges with a pronounced narrowing in the bed part, a plug is arranged in the lower part – a massive concrete structure, which in terms of its rigidity can be calculated in terms of a rock foundation; for example, Chirkei dam (Figure 8.41).

In some cases, piers are made in the form of a small arch dam: Santa Lucia (Portugal, Figure 8.43) and Cambamba (Angola).

According to the SSS conditions, it is advisable to separate the plug from the arch part of the dam with a constructive joint as was done at the arch dam of the Miatla HEP (Figure 8.44).

If there are weakened rocks on the upper sections of the gorge banks, the strengthening or removal of which is irrational for one reason or another, the so-called "diving" crest can be constructed.

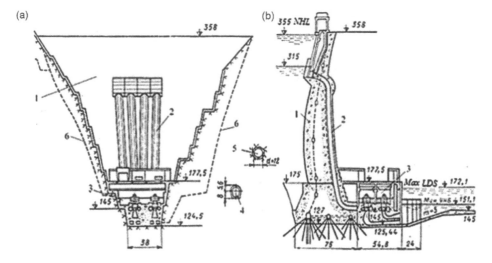

Figure 8.41 Chirkei arch dam: 1 – dam, 2 – HEP, 3 – building of the HEP, 4 – construction tunnel, 5 – spillway, and 6 – circuit cutting.

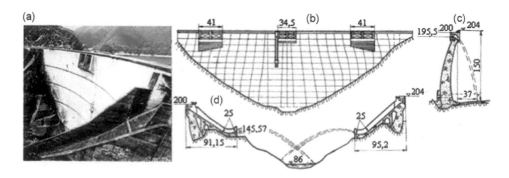

Figure 8.42 Hitoshuse arch dam (Japan) with three weirs in the central part and in the bank piers: (a) view from left bank and downstream; (b) view from the upstream; (c) section along the center; and (d) view from the downstream.

Figure 8.43 Santa Lucia two-arch dam (Portugal).

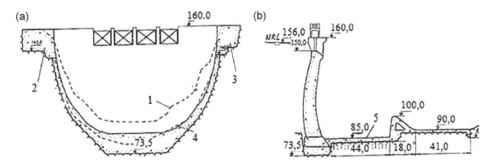

Figure 8.44 Miatla arch dam: (a) view from the upstream and (b) section along the center console: 1 – natural surface, 2 and 3 – left and right bank piers, 4 – perimeter joint, and 5 – water well.

A "diving" crest is formed by leaving cemented construction joints in the upper part of the bank sections of the dam uncemented so as to direct thrust forces below the crest. Nonmonolithic side sections of the dam operate as gravity. A certain constructive solution was implemented on the Kurobe-4 dam (Japan, see Figure 8.20), where the rock at the upper elevations had a deformation modulus of more than four times lower than the rock lying in the lower part of the gorge.

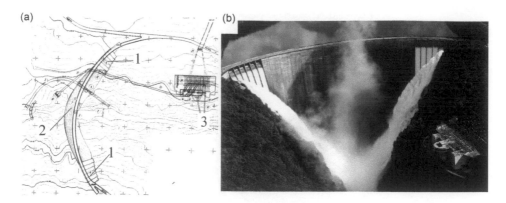

Figure 8.45 **Contra arch dam (Switzerland): (a) plan and (b) view from the downstream.**

Under adverse natural conditions (wide river valleys with a developed bed part, as on the Kontra dam (Figure 8.45), very hard rock foundations, etc.), when it is not possible to get rid of the tension in the dam zones adjacent to the foundation, by optimizing the shape of dams, it is recommended to consider the device in the contact section of the joint notch, hinges, sliding joints, elastic gaskets, etc.

A horizontal joint notch was made in the four central sections of the Lauza arch dam (Figure 8.46) 50 m high and 285 m long along the crest. In the 63.5-m-high arch dam Sibing built in rolled concrete constructed in China (1995), to reduce tension in the lower third of the dam height, a joint incision is made on the upstream face of the dam [83,207].

When installing a joint notch to relieve vertical tensile stresses in a dam above contact with the foundation, it should be noted that with a high location of the joint notch, it can be ineffective due to the rapid decrease in tensile stresses with distance from the contact. At the same time, with a low location of the joint notch in the part of the dam, significant tensile stresses arise under it, which can lead to its separation from the rest of the dam [35].

Horizontal hinges were used in the Moulin-Ribou dam (Figure 8.47) in France with a height of 16.2 m with a crest length of 162 m [65].

It should be noted that in arched dams with a height of more than 150–200 m, in relatively wide sections, $\ell/h > 2.5$, it is usually not possible to avoid significant tensile stresses along the supporting contour of the dam from the side of the upstream face, as on the Mauvoisin arch dam, which affects the reliability of the dam (Figure 8.48).

In order to prevent rupture of the cementation curtain in the contact zone of the rock foundation (caused by tensile stresses under the upstream face of the dam), it can be carried into the upstream by means of a concrete apron (see Section 4.4.3). Such a solution was used on a number of high dams, for example, Kelnbrein (see Figure 8.34), Zillergründl, and Bor.

The effectiveness of the apron device should be justified by computational studies of the SSS and the filtration regime of the "apron-dam-foundation" system. If necessary, the apron is prestressed. Particular attention should be paid to the justification

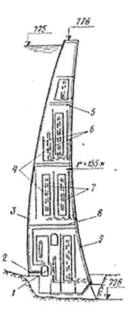

Figure 8.46 Lauza arch dam (France): I – gallery, 2 – uncemented notched joint with two dowels, 3 and 9 – dowels, 4 – recementation valves, 5 – drainage grooves, 6 – grooves of the Ist cementation stage, 7 – 2nd cementation stage, and 8 – horizontal keys.

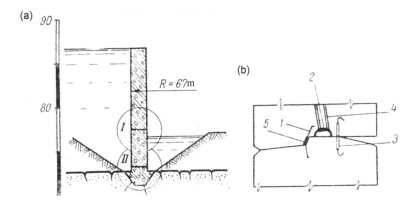

Figure 8.47 Mulen Ribou arch dam (France): (a) vertical section and (b) joint detail II: I – sheet, 2 – bitumen, 3 – dowel, 4 – well, and 5 – step.

of the water tightness and durability of the apron interface with the upstream face of the dam, on which the reliability of the dam depends.

Another possible way to increase the reliability of high and ultrahigh arch dams is to install a vertical slit-seam in the foundation under the upstream face of the dam with a depth of 5%–10% of the dam height, as was used at the Schlegeis Dam in Austria.

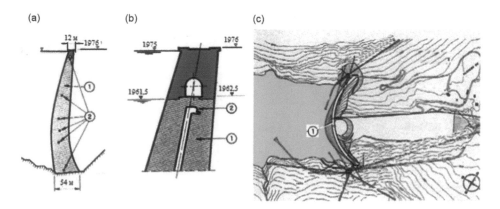

Figure 8.48 Mauvoisin arch dam (Switzerland): (a) section along the center console, (b) a crest, and (c) a plan.

The Schlegeis arch-gravity dam, 131 m high, 725-m-long crest, and $\ell/h = 5.5$ (the thickness along the crest is 9 m and the bottom is 34 m), was erected in 1971 in order to generate electricity. At the last stage of filling the reservoir, the filtration costs at the foundation increased sharply to 251 L/s with 90% of the filtration concentrated in the 150 m bed part of the dam. Piezometers showed 90%–100% of pressure in this zone and strain gauges indicated the opening of the contact between the dam and the foundation under the upstream face; they also provided information that tensile stresses propagate deep into the base to only 5 m.

Computational studies have shown that the most radical measure to relieve tensile stresses at the contact and under the dam turned out to be a vertical slip in the foundation with a depth of only 5 m arranged at the bottom of the viewing gallery. After testing the technology for creating a slip in the experimental section, such a slip was created in the entire 150 m section where increased filtration was observed (Figure 8.49). After filling the reservoir, the filtration discharge decreased by 10 times – from 251 to 25 L/s.

The possibility of applying such new solutions should be justified by appropriate SSS calculations of the "dam foundation" system and an analysis of the design features and its reliability in specific natural conditions.

8.3.3 Preliminary definition of arch dam parameters and inscribing

The process of designing an arch dam begins with a preliminary selection of its main dimensions, which is carried out using empirical formulas, statistical relationships, analogs, and approximate calculation methods.

The statistical dependences of the Bureau of Reclamation of the USA obtained on the basis of processing the parameters of more than a hundred dams [142] are the most popular as a tool for the operational determination of approximate geometry parameters of arch dams. The construction of the center console (Figure 8.50) is

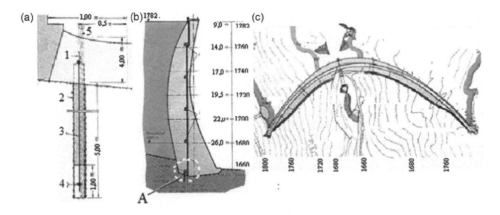

Figure 8.49 Arch dam Schlegeis (Austria): (a) detail A: I and 4 – metal sheet, 2 – well (slit), 3 – PVC sheet, and 5 – tube; (b) vertical section, and (c) plan.

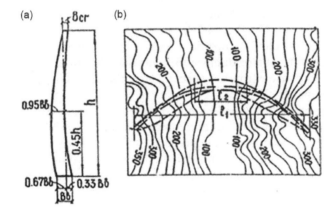

Figure 8.50 Construction of the center console (a), (b) plan.

performed according to the given maximum construction height h, the lengths of the chords of the crest arch ℓ_1 and arch ℓ_2, located at a height of $0.15h$ from the base of the console.

The thickness of the dam on the crest is

$$b_{cr} = 0.01(h + 1.2\ell_1) \tag{8.3}$$

To determine the thickness of the dam on the crest, taking into account its longitudinal stiffness and placement on the crest of the road, the adjusted formula (8.3) is recommended:

$$b_{cr} = 0.0145(h + 2r_{cr}) \tag{8.4}$$

where
r_{cr} is the radius of the axis of the crest arch.

The thickness at the base is determined by the formula

$$b_b = \sqrt[3]{0.0012\, h \times \ell_1 \times \ell_2\, (h/122)^{h/122}} \tag{8.5}$$

The concrete volume of the dam body is determined by the dependence

$$V = 0.000171\, h^2 \ell_2 \frac{(h+0.8\ell_1)^2}{\ell_1 - \ell_2} + 0.0108h \cdot \ell_1 (h + 1.1\ell_1) \tag{8.6}$$

The above dependences are valid for

$$30 \le h \le 365 \text{ m}; \ 30 \le \ell_1 \le 1{,}830 \text{ m}; \ 4.5 \le \ell_2 \le 365 \text{ m} \tag{8.7}$$

As an analogue, it is recommended to use one of the constructed or designed dams with close heights, the ratio of the length of the crest to the height ℓ/h, the coefficient of shapeliness $\beta = b_b/h$, the shape of the gorge, engineering-geological conditions, etc. Having accepted for the designed dam, taking into account the geometric scale and the dimensions of the analog dam (radii, central angles, thicknesses, deviations of the axis of the center console from the vertical, etc.), the dam is inscribed into a given site, adjusting the geometry in place.

Figure 8.51 shows *R. Schroeder's* nomogram for preliminary determination of the shapeliness coefficient $\beta = b_b/h$ of arch dams of cylindrical shape (C), with a constant central angle (CA) and dome (D): depending on the relative section width ℓ_1/h (ℓ_1 – the width of gorges along the arch chord at the level of the dam crest), the shape of the valley profile in the dam site $\psi = s/h$ (s – the height from the bottom to the center of gravity of the gorge cross section to the mark of the dam crest), the central angle of the arch on the crest 2α. This does not take into account differences in the deformability of rock and dam concrete, the height of the dam, and the vertical component of hydro-static pressure on the dam.

The ratio β/β_b is given on the nomogram, where $\beta_b = 0.65$ is the value of β adopted by R. Schroeder at $\alpha = 0$ (i.e., for a gravity dam).

Figure 8.51 shows examples of the determination of β for three dams:

- type D (dashed line) with $\ell_1/h = 2.09$; $\psi = 0.64$; $2\alpha = 87° - \beta/\beta_0 = 0.245$; $\beta = 0.65 \times 0.245 = 0.16$;
- type CA (dash-dot line) with $\ell_1/h = 2.08$; $\psi = 0.63$; $2\alpha = 126° - \beta/\beta_0 = 0.46$;
- $\beta = 0.65 \times 0.46 = 0.30$;
- type C (dash-dot line) with $\ell_1/h = 3.3$; $\psi = 0.652$; $2\alpha = 100° - \beta/\beta_0 = 0.9$; $\beta = 0.65 \times 0.9\ 0.57$.

The asymmetry of the profile is taken into account as follows: the values are calculated: $2\ell_1'/h$ and $2\ell_1''/h$ (Figure 8.51) and the corresponding values of ψ' and ψ'' for

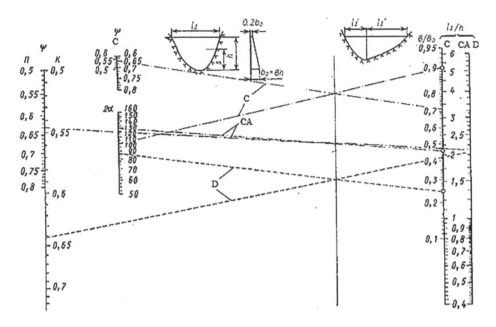

Figure 8.51 Nomogram for preliminary determination coefficient of arch dam shape – lines β for cylindrical (C), with a constant central angle (CA) and dome (D).

symmetrical valleys with a width at the level of the dam crest of $2\ell_1'$ and $2\ell_1''$, respectively; then the values of β' and β'' are determined from the nomogram, and then either the most cautious variant of the profile is accepted (the maximum of the obtained two values of β) or the average of the obtained values of β.

For an approximate determination of the concrete volume of the dam, the graph shown in Figure 8.52 can be used with which the relative width of the dam ℓ/h determines the relative concrete volume v m^3 per 1 MN load. By multiplying υ by the value of the hydrostatic load acting in the site of the dam, an approximate volume of concrete of the dam that meets the current level of development of arch dam engineering is obtained [126].

After constructing the profile of the center console, the dam is inscribed horizontally on the surface of a healthy rock mass, and the depth of cutting the rock foundation is determined.

When inscribing, one should adhere to the principles of smoothness and continuity of all geometric parameters of the dam in height which are fundamental for arch dams, because any irregularities, in particular, jumps in arch or cantilever curvatures and fractures of the supporting surface, give rise to stress concentration.

In this sense, graphs of changes in geometric parameters along the height of the dam are useful. The formula can be used to determine the radius of the axis of the crest arch

$$r_{cr} = 0.6\ell_1 \tag{8.8}$$

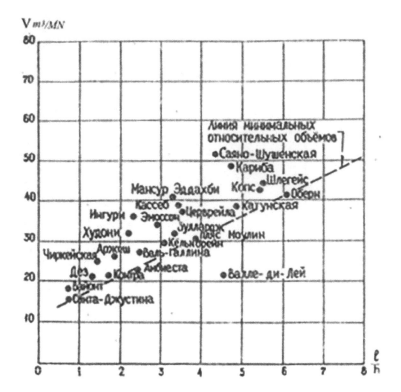

Figure 8.52 Graph for determining relative volume *v*.

The central angle of the crest arch from the conditions of ensuring a favorable stress state of the dam and the stability of the bank rock abutments is recommended to be taken not less than 90° and not more than 120°.

The inscription is made along several arch belts (slice plans) the amount of which is usually taken in the range of 6–12 depending on the height of the dam. The smoothness of the geometrical parameters of the dam is checked and if necessary, the radii, thicknesses in the key and heels, the central angles as well as the sweep of the upstream and downstream faces of the dam on a vertical plane are corrected for height.

To a first approximation, the stability of the bank rock abutments is ensured if the straight line drawn at an angle of 30° from the tangent to the downstream face of the arch at the heel to the downstream remains inside the bank massif at the level considered.

8.3.4 Arch dams form selection

The relative width and configuration of the site significantly affect the selection of the shape of arch dams [126]. Figure 8.53 shows the sweeps of the contact between the downstream face and the foundation, combined in one conventional site of the reduced

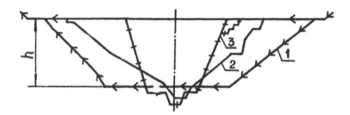

Figure 8.53 Sweeps of the upstream faces of arch dams combined in a conventional site of a reduced height *h*: I – Sayano-Shushensk, 2 – Inguri, and 3 – Chirkei.

height h for three arch dams: Chirkei, Inguri, and Sayano-Shushensk. As follows from the figure, the Chirkei dam is located in a relatively narrow site, its support contour has a shape close to a U-shaped one (see Figure 8.41) with the ratio of the site width at the level of the dam crest to the height $\ell_1/h = 1.5$ and the ratio of the crest and lower arch lengths $\ell_{cr}/\ell_b = 1.6$. The Inguri dam was erected in a site of medium width with $\ell_{cr}/h = 2.3$, which has a shape close to parabolic (see Figure 8.23). The Sayano-Shushensk dam was built in a very wide trapezoidal site (see Figure 1.28) characterized by $\ell_{cr}/h = 4$ and $\ell_{cr}/\ell_b = 2.2$.

According to the characteristics of the site, the Chirkei dam was designed with circular-shaped arches of constant thickness and a slightly curved profile of the center console (see Figure 8.41). The Inguri dam has arches of variable curvature (with maxima in the key and heels), variable thickness, and a substantially curved profile of the center console (see Figure 8.23). The Sayano-Shushensk dam is an arch-gravity type with three-center arches of constant thickness and a massive section of the center console with a vertical upstream face slightly cropped in the lower part (Figure 8.29).

During the designing, the shape of the dam is refined through its sequential improvement based on the criterion of minimizing the volume of concrete and rock cutting, while satisfying the conditions of a favorable stress state and stability of the "dam foundation" system.

The vertical curvature of the dam is checked and if necessary corrected based on the conditions of autostability (self-stability for slide and turning over) of columns and sections during the construction period.

When analyzing the slide and turning over stability of individual dam elements during the construction period, it should be in mind that

1. the generally accepted helicoidal (propeller) shape of intersectional joints and vertical grooving of their surfaces impede the independent operation of individual sections regardless of whether cementation is performed or not; the turning over stability several sections to the upstream is increased by shifting the common center of gravity to the downstream;
2. sections of arch dams erected on an oblique are often concreted into a spreader with an uphill slope of the pit, which in this case plays the role of a natural stop (Figure 8.54);
3. high dams as a rule are erected monopolized and loaded in stages; therefore, it is usually not necessary to ensure the stability of the dam sections when erecting

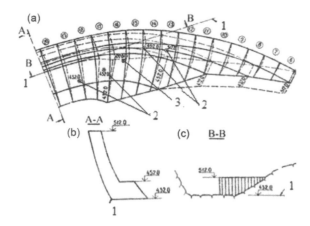

Figure 8.54 Scheme for testing for turning over of the first stage of the Katun dam [120]: (a) plan, (b) section along A-A, (c) section along B-B; 1 – axis of turning over and 2 – centers of gravity of arch sections on elevation 432.0, 452.0, 492.0, 512.0; 3 – center of gravity of the 1st stage.

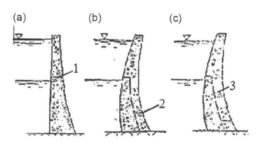

Figure 8.55 Scheme for the staged construction of arch dams: 1 – joint between the stages and 2 and 3 – longitudinal nonmonolithic and monolithic joint.

them to full height; sequential cooling and monolithic sections as the dam grows eliminate the problem of autostability (Figure 8.55).

Figure 8.55 shows three schemes for the staged construction of the arch dam: (a) the construction of the second stage without increasing the thickness of the first stage; (b) and (c) the construction of the second stage with a nonmonolithic and monolithic longitudinal joint, respectively

The arch dam Les Toules with a height of $h = 86\,m$ and $\ell/h = 5.35$ was erected in two stages. The first stage of the dam, 20 m high, served as a supporting structure for the dam of the second stage and was erected in 1958 (Figure 8.56).

The second stage was erected in 1963; cavities were left in the joint between the two dams which were cemented after filling the reservoir. However, the dam parameters (crest thickness 4.50 m, greatest thickness 9.90 m, and crest length 460 m) were too bold, and in 2011, the dam was strengthened: nine right- and eight left-bank sections

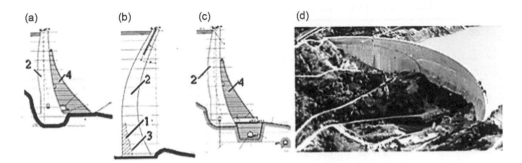

Figure 8.56 Les Toules arch dam (Switzerland): (a–c) the left-bank, central, and right-bank consoles; (d) view from the downstream: 1 – 1st-stage dam, 2 – 2nd-stage dam, 3 – cemented joint, and 4 – project reinforcement dam.

were thickened. The volume of thickening of the dense amounted to $70,000 \, \text{m}^3$ with the volume of the dam before strengthening being $235,000 \, \text{m}^3$. With insufficient autostability of the dam sections, various supporting structures including temporary ones can be used to increase their stability during the construction period (see Figure 8.12).

Changing the thickness of the arch along its axis is most often achieved by constructing its upstream and downstream lines from different centers (Chirkei, Kirdzhali, Frera, Piave di Cadore, Euni, Mauvoisin, etc.). The law of changing the thickness of such arches is usually very complex.

The favorable stress state of the arch dam should be considered in which

- strength conditions are satisfied for all points of the dam body;
- tensile stresses in the dam body are absent or have local distribution;
- the values of the main compressive stresses are close to the maximum permissible, which ensures the greatest degree of use of the strength properties of the dam concrete;
- distribution of arch compressive stresses in the dam body evenly.

The first condition meets the requirements of the SR [14], and the remaining conditions are necessary.

The conditions of the strength and stability of the rock foundation of the arch dam are required.

The most favorable stress state of the arch dam can be ensured by operating it as a moment-less shell. The shape of the middle surface of the dam in this case should be close to the shape of the pressure surface from the existing forces. However, for the vast majority of arch dams, the implementation of moment-less schemes can be performed as a rule only approximately.

When selecting the outlines of the cantilever elements, it is tended to maximally neutralize tensile stresses caused by hydrostatic loading and the dead weight of the section. To this end, the cantilever elements are cut in the lower part from the upstream side and tilt toward the downstream in the upper part. The crest part of the

Table 8.1 Description of the shape of the arches

Arch shape	The equation of the curve describing the arch	Dam parameters (country, year of completion, height (m), and ratio of crest length to height)
Circular	$\rho = const$	Chirkei (Russia, 1975, 233, 1.29)
		Vaiont (Italy, 1960, 262, 0.73)
Multicenter	$\rho = \rho_1, \varphi_1 \leq \varphi \leq \varphi_2$	Sajano-Shushensk (Russia, 1996, 245, 4.54)
	$\rho = \rho_2, \varphi_2 \leq \varphi \leq \varphi_3$	Glen Canyon (USA, 1964, 216, 2.12)
Parabolic	$\rho = p/(1 + e \times Cos\varphi)$	Muari (Switzerland, 1956, 148, 4.12)
		Limmery (Switzerland, 1963, 145, 2.57)
Elliptic	$\rho = p/(1 + e \times Cos\varphi)$	Schlegeis (Austria, 1972, 131, 5.53)
		Les Toules (Switzerland, 1963, 86, 5.35)
Pascal's snail	$\rho = 1 + a \times Cos\varphi$	Tolla (France, 1961, 88, 1.36)
		Biou (France, 1936, 56, 2.79)
Logarithm spiral	$\rho = k \times e^{\alpha\varphi}$	Vouglans (France, 1969, 138, 2.9)
		Henrik Verwoerd (South Africa, 1972, 90.5; 10.5)

dam thrown into the downstream is also advisable from the point of view of perceiving tensile seismic stresses at low water levels in the reservoir and the maximum deviation of the stream when a surface weir is installed on the crest of the dam.

Table 8.1 presents the equations describing the shape of the arches used in the construction of a number of arch dams.

A characteristic feature in the design of arch dams at the present stage is the representation of the shape of the dam (meaning the middle surface of the dam and the distribution function of thicknesses) in an analytical form due to the use of computers for computational research.

In the process of detailed design of the Inguri arch dam at JSC Hydroproject Institute, a computer program package was developed with the help of which computational studies of the dam strength were performed, and the main tasks of the structural geometry of the arch dam were realized: construction of dam sections (along arches, along consoles, and general view); construction and analytical description of intersectional, perimetral, and longitudinal seams; calculation of geometric characteristics of concrete blocks and elements of their removal in kind; and calculation of concrete volumes in the dam body.

8.4 Construction solutions of arch dams

8.4.1 Concrete zoning

The greater the heterogeneity of the stress state of the arch dam and the greater its volume, the higher the efficiency of concrete zoning. Zonal distribution of concrete is carried out by selecting the required strength for the set of stress fields obtained on the basis of computational studies for various combinations of loads and effects, while the requirements for water resistance and frost resistance for different zones must be met.

For example, with the zonal distribution of concrete in the arch-gravity dam of the Sayano-Shushensk HEP, the main volume of the dam is laid out of concrete of four classes: M250, W8, F100; M300, W8, F100; M300, W8, F200; and M400, W12, F500.

Rational concrete zoning made it possible to increase the use of the strength properties of concrete, improve the temperature regime of concrete masonry, reduce cement consumption by 300 thousand tons, and get a significant economic effect.

8.4.2 Intersectional, perimetral, and longitudinal joints

Intersectional joints. The modern technology of erecting arch dams provides in order to combat temperature cracking and uneven settlements cutting the structure with transverse (intersectional) joints 12–20 m long along the dam, which are cemented after cooling of the dam body to the design temperature.

The cutting step is determined by calculations of the thermal stress state of the dam sections during the construction process and during operation as well as structural and technological considerations.

Dams of small height as well as those having a shape close to cylindrical intersectional joints are designed flat and vertical. Intersection joints of arch dams must satisfy the following structural and strength requirements:

1. the lines of intersection of the surface of the joint with horizontal arches should be orthogonal to the axes;
2. the trajectories of the joints on the middle surface of the dam (the middle lines of the joints) should be as close as possible to the vertical;
3. the deviation of the joint from the vertical plane should be minimal;
4. the intersection of intersectional joints with a perimeter joint and contact with the foundation should be orthogonal or close to one.

Deviations of the trajectories of intersectional joints of the Inguri dam from verticals drawn through the joint points on the crest do not exceed 1–2 m at a column height of up to 230 m. At the same time, the traditional system of cutting using a center cylinder, for example, for the Tashan dam (Taiwan) designed by the company "Electroconsult" (Italy) led at lower column heights (160 m) to deviations of up to 4 m.

The perimeter joint previously used mainly to give the dam symmetry and a smooth outline of contact with the foundation (Italian dams Guzana, Ambiesta, Bearegard, etc.) has recently been arranged mainly to relieve tensile stresses that may occur at the contact of the dam with the foundation (due to seismic effects or deformation heterogeneity of the rock foundation) and lead to rupture of the antifilter contour of the dam.

The device of a massive saddle separated from the dam body by a perimeter joint allows reducing the stresses transmitted by the dam to the foundation, by increasing the width of the supporting surface to extinguish stress concentrations associated with unevenness and heterogeneity of the rock foundation within the saddle. In addition, it becomes possible to quickly close the rock excavation with concrete in order to protect the rock from decompression and erosion which, in turn, avoids additional rock excavation.

The Inguri (see Figure 8.23) and Myatla (see Figure 8.44) arch dams located in the area of high seismic activity were made with a perimeter joint. In some cases, the saddle may include bank abutments and plug as components (for example, the Pontesey arch dam, 93 m high, 150.2 m crest length, and $\ell/h = 1.61$, Figure 8.57).

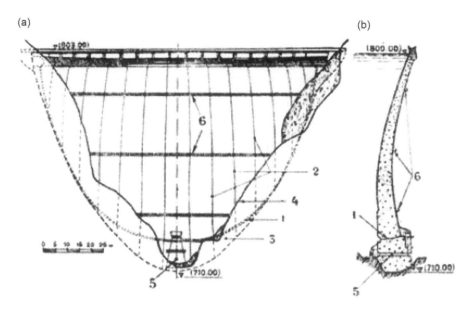

Figure 8.57 Pontesey arch dam (Italy): (a) view from the downstream (sweep) and (b) section along the spillway; I – perimeter joint, 2 – intersectional joints, 3 – plug, 4 – natural surface, 5 – drainage, and 6 – bridges.

The methodology for constructing the perimeter joint developed during the design of the Inguri dam in contrast to the known approaches used on the geometric representation is based on taking into account the force interaction of the dam and the saddle.

Cross sections of the perimeter joint are set in the form of an arc of a circle, which provides virtual rolling of the dam body along the surface of the saddle. Unknown circle parameters – radius r and angular displacement α of the center of the arc n from the unit vector γ tangent to the middle surface of the dam (Figure 8.58) are determined from the following conditions:

1. the normality of the arc to the lines of the upstream and downstream faces in the considered cross section;
2. passing through the center of the arc of the main vector of forces $\bar{R}$ transmitted from the dam to the foundation.

Longitudinal joints are used when erecting a dam in several columns, which can be caused by the requirements for crack resistance of concrete masonry during the construction period or by the desire to reduce the starting volume of concrete during construction of the dam by stages. Longitudinal joints are usually cemented although there are dams with an uncemented longitudinal joint, for example, Frera dam (see Figure 8.36). In the case of the construction of a dam in two columns, the longitudinal seam as a rule is combined with the middle surface of the dam.

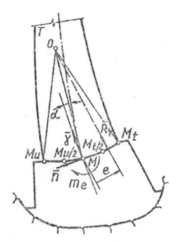

Figure 8.58 **Scheme to construction perimeter joint.**

8.4.3 Dam body drainage

The drainage of the body of arch dams is arranged similarly to drainage in gravity dams (see Section 7.3.4) in order to organize the interception and drainage of water filtered through the dam body and to prevent its entry to the downstream face of the dam. Drainage is carried out along the upstream face of the dam in the form of vertical wells (drains) with a diameter of 10–30 cm with a distance between the axis of the drains of 2–3 m. Drains have access to the longitudinal horizontal galleries from which the filtered water is discharged to the downstream.

The distance from the upstream face of the dam to the drainage axis a_{dr}, as well as to the upstream face of the longitudinal gallery, is assigned at least 2 m subject to the conditions [71]:

$$a_{dr} \geq \frac{H_d \gamma_n}{I_{cr,m}} \tag{8.8}$$

where
   H_d – the head at the design section;
   γ_n – the reliability coefficient for the purpose of the structure;
   $I_{cr,m}$ – the critical average head gradient, the value of which for arch and arch-gravity dams is taken equal to 50.

Concrete of the appropriate class for water resistance, strength, and frost resistance is laid from the upstream face of the dam to drainage.

A number of arch dams especially thin ones were erected without drainage of the dam body, which in some cases is considered undesirable: it may contribute to leaching of concrete due to large filtration gradients and reduce the effect of concrete swelling usually positive. So, the arch dam Valle di Ley with a height of 143 m and a length along the crest of 690 m was built in 1960 in Switzerland with the ratio $\ell/h = 4.82$ (Figure 8.59).

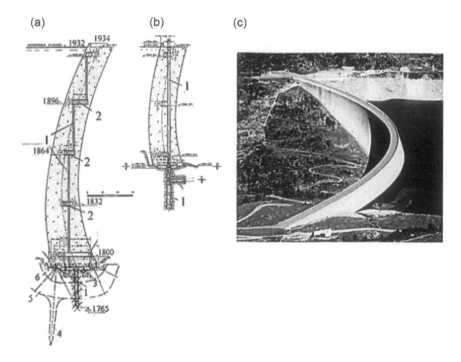

Figure 8.59 **Valley de Ley dam (Italy and Switzerland): (a) and (b) sections along the central and lateral consoles; (c) view from the left bank; I – plumb shaft, 2 – gallery, 3 – pump, 4 – cement curtain, 5 – zone high-pressure cementing, and 6 – the same low and medium pressure.**

8.4.4 Construction solutions of dam faces

The design solutions of the faces of arch dams as in gravity dams (see Section 7.3.5) are determined by climatic conditions, concrete zoning, construction technology, and operating conditions. Typically, zoning of concrete in the body of an arch dam provides normal conditions for its operation. However, under certain conditions, protective coatings of the faces of arch dams are used. On the upstream face of the arch dams, watertight screens are used under the same conditions as in gravity dams and are made of polymeric materials, reinforced concrete, asphalt concrete, asphalt-polymer concrete, etc.

On the downstream face of arch dams, in severe climatic conditions, thermal insulation is provided in the form of a heat-shielding zone of foam-epoxy compounds and other materials or a special heat-shielding wall forming a closed cavity with air heating. Thermal protection of the downstream faces of arch dams is extremely rare. So, heat shielding walls were applied on several low arch dams in Norway. There is a proposal to improve the SSS of arch dams in severe climatic conditions to regulate the temperature on the downstream faces by circulating heated air in a special closed casing. The performed studies of the SSS of the arch dam in a wide site showed that heating the downstream face by 23°C allows increasing compressive stresses on the upstream face of the dam by −1.5 MPa [35].

8.4.5 Reinforcement of dams

In arch dams, reinforcement is commonly used

- in areas where local stress concentrations can occur (around spillways and other openings, galleries in the dam body, at the mouths of nonthrough joints, in the places of profile fracture, etc.);
- on the outer surfaces of the dam to limit the size of temperature cracks (for example, in the form of a grid of reinforcement of class AII with a diameter of 16–25 mm to 4 rods per 1 m);
- at surfaces of contour joints;
  in the upper arches of the dam for the perception of seismic impact; for example, an antiseismic reinforcement belt is made in the upper quarter of the Inguri dam;
- at the bottom of the dam to prevent cracking due to stress concentration due to heterogeneity of the foundation.

In some cases, dispersed reinforcement is used, which increases the elongation of concrete and prevents the formation of cracks. Such reinforcement was used when the saddle of the Inguri dam was installed.

In arch dams, usually the reinforcement consumption does not exceed 5 kg per m^3 of concrete; sometimes, it reaches 10–15 kg m^3.

8.4.6 Spillways, water intakes, and head water conduits

The installation of spillway, water intake, and head pipelines of the dam HEP in arch and arch-gravity dams is their most important advantage, which allows reducing the total cost but can negatively affect their SSS and complicate the design of dams and the technology of their construction. Therefore, when choosing constructive-technological solutions of such dams, it is necessary to minimize the negative impact of spillways and HEP water intake on dams and ensure favorable construction conditions.

The passing of discharge in HPs with arch dams can be carried out through open spillways such as weirs or deep spillways combined with the body of the dam as well as through open or closed spillways located on the banks.

The following are used as the main schemes when combining spillways with the body of an arch dam as in gravity dams (see Section 7.3.7):

a. free overflow of jets into a downstream through a surface spillway with a discharge flow along the downstream face: Salime dam, $h = 139$ m high and spillway capacity $Q = 2,000 \, m^3/s$, see Figure 2.26; Sausels, $h = 83$ m, $Q = 11,200 \, m^3/s$ in Spain; the Ova Spin dam, $h = 73.5$ m (1968) with a spillway above the HEP building in Switzerland; (see Figures 1.35 and 2.27) and others;

b. throwing the jet to a considerable distance from the dam using a special vertex on the crest: the Inguri dam, $Q = 2,200 \, m^3/s$ (see Figures 1.27 and 8.23); Mossy-rock, $h = 184$ m and $Q = 2,200 \, m^3/s$ (Figure 8.60); and others or springboard toe: Aldeadavila dam, $Q = 10,000 \, m^3/s$ (see Figure 8.5), including a flow pass over the

(a)   (b)

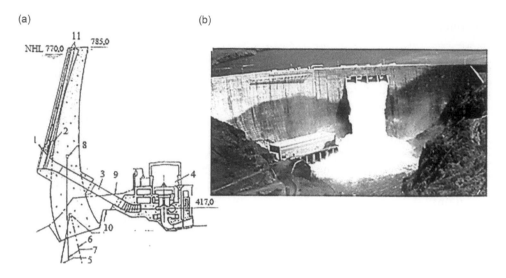

Figure 8.60 Mossyrock arch dam (USA): (a) vertical section; (b) view from the downstream; I – lattice, 2 – working gate, 3 – water conduit, 4 – building HEP, 5 and 6 – cementation and drainage curtains, 7 – control well, 8 – gallery, 9 – surface of the rock, I0 – foundation pit, and II – room gates.

HEP building: Picoti dam, h = 100 m and Q = 10,400 m^3/s (see Figure 8.32) and the French dam Aigle, h = 95 m and Q = 4,000 m^3/s;

c. throwing water from deep spillways: Sayano-Shushensk dam, Q = 13,600 m^3/s (see Figure 8.29); Kariba dam, Q = 9,500 m^3/s (see Figure 1.33), etc.

It is especially advisable to use jet throw directly from deep spillways in the case of thin arch dams.

At the same time, deep spillways are short and can perform both horizontal and inclined, for example, the Boundary dam in the USA, h = 104 m and Q = 7,100 m^3/s (see Figure 2.28).

The quenching of the energy of the waste stream can be carried out directly in the bed or in a water well.

Combining a spillway with an open face with the central part of the dams is recommended to use only in the case of thick (arch-gravity) dams, for example, the Sayano-Shushensk dam (see Figure 8.29).

In the case of thin arch dams, open spillways can be arranged on the crest in the form of a spillway vertex with a free fall of water into the downstream, for example, Portugal's arch spillway dams: Busan with a height of h = 65 m, Salamondi with h = 78 m (Figure 8.61), Kurobe-4 dam (see Figure 8.20), and Vaiont (see Figure 1.29). It is recommended to arrange such spillways (regulated or unregulated) in the central part of the dam although there are known cases of their placement in the lateral abutments of the dam: Hitoshuse (see Figure 8.42), New Bullards Bar in the USA, h = 195 m and Q = 4,530 m^3/s (Figure 8.62); Novilla dam in Mexico, h = 135 and Q = 14,200 m^3/s, etc., or the dams: Rail in Chile, h = 109.5 m and Q = 11,300 m^3/s, Santa Eulalia (see Figure 8.21), and others.

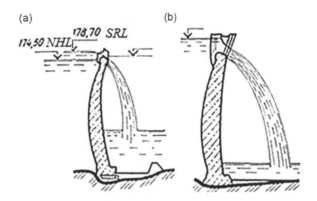

Figure 8.61 Portugal arch spillway dams: (a) Busan with *h* 65 m and (b) Salamondi *h* = 78 m.

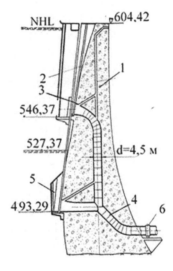

Figure 8.62 New Bullards Bar arch dam (USA): 1 – dam, 2 – air tube, 3 – gate, 4 – head water conduit, 5 – grate, and 6 – disk gate.

Shore spillways of arch dams may be

- tunnel (ChirkeI dam with $Q = 3,190\,\text{m}^3/\text{s}$; Glen Canyon Dam with $Q = 7,820\,\text{m}^3/\text{s}$ (see Figure 1.36a); Dez dam (Iran) with $h = 191\,\text{m}$ and $Q = 5,900\,\text{m}^3/\text{s}$);
- open dam Karaj (Iran) with $h = 180\,\text{m}$ and $Q = 1,480\,\text{m}^3/\text{s}$; dam Yagisawa (Japan) with $h = 131\,\text{m}$ and $Q = 1,300\,\text{m}^3/\text{s}$, etc.).

When designing spillways located on the banks, the following points are recommended:

- pay special attention to the design of the input sections where there is a significant increase in specific discharge and a significant deformation of the flow;

- spillway paths to perform straightforward (turning the path is allowed only with the appropriate feasibility study);
- to use special-shaped springboard that allow you to change the direction of the throwing and ensure that the jets fall on a given section of the bed (scattering springboard-bends, spillways with side flow, vertical springboard socks, etc.);
- on open spillways with several spans, arrange separate walls providing a reduction in the range of changes in specific water discharges and therefore maintaining the required throwing of jets;
- in the case of the conjunction between the downstream and the jets, arrange (to eliminate dangerous erosion in the event of discharges significantly lower than the designed) fastening near the structure, perform a springboard toe with a small angle at the exit, or limit the minimum discharges;
- if there are two spillways on different banks and there is a sufficient angle between the axes of the discarded jets (not more than 45°), consider the possibility of using the effect of collision of jets (Hitoshuse dam, Japan, see Figure 8.42).

When designing spillways of arch dams, one should not allow the discharge flow to fall on the slopes of the gorge above the downstream; water spray is allowed. Failure of taking into account the geological structure of the rock mass on which the spillway was located on the Karun-1 arch dam led to the collapse of the ski jump sock.

An arch dam of double curvature, 200 m high and 6 m thick on the crest, had a crest length of 380 m, the ratio $\ell/h = 1.90$ (Figure 8.63a), and a concrete volume of 1.35 million m^3. In the downstream, the river sharply turned to the left forming a rock spur, on which the dam rested by the left side; on the back slope of the spur, a surface spillway with a spring-toe at the end was constructed. It should be noted that the designers (US design firm HARZA) successfully used the topography of the area. But during the flood in May 1993, the end section of the spillway with the toe-springboard was destroyed (Figure 8.63b).

As it turned out later that the bedding on the left bank was parallel to the back slope of the spur creating anisotropy in slide strength, in the absence of drainage, water penetrated the formation cracks, and slide along the cracks occurred.

The spillway was restored in the late 1990s, and the dam is still in operation.

(a)  (b)

Figure 8.63 Karun-I arch dam (Iran): (a) after the accident at the bank spillway and (b) after the restoration of the spillway.

The presence of a spur on the left bank caused uneven compliance of the banks (the right bank is parallel to the river channel), which led to unusual movements: the arch dam turned in plan while the movements of the left-bank were directed to the downstream and the right-bank to the upstream.

On many HPs with arch and arched gravity dams, HEP buildings are carried out by the dam. In this case head pipelines supplying water from the water inlets to the HEP building, as in gravity dams (see Section 7.3.7), can be placed

- on the upstream face of the dam (for example, Nagowado in Japan, $h = 156\,\text{m}$ (Figure 1.36b), passing through the dam in the lower less-stressed part or in the upstream outside the dam, due to which their impact on its SSS is reduced and more favorable construction conditions are also ensured;
- in the body of thick arch and arch-gravity dams, more in the stressed middle part of the dams, which also complicates the construction conditions, for example, in the Glen Canyon arch-gravity dam, $h = 216\,\text{m}$ (see Figure 1.36a), the thick domed New Bullards Bar (see Figure 8.62), and Mossyrock (Figure 8.60);
- on the downstream face of the dam, due to which their influence on the SSS dam is sharply reduced and more favorable conditions for the construction of the Sayano-Shushensk (see Figures 1.28 and 8.29) and Chirkei (see Figures 1.32 and 8.41) are created in Russia.

8.4.7 Dams from two arches

In wide sections with a specific shape and geological conditions, an arch dam of two arches with an intermediate pier may be effective.

The Hongrin arch dam was erected in 1969 in Switzerland to create the upper basin of the SPP and consists of two dams (see Figure 8.22). The volume of the basin created by the dams is 52 million m^3. The north and south arch dams are based on a common bank abutment and in the upper part on a common gravity pier, 26 m high. The height of the northern dam is 123 m and of the southern dam is 95 m; the crest of the northern dam is 325 m and of the southern dam is 272 m; the thickness of the northern dam along the crest 3 m and along the bottom is 25 m, and those of the southern dam are 3 and 18 m, respectively; the total volume of concrete is 345 thousand m^3. Both dams are thin and of double curvature. Limestones lie at the base. The spillway built in the form of a "daisy" near the central abutment is designed for a flow discharge of 100 m^3/s and is intended for sanitary releases. Both dams are equipped with bottom outlets for emptying the reservoir.

Field observations of dam displacements show that both dams operate in design mode despite the cyclic load on the dam and the foundation caused by the drawdown and filling of the upper basin during the operation of the SPP [220].

The Okura arch dam built in 1961 in Japan to regulate runoff, irrigation, and water supply and generate electricity with a reservoir volume of 26 million m^3 also consists of two arch dams: left- and right-bank (Figure 8.64), based on a common concrete gravity pier.

The right-bank dam, 42 m high, has a crest length of 161 m, and the ℓ/h ratio is 3.83; the left-bank, 82 m high (of which 40 m is a plug), has a crest length of 162 m, and the ℓ/h ratio is 1.98; the thickness along the crest and at the base (for the left-bank dam at the

Figure 8.64 Okura arch dam (Japan). View of downstream.

top of the plug) of both dams is the same and is 8.5 and 17 m, respectively; the thickness of the plug of the left-bank dam on the top is 22 m and on the bottom is 28 m; the total volume of the dams is 226 thousand m³.

The dam foundation is piled with tuffs with andesite dikes. In these topographic conditions characterized by the presence of a terrace (apparently, the old riverbed) near the river valley, the dam variant in the form of two arch dams based on a common gravity pier turned out to be the most economical of the considered variants. For the purpose of symmetrical loading of the pier by forces from arch dams, the spans of the latter were designed almost the same, and in order to ensure that the heights of the dams were the same, a plug 40 m high was cut in the bed, cut off by a joint from the arched part of the left-bank dam. The arched dam Khudoni is designed on the river. Inguri is 32 km above the Inguri arch dam. According to the project, the height of the dam was to be 201.5 m (of which 171.5 m was an arched part and 30 m was plug), the length of the crest was 525 m, the thickness of the crest was 6 m, and the bottom was 16 m (Figure 8.65). The volume of concrete of the dam is 1.35 million m³.

The construction of the dam was started in the late 1980s of the last century, a pit was constructed under the dam and rock blocks were laid, and work was completed under the underground building of the HEP. After the collapse of the USSR and the independence of Georgia, construction ceased, and construction is currently frozen.

The section of the dam is a V-shaped gorge with a width of 15–20 m along the bed. The slopes of the gorge are composed mainly of tuff breccias and tuff shales; in the left-bank adjoining, there is a thick (up to 50–60 m) thickness of the quaternary deposits; the designed seismicity of the dam area is eight balls.

At the upper elevations of both banks, the dam connection with the foundation by means of gravity piers passing below into a massive saddle was cut off from the arched part by a perimeter joint. The right-bank pier abuts against a low-strength rock, and the left-bank pier at an altitude of 60 m has no thrust on the rock foundation.

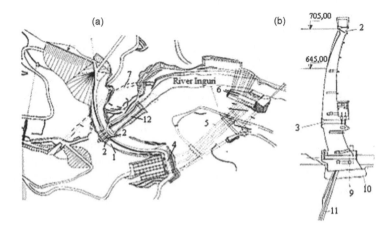

Figure 8.65 Khudoni arch dam (project): (a) plan and (b) a section along the center console: 1 – arch dam, 2 – surface spillway, 3 – deep spillway, 4 – water intake, 5 – head pipelines, 6 – underground building of HEP, 7 – spillway openings of the construction period, 8 – level drawdown, 9 – strengthening cementation, 10 – drainage curtain, 11 – cementing curtain, and 12 – damping plunge pool.

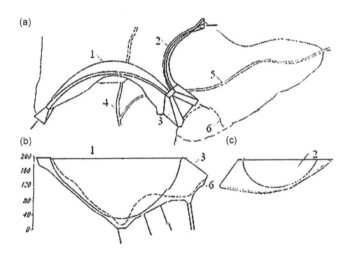

Figure 8.66 Two-arch variant of the Khudoni dam: (a) plan, (b) and (c) view from the downstream of the large (1) and small (2) arch dams; 3 – gravity pier, 4 – river Inguri, 5 – side tributary, and 6 – loose deposits.

The lack of thrust on the rear face of the pier due to loose deposits led to the impressive pier dimensions: the length along the crest will be 50 m, the height 61.5 m, and the bottom thickness 67 m. A head front on the left bank within the loose sediments is planned to be formed using retaining wall adjacent to the concrete pier and a flexible screen.

The quaternary sediments in the left-bank abutment and the lateral canyon on the left-bank upstream prompted to propose a variant of a two-arch dam (Figure 8.66), in which the main arched dam 200 m high with the right side rests on the right bank of the river Inguri and the left bank – in the lower part to the left bank and in the upper part, within 50 m to the gravity pier having a "diving" rear face; a small arch dam 80 m high rests on the same pier by the right side, and on the left side, a small dam rests on the right bank of the side canyon. Both arch dams are of double curvature and have a perimeter seam.

Loose deposits remain in the downstream, and since they are not soaked, there is no need for measures to stabilize them, just as there is no need to carry out special work to create a head front in them.

The abutment is located in the downstream and is not affected as usual by a hydrostatic load, which simplifies its working conditions, but its rigidity and position significantly affect the SSS of both arch dams. The optimal position and design of the pier in the form of a "fork" were selected using computational studies using the FEM under the action of only hydrostatic pressure on the dams.

Calculations showed the viability of a nontrivial dam variant.

8.5 Arch dams constructed by stages and built up

The construction of high arch dams with stages and phased filling of the reservoir allows increasing their economic efficiency due to the commissioning of the HEP during the construction period with minimal workloads and capital investments.

When erecting arch dams with stages or building them up, it is necessary to solve a set of issues related to the conditions and technology for erecting subsequent stages, their joint work, phased application of loads, and the formation of the SSS of the "dam foundation" system as well as ensuring normal operating conditions for the HEP and spillways in the ongoing construction process.

The main applicable schemes for constructing dams in stages and building dams without drawdown of the reservoir are shown in Figure 8.55 [65].

The second and third schemes allow you to limit the volume of concrete of the first-stage dam necessary for the perception of water pressure of the reservoir of the first stage and accordingly reduce the cost of the first stage of the dam (Figure 8.55b and c).

In the first simplest scheme, which is widely used in practice, the construction of the second stage is carried out without increasing the thickness of the first-stage dam equal to the thickness at the full height of the dam (Figure 8.55a). The concrete volume of the first-stage dam is greater than the necessary volume for the perception of water pressure at an intermediate level of the first stage reservoir. The first scheme was applied on the arch dams of Nagowado in Japan and Ross in the USA (Figure 8.67), which was built in four stages: the first with a height of 88.4 m (1940), the second with a ramp up to 133.7 m, the third with a ramp up to 164.6 m, and the fourth with a ramp up to 201.5 m (1970) [142].

According to the same scheme, the Emosson arch-gravity dam (Switzerland) was built with a height of 45 m and a concrete volume of 62 thousand m^3 in 1955. The upper part of the existing dam with a volume of 15 thousand m^3 was dismantled; after a ramp up of its height, an arch dam was formed in two-curvature with a total volume of 97 thousand m^3 (Figure 8.68) [204].

With such a construction of the dam, the mortification of capital investments is reduced, and an additional economic effect is obtained. According to the second scheme, the monolithic longitudinal joint between the first and second phases is not required,

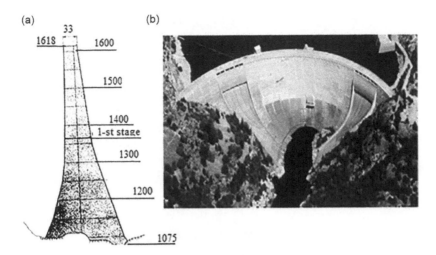

Figure 8.67 Ross arch dam (USA). Marks in feet: (a) section along the center console and (b) view from the downstream.

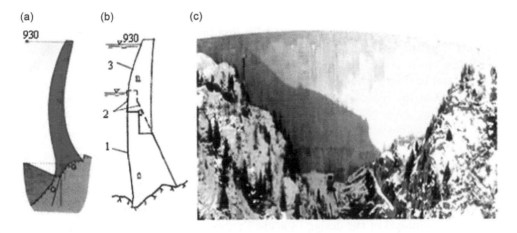

Figure 8.68 Emosson arch dam (Switzerland): (a) and (b) vertical sections: 1 – initial dam, 2 – sections of the workable part, and 3 – extension dam; (c) view from the downstream.

which simplifies the work during the construction of the second stage of the dam. Such a scheme was applied in the construction of the Frera dome dam (height 138 m, length along the crest 315 m, and concrete volume 400 thousand m³) in Italy in 1961 (see Figure 8.36).

The third scheme requires a monolithic longitudinal joint between the first and second phases, which requires special measures; the construction of the second stage of the dam is somewhat more complicated. Such a scheme was applied in the construction of Kankano-II dams (Figure 8.69), Les Toules (see Figure 8.56).

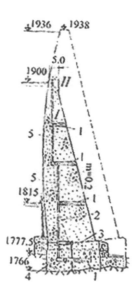

Figure 8.69 Kankano-II arch dam (Italy): I – Ist stage and II – 2nd stage; I – gallery, 2 – drainage, 3 – perimeter joint, 4 – saddle, and 5 – plumb bore.

During the phased construction and loading of the SSS dam, it will differ significantly from the SSS of the dam, built in one turn and filling the reservoir to the NHL.

When constructing dams in stages, in most cases, the longitudinal joint is monolithic, and the dam is monolithic.

Studies of the SSS of a number of dams with a nonmonolithic longitudinal seam showed that under certain conditions, their SSS can be more favorable than in the presence of a clutch.

8.6 Arch dams from rolled concrete

The positive experience of the widespread use of RCC technology in the construction of gravity dams became the basis for the use of RCC for arched dams, which, combining the advantages of the design of arched dams and RCC technology, increased the efficiency of arch dams.

In arch dams, both ordinary and RCC, in order to ensure solidity and transfer of the load in the arch direction to the rocky foundation, it is necessary to cement transverse temperature-shrinkable joints. Before grouting, the average temperature of the dam should not be higher than the average annual temperature. Given that RCC is less sensitive to cracking than ordinary [208] concrete, the distance between transverse heat-shrinkable joints increases from 30 to 90 m.

Transverse joints can be formed, for example, from prefabricated elements 1 m long, 0.3 m wide, and 0.3–0.6 m high (equal to the height of one or two concrete layers), in which devices for cementing joints are embedded. Such a system was used on the dams Shapai (see Figures 1.34 and 8.70b), 132 m high, and Pudding, 75 m high (Figure 8.71).

The Shimenzi arch dam (Figure 8.70a), 110 m high, was constructed with one transverse joint in a conventional formwork with a cementation system as in ordinary arch dams. In low arch dams, which are constructed from RCC in one season at a low temperature period with measures for concrete cooling, it is possible to refuse transverse joints with the corresponding calculation justification [35,160].

The consumption of the binder (cement and fly ash) for RCC in arch dams ranges from 150 to 230 t/m^3. In high arch dams, two or more compositions of RCC can be used.

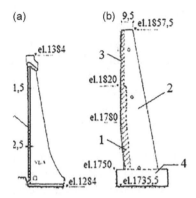

Figure 8.70 Shimenzi. arch dam (China) with a height 109 m (a) and Shapai (China) with a height 132 m (b).

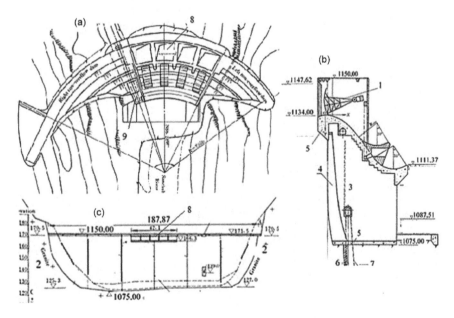

Figure 8.71 Puding arch dam (China): (a) plan, (b) spillway section, and (c) view from the downstream; 1 – gate, 2 – granite, 3 and 4 – RCC of 3 and 2 compositions, 5 – vibrated concrete, 6 – cement curtain, 7 – drainage, 8 – weir spillway, and 9 – deep spillway.

For the most stressed zones of dams, a concrete composition with a cement flow rate similar to that for conventional arch dams with high compressive strength of concrete is used [262].

Considering that the slide strength in construction joints of arch dams made of RCC should be higher than in gravity dams, in order to increase the slide strength, in some cases, cement mortar is laid before the next layer of concrete is rolled.

A serious problem in the construction of arch dams from RCC especially high with increasing thickness is the dissipation of excess heat to lower the temperature of the dam concrete before cementing heat-shrinkable transverse joints. For this, pipe-cooling was used at the Shuikou dam [194]. In China, a number of RCC arch dams used concrete with controlled swelling due to the addition of a high percentage of magnesium oxide MgO: Shimenzi dam (Figure 8.70a) and Longshou-1. However, the effectiveness of such an additive on the long-term strength of RCC in arch dams has not been investigated.

The first two dams of RCC Wolverdans and Knellport, 70 and 50 m high, respectively, built in South Africa in 1990 were arch-gravity with a vertical upstream face and a downstream face of 0.5; in 1994, in China, the arch dams Puding (Figure 8.71) and Shuoshai, 75 m high, were built.

Currently, over 30 arch dams from RCC have been built in China with a height of more than 75 m, including 10 dams with a height of more than 100 m (Table 8.2) [93,162,194]; the height of the arch dams Dahuashui and Yunlonghe-III reached 135 m. In Pakistan, the arch-gravity dam Gomalzan was erected with a height of 133 m.

One of the highest dams, the Shapai Dam (Figure 8.70b), with a height of 132 m and a crest length of 250.2 was built in 2002 in a highly seismic area. Preserved granite-diorites lie at the foundation of the dam. A concrete plug is arranged at the bottom of the dam. The horizontal sections of the dam are made in the form of three-center arches (the maximum central angle is 92.5°), and the vertical axial section has a

Table 8.2 Arch dams with a height of 75 m or more from RCC built in China [194]

No	Name	Years of construction	Parameters (m)		Volume (thousand m^3)		Discharge flow rate (m^3/s)
			Height	Length	RCC	Overall	
1	Suoshai	1989–1994	75	195.7	-	88	560
2	Puding	1989–1994	75	195.6	103	137	2,510
3	Oilinguan	2005–2007	77	140	45	55	155
4	Xuanmiaoguan	2003–2005	79.5	243	-	95	2,743
5	Longshoul	1999–2001	80	258	187	210	3,090
6	Yujianhe	2002–2005	83	167	-	155	231
7	Longqiao	2005–2007	95	155.6	-	160	3,580
8	Linhekou	1999–2003	100	311	229	293	3,480
9	Bailianya	2006–2007	104.6	422	-	-	-
10	Zhaolaihe	2003–2005	107	206	166	255	1,710
11	Shimenzi	1999–2002	110	176.5	-	211	-
12	Huanghuazhai	2006–2008	110		-	-	-
13	Tianhuaban	2006–	113	159.8	-	360	5,046
14	Silin	2005–2012	117	310	-	1100	32,922
15	Shapai	1997–2002	132	250.2	362	392	-
16	Dahuashui	2005–2010	134.5	306	560	650	-

constant laying of the downstream face of 0.21, the vertical upstream face with a reverse slope of 0.1 in the lower part, and the thickness at the base of 28 m. The dam is divided into five sections by two vertical seams and two joint notches. The thickness of the layers of rolled concrete was 0.3 m; the compressive strength of concrete at the age of 90 days is 20 MPa. On the upstream face, a protective layer is made of ordinary vibrated concrete and a screen of 2 mm thick synthetic film.

According to the results of calculations with an unusual combination of loads taking into account seismic effects (at an acceleration of 0.138 g), the maximum compressive stresses on the downstream face were −5.4 MPa and the maximum tensile stresses on the upstream face were about +1 MPa [194]. The dam suffered a major earthquake without damage on May 12, 2008, with an acceleration of about 0.8 g with an epicenter 12 km from the dam

The Shimenzi Dam (Figure 8.70a), 110 m high and 176.5 m long, was built in 2001 in the north-west of China in a seismic region in severe climatic conditions (with an average annual temperature of + 4.1° C and a minimum winter temperature of −30°C). Conglomerates with a low elastic modulus of 4,000 MPa lie at the foundation, which caused an increase in the thickness of the dam in the foundation to 30 m. In the upper part of the bank abutments of the dam, intensified cementation was performed, and flexible arch belts were constructed in the part of the dam capable of absorbing significant deformations of weak rock. A protective layer of ordinary concrete and a layer of polymer sealant are made on the upstream face.

In the five winter months when the temperature was below 0°C, the laying of RCC was stopped and a layer of sand and crushed stone 3 m thick was laid on its surface to protect it from freezing [93].

One of the highest arch dams made of RCC, the Dahuashui Dam, 134.5 m high and 306 m long, with a concrete volume of 650 thousand m^3 (including rolled 560 thousand m^3) was built in China in 2010. The dam of double curvature rests on powerful gravity pier on the left bank 37 m high and 71 m wide at the base. The width of the dam along the crest is 7 and 28 m at the base. Surface and deep spillways are constructed in the dam.

The thinnest arch dam made of RCC, Xibingxi, 63.5 m high, 93 m long, and concrete volume 33 thousand m^3, built in China in 1995 in a narrow canyon with a bed thickness of 12 m is characterized by a thickness to height ratio of 0.19. At the foundation of the dam are weathered sandstones. The thickness of the zone of vibrated RCC at the upstream face was 0.5 m, and it was enriched with cement mortar (Figure 8.72).

A transverse joint is made in the upper part at 2/3 of the height of the dam. Binder consumption per 1 m^3 of concrete was cement 80 kg and fly ash 105 kg. To reduce tensile stresses in the dam, in the heel sections of the arches, joint notches were made overlapped by a seal [93,194].

8.7 New technical solutions

8.7.1 Improved arch dams

In recent decades, in the design and construction of arch dams, a number of progressive structural and technological solutions have been developed and improved that have provided increased reliability and significant savings in cost, material, and labor resources.

(a)   (b)

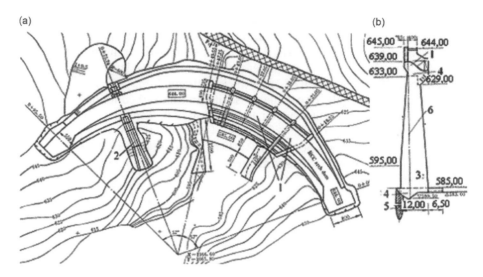

Figure 8.72 Xibingxi arch dam (China): (a) plan and (b) a section along the spillway;
I – weir spillway, 2 – bottom spillway, 3 – RCC, 4 – vibrated concrete,
and 5 – cement curtain.

An analytical description of the form (see Section 8.3.4) implemented in the working draft of the Inguri arch dam ensured the smoothness and continuity of the faces of the dam and the absence of stress concentrations and as a result, reduced the volume of concrete in the dam compared to the project of about 100 thousand m^3 [32,126]. A polynomial form was also used in the design of the Khudoni arch dam.

"Diving" *piers* do not rely on the rock mass in their upper part and are oriented in the direction of the main stresses in the arch dam. The head front of the dam on the inclined section of the pier is closed using the retaining wall-wing, which is part of the pier due to which the pier is freed from the action of hydrostatic load and filter uplift, which greatly facilitates the conditions of its static operation.

The piers in the Inguri arch dam made it possible to significantly reduce the length of the crest arch and to avoid the transfer of forces from the arch dam to the intensely fractured rock formations of both banks; the total volume of piers was very significant –175 thousand m^3 (see Figure 8.23).

The design of the saddle with a joint system is designed to neutralize the effect on the arch dam of differentiated tectonic displacements at the foundation, which are possible when filling the reservoir, and seismic influences. The height of the saddle section above the tectonic disturbance increases to its double width and is cut by two joint systems, one of which is oriented parallel to the middle surface of the dam and the other approximately along the plane of the fault. At the same time, a system of blocks is formed in the saddle with a certain freedom of movement relative to each other, which makes it possible to suppress relative displacements in the saddle due to tectonic disturbance.

As shown by model studies in VNIIG, the design of the saddle (see Figure 8.37) neutralizes displacement along a fault of up to 10 cm with minimal changes in the stress state of the dam body [20]. The loss of the bearing capacity of the dam occurred with an increase in displacements of up to 24 cm.

A water well of cylindrical shape was designed to protect from erosion of the bed and the sides of the gorge in the downstream during operation of high-head spillway devices. As you know, the traditional design of a water well is longitudinal walls and a massive flat plate which accepts alternating loads caused by the impact of a falling stream of water and the weighing forces of the filtration flow in the foundation under the plate. On rock foundation, the slab in order to facilitate it can be anchored to the foundation (like the Sayano-Shushensk HEP).

To reduce the amount of work, a spatial waterproof design of water well was developed including bank concrete walls and a water plate made of separate blocks elastically connected between each other and laid directly on alluvial soils in the form of a return arch, expanded into the side wall piers. When quenching the energy of a falling water jet, the bottom of the well works like a cylindrical plate on an elastic base; when perceiving the weighing load of the filtration flow at the foundation, it works as a vault, which transfers a considerable part of the load to the bank massifs due to thrust, which allows reducing the thickness of the plate and abandoning its anchoring to the rock foundation.

A similar design of water well was introduced at the construction of the Inguri arch dam; under difficult geological conditions in the downstream and tight construction periods, this design allowed us to reduce the volume of earth-rock and concrete work [33]. The width of the well was 50 m, and the total length was 270 m, including 42 m – the inclined threshold (Figure 8.73).

The riverbed below the well is protected from erosion by a 50-m-long apron made of $1.7 \times 1.7 \times 1.7$ m concrete cubes laid in two layers. Blocks of a water plate, 10 m wide, are each connected to each other along the bottom by reinforcement installed in the direction along and across the stream, according to the principle of dispersed reinforcement in the form of frames of four rods with a diameter of 40 mm with a step of 1.5 m.

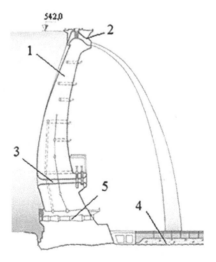

Figure 8.73 Inguri arch dam. Key console section: 1 – dam, 2 – surface spillway, 3 – deep spillway, 4 – water well, and 5 – construction openings.

The antiseismic reinforcement belt is designed to ensure the reliability of the dam under the action of seismic loads. When the reservoir is filled, arch tensile stresses arising in the upper part of the structure during the seismic are compensated by compressive arch stresses from the hydrostatic load of the main combination. During reservoir drawdown, the upper zone of the dam is practically uncompressed. In case of seismic impact, the upper part of the dam turns into separate consoles. As shown by experimental studies at GruzNIIEGS, under the action of relatively low accelerations, significant opening of intersectional joints occurs, accompanied by the formation of thorough horizontal cracks in the dam.

The increase in seismic resistance during seismic effects can be achieved by the device in the upper part of the arch dam of the antiseismic belt from horizontal reinforcement with its passage through intersection joints.

The antiseismic belt of the Inguri arch dam is made on the basis of the principle of dispersed reinforcement in the form of four core frames of reinforcement Ø40 mm, laid in 2–3 rows on inter-block surfaces (1.5 m in height) along the upstream and downstream faces. The frames are equipped with damping insert ties made of reinforcement Ø70 and 50 mm, 9.2 and 7.2 m long, respectively, passed through intersection joints. The rods with a length of 3.2 m are enclosed in polyethylene pipes, which made it possible to open the joints during pipe cooling of concrete before subsequent monolith. The number of horizontal antiseismic reinforcement was 11,400 tons. The vertical antiseismic reinforcement of the dam was also made according to the principle of dispersed reinforcement in the form of frames of 4–8 rods Ø40 and 50 mm installed along the outer faces of the dam with a pitch of frames of 1.5–0.75 m. The number of vertical antiseismic reinforcement was 12,500 tons.

The tiered technology for the construction of arch dams is aimed at increasing the rate of concreting and preventing cracking of concrete. Distinctive features of the technology are the uniform construction of the dam in tiers concreted from bank to bank without lagging and leading sections, organization of all technological operations on the concreting tier according to the flow diagram, and use of intensive concrete supply methods.

The location of concreted blocks at one level facilitates the management of the production process, allows rational organization of the process flow, and increases labor safety. The technology of tiered concreting was first used in the world practice at the Myatla arch dam [92]. The height of the concreting tier was 1.5 m; the designed time for concreting one tier was seven days. For the formation of intersectional joints, prefabricated reinforced concrete panels left in concrete were used. The main volume of concrete was laid by two cranes located along the sides of the gorge; to compact the concrete mixture, a modernized self-propelled manipulator with a package of vibrators was used. At the same time, the average monthly intensity of concreting was 8.9 thousand m^3/month, and the average growth rate of the dam in height was 7.7 m/month. Simultaneous concreting of adjacent sections and continuous pipe cooling ensured the opening of intersectional joints by 5–7 mm, which made it possible to reliably monolize the dam (see Figure 8.44).

Arch dams were erected on the bridge in the presence of a deep and narrow gap in the canyon, filled with alluvial and other deposits. Thus, the arch dam Zhaixiangkou

(a)   (b)

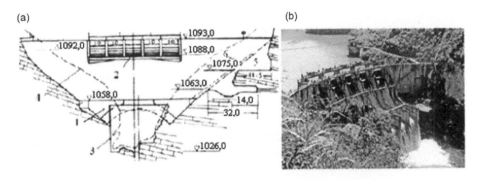

Figure 8.74 Zhaixiangkou arch dam (China): (a) scan on the downstream face; I – arched bridge, 2 – spillway, 3 – wall in the ground, 4 – natural surface, and 5 – concrete.

(Figure 8.74) in China was erected in 1975 on a curved bridge in plan, working mainly in the arch direction.

Given the great difficulties associated with the removal of alluvium and excavation of the pit in a narrow bed part after considering various variants, the variant with an arch dam on the bridge was chosen. The height of the dam above the bridge is 39.5 m, the length of the crest is 151.9 m, and the thickness along the crest is 3 m and the bottom is 8.7 m; the concrete volume of the dam was 62.8 thousand m^3. In alluvium, two reinforced concrete walls 1 m thick at a distance of 1 m from each other and 31.4 and 28.3 m high were erected using the wall-in-ground method.

To pass the flood on the dam, a surface spillway with a springboard is arranged.

The idea of construction of a dam on the bridge was used in the design of the Kowsar gravity dam in a narrow canyon in Iran (see Figure 2.22) and was successfully implemented during construction.

An arch dam was designed with a concrete apron and a vertical slit-seam in the foundation under the upper face to reduce uplift on the dam bottom and to neutralize the rock mass of the dam foundation from crack formation (see Sections 4.4.3 and 8.7.4). This decision was made on the Schlegeis arch dam (see Figure 8.49).

8.7.2 Suggestions for new decisions

Longitudinal noncemented joints were arranged in the lower part of arch (arch-gravity) dams in relatively wide sites for the purpose of more favorable distribution of hydrostatic load between arches and consoles (Figure 8.75).

It should be noted that the Frera arch dam, 138 m high, is operated with an uncemented longitudinal seam more than half the dam height (see Figure 8.36).

The arch dam was interacted with the foundation by a flexible apron by means of a device between a concrete apron (see Sections 4.4.3 and 8.3.2) and the dam and a trapezoidal cavity expanding to the top and filled with molded asphalt concrete characterized by high water resistance and durability. Under the influence of water pressure

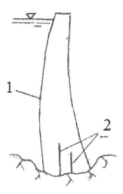

Figure 8.75 Arch dam with uncemented longitudinal seams (offer): I – dam and 2 – joints.

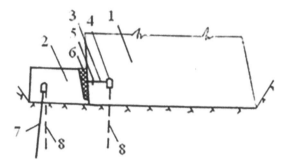

Figure 8.76 Interaction of the arch dam with a foundation by flexible apron 2: I – dam, 3 – notched joint, 4 – gallery, 5 – dowel, 6 – bitumen, 7 – cementation curtain, and 8 – drainage curtain.

on the surface of the asphalt concrete, its compression is ensured with the formation of a waterproof contact of the asphalt concrete with the apron and with an arch dam in the area of the joint notch. A gallery is arranged in the concrete apron from which cementation and drainage curtains are made (Figure 8.76) due to which the filtration pressure on the bottom of the arch dam decreases sharply.

The connection of a concrete apron with a dam can be performed using arch ceilings forming a single structure with a dam and an apron (Figure 8.77).

In this design, the apron is made in the form of a concrete slab, which if necessary, is anchored to the foundation, and cementation and drainage curtains are arranged from the cavity formed by the arch ceiling. An additional drainage curtain can be made from a gallery arranged in an arch dam, where a joint cut can also be made, if necessary, in a part below the arch ceiling. In order to provide the necessary flexibility and favorable stress state with compressive stresses in the arch ceiling, the central angle of the arch ceiling should be about 150° with an average thickness ratio of 0.15–0.2

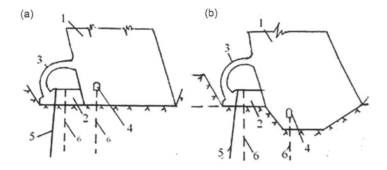

Figure 8.77 Sections on the connection dam: (a) on the bed part and (b) on the bank part: I – dam, 2 – apron, 3 – arch ceiling, 4 – gallery, 5 – cement curtain, 6 – drainage curtain.

for its thickness to radius; in addition, it is advisable to perform arch ceilings with a thickening to the heels and of high-strength concrete.

There is no hydrostatic pressure on the upstream face part of the arch dam connected to the foundation since this part is protected by arch ceiling.

The pressure of the water load on the arch ceiling causes compressive forces connecting to the arch dam and the apron plate and also compresses the foundation under the plate in the area of the cement curtain, ensuring its reliable operation.

In a composite arch dam consisting of upstream arch perceiving hydrostatic pressure and downstream arch erected from RCC, the high compressive strength of ordinary concrete and the manufacturability of RCC are successfully used.

The upstream arch is connected to the downstream arch by the prefabricated reinforced concrete elements, which ensure the transfer of load from the upstream to the downstream arch and uniform loading of the entire system. Prefabricated elements should be flexible enough to minimize transmission of slide forces. At the same time, it is advisable to perform them in coil form with expansion at the ends that form the inner surface of the arches (Figure 8.78) [91,223,225].

The increased flexibility of the composite arch dam and the combined "spring" operation of the upstream and downstream arches can reduce the dynamic effect and create the prerequisites for increasing its seismic resistance.

The construction of a composite dam is better adapted to the difficult geological conditions of the site since in the presence of weaknesses in rock abutment, first of all, the redistribution of forces in a stiffer downstream arch will occur, and to a much lesser extent, this will affect a more flexible upstream arch. It is also less affected by deformations of the sides and bed of the valley, which occur when the reservoir is filled and, in some cases, leads to the formation of tensile stresses in the foundation under the upstream face of high dams.

Due to the absence of a decompression zone in the foundation under the upstream face over the entire contour of the dam abutment, the filtration conditions in the foundation and sides improve, the filtration pressure decreases, the abutment stability increases, and the overall reliability of the "dam foundation" system increases.

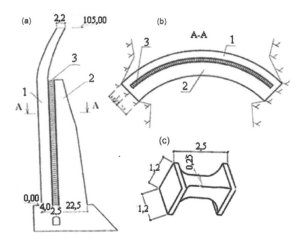

Figure 8.78 Composite arch dam: (a) cross section; (b) section along AA; (c) prefabricated reinforced concrete element; 1 – upstream dam from ordinary concrete, 2 – downstream RCC dam, 3 – prefabricated reinforced concrete elements.

Conclusion

The human mind is greedy.
He cannot stop
Not to be at rest
And strives further

Francis Bacon (1561–1626),
English philosopher

The history of the development of dam engineering shows that it is the constant competition between concrete and soil dams as well as competition between different types of concrete dams that led to the revision of the criteria for optimality and overcoming the stereotypes of traditional solutions that brought to life new effective ideas.

Currently, gravity, massive, and arch dams are one of the most common and effective types of dams; their construction is carried out all over the world in various natural conditions.

The authors sought on the basis of analysis and generalization of accumulated experience, current trends and new structural-technological solutions of gravity massive and arch dams to show not only the achieved level, but also possible directions for improving layout and structural solutions, construction technology, computer analysis, and more complete consideration of natural features of construction sites.

Constructive solutions and technology for the construction of concrete dams are closely interconnected, which is one of the most important factors in their design.

The experience in the construction and long-term operation of massive gravity and arch dams has shown their effective operation and high reliability in a variety of natural conditions, including complex engineering-geological conditions, severe climate, and high seismicity. A characteristic trend is the further movement of their construction to foothill and mountainous regions, in many cases characterized by extremely difficult natural conditions and increase in the height of the dam, and their safety remains the top priority in the design, construction, and operation of dams.

Significant progress in the construction of concrete dams was achieved, thanks to the introduction of RCC technology, which provided a sharp reduction in labor costs and construction time and cost, increasing their competitiveness and efficiency.

Gravity, massive, and arch dams have significant reserves for improving design solutions, the implementation of which will ensure a further increase in their reliability and efficiency, including the following:

- the choice of an effective layout of the HP based on the natural conditions of the site in conjunction with the structural-technological solutions of the structures;
- further improvement of dam structures and the technology of their construction in various natural conditions;
- increasing the seismic resistance of dams due to the targeted change in their dynamic characteristics;
- optimization of solutions for the construction of dams in stages;
- ensuring the spatial character of the operation of massive gravity dams in relatively narrow sites;
- optimization of the shape and design of arch dams aimed at a more complete use of the strength properties of concrete and rock foundations as well as improving the manufacturability of the construction.

Appendix 1

Links to sources from which photographs of dams and HEPs are borrowed

Photo in Figure	Source
Figure 1.1. Proserpina	Arenillas M. The historical development of Spanish dams. The International Journal on Hydropower and Dams, № 3 и 4, 2006
Figure 1.2. Shāh Abbās	sanaei@livejournal.com 1. Mehdi Sanai Ambassador Extraordinary and Plenipotentiary of the Islamic Republic of Iran December 2013.
Figure 1.3. Almansa	© ProEnergo. Blogspot.Com
Figure 1.4. Tibi	Web oficial del Ayuntamiento de Tibi.
Figure 1.5. La Sierra de Elche	Rozanov N. S. et al. Accidents and damage to large dams. Energoatomizdat, M., 1986
Figure 1.8. Dnepr HEP	Landau, Yu.A. et al. Hydropower and the environment. Kiev, Libra, 2004
Figure 1.11. Hoover Figure 8.28. Nambe Falls Figure 8.60. Mossyrock Figure 8.67. Ross	Development of Dams Engineering in the United States. Preparing in Commemoration of the Sixteenth Congress of International Commission on Large Dams by United States Committee on Large Dams, ©.1988, Pergamon Press, 1072 стр.
Figure 1.21. Krasnoyarsk Figure 1.22. Bratsk Figure 1.23. Toktogul Figure 1.28. Sayano-Shushensk Figure 1.32. Chirkei Figure 1.42. Zeya	HEP of Russia. Renewable energy. 2018, 223 pp. ©. RusHydro, ©. NTF Energoprogress, ©. Association "Hydropower of Russia", ©. St. Petersburg Polytechnic University of Peter the Great.
Figure 1.24. Mieyagase Figure 8.20. Kurobe-4	Dams in Japan, ICOLD, 2012. http://www.jcold.at.jp.
Figure 1.25. Porce-II	Tejada L. C. and other. The design and construction of Porce II RCC Dam in Colombia, the International Journal on Hydropower and dams, Issue 3, 2001
Figure 1.26. Gergebil	© Branch of PJSC RusHydro-Dagestan Branch, 2009–2019
Figure 1.27. Gunib	Description of the Gunib hydroelectric station on the website of the Dagestan branch of JSC RusHydro. RukiVNogi.com 2012–2019
Figure 1.30. Inguri Figure 8.23. Inguri	
Figure 1.34. Shapai Figure 1.45. Three Gorges	Large Dams in China. A Fifty-Year Review. China Water Press. Beijing, 2000, 1029 pages©.

(Continued)

Photo in Figure	Source
Figure 1.35. Ova Spin Figure 8.22. Hongrin Figure 8.25. Schifferen Figure 8.30. Punt dal Gall Figure 8.40. Almendra Figure 8.45. Contra Figure 8.56. Les Toules Figure 8.59. Valle di Ley Figure 8.68. Vieux Emosson	Dams in Switzerland. Source for Worldwide Swiss Dam Engineering. Swiss Committee on Large Dams, 2000, 277 CTP. www.swissdam. Ch
Figure 1.38. Roseland Figure 1.39. Daniel Jonson	Pinterest.ru. The Roseland Dam in French Alps. ©. 2007 Hydrotechnics.ru.
Figure 8.43. Santa Lucia Figure 1.43. Itaipu Figure 2.1. Tucuru	Main Brazilian Dams. III Design. Construction and Operation. Brazilian Committee on Large Dams, 2009. 496 pages.
Figure 2.2. Dvur Kralove nad Labem	© 2006–2019 Pardubický kraj & Destinační společnost Východní Čechy
Figure 2.36. Vaiont Figure 2.22. Kowsar Figure 8.67. Karun-I	Photo of Mgalobeliv Ju, B.
Figure 2.10. Son La Figure 7.26. Boguchansk HEP	Photo of Volinchikov A, N.
Figure 2.11. Salto Caxias	Toniatty N. B., Mussi J. M. Salto Caxias: Responding to Electricity Demands in Brazil, HRW, May, 1999
Figure 2.16. Revelstoke	Brunner W. Cracking of the Revelstoke concrete gravity dam mass concrete. Proc. 15th ICOLD, Lausanne, V.2., Q.57.R.1, 1985
Figure 2.18. Yeywa	Win Kyaw U. and other, Myanmar's Yeywa RCC dam. The International Journal Hydropower and Dams, Issue 4, 2007
Figure 3.5. Installation Figure 3.7. Installation Figure 3.9. Spice Figure 3.12. View installation Figure 3.16. Determination of strength Figure 3.42. La Amistad dam Figure 3.43. Mont Toc Figure 3.48. Uites Figure 3.55. Fernando Irnart Figure 3.62. Rock block Figure 3.63. Vahsh Figure 3.65. Pilon blocks Figure 3.66. Strengthening	Photo of Gaziev E.G.
Figure 4.14. El Atazar Figure 8.56. Aldeadavila	Dams in Spain/Spanish National Committee on Large Dams, 2006, ©. Colegio de ingenieros de Comines. Canaris y Puartos Almago, 430 CTP.
Figure 4.15. Grandcharevo	Stojic P. Bearing Capacity of abutment and improvement of stability of left slope on Grancharevo dam. 9th Congress on Large Dams, vol.I, Q32-R40, Istambul, 1967
Figure 7.37. Cindere	ru.knowledgr.cor
Figure 8.1. Zola	Wikidata.org ARTICLE Notre Provence Group
Figure 8.2. Psirha	©1999–2018 © "В ОТПУСК.РУ"
Figure 8.3. Bear Valley	British Library HMNTS 10410. pp.13

(Continued)

Photo in Figure	Source
Figure 8.4. Salmon Creek	Chanson, Hubert; James, D. Patric (2000). "Historical Development of Arch Dams. From Cut-Stone Arches to Modern Concrete Designs (A revised history of arch dams)". University of Queensland, Australia. Retrieved 2010–07–15
Figure 8.6. Moye	Wegmann, Edward (1918). <u>The Design and Construction of Dams</u>. John Wiley & Sons, Incorporated. pp. 54–55
Figure 8.7. Lake Lanyo	Kayla Robins, "Lake Lanier levels concerning for drought" Archived 2016-08-02 at the <u>Wayback Machine</u>, *Forsyth County News*, 29 June 2016; accessed 25 July 2016. From Wikipedia, the free encyclopedia
Figure 8.8. Saint Francis	"St. Francis Dam Disaster Site". Office of Historic Preservation, California State Parks. Retrieved 2012-10-08. From Wikipedia
Figure 8.10. Mareges	Marèges Dam Saint-Pierre/Liginiac, 1935 Structurae
Figure 8.11. Vouglans	Wikipedia.org/wiki/Вуглана
Figure 8.14. Xiluodu	Current Activities of Dam Construction in China.
Figure 8.15. Xiowan	2008, China National Committee on Large Dams,
Figure 8.16. Jinping-1	202 стр. Email:chine@iwh.com.
Figure 8.74. Zhaixiangkou	
Figure 8.42a. Hitotsuse	Current Activities on dams in Japan, 2003, Japan
Figure 8.64. Okura	National Committee on Large Dams, 422 pages. Email:secretariat@jcold.ld.at.jp. Copyright:JCOLD
Figure 8.34a. Kelnbrein	<u>CAWater-Info/International Organizations/ International Commission on Large Dams</u>
Figure 8.39. Mansur Eddahbi	http://blog.rushydro.ru visualrian.ru sputnikimages. com
Figure 3.44. Malpasset	Commission d'enquête du barrage de Malpasset.
Figure 3.46. Malpasset	Rapport définitif, ministère de l'Agriculture, 1960, 4 vol. J. Bellier, « Le Barrage de Malpasset », dans Travaux, n° 389, juillet 1967, pp. 3–63.

Accepted abbreviations

BP	Basic provision
BR	Building regulations
CME	Control measure equipment
CSGOES	Center for the Service of Geodynamic Observations in the Energy Sector
DE	Design earthquake
DGRF	Decree of the Government of the Russian Federation
DVL	Dead Volume Level
EC	Energy (Power) construction magazine, Russia (former USSR)
FDM	Finite difference method
FEM	Finite element method
FL	Federal Law
HC	Hydrotechnical construction magazine, Russia (former USSR)
HEC	Hydropower construction magazine, Russia (former USSR)
HECA	Hydropower construction abroad magazine, Russia (former USSR)
HEP	Hydroelectric plant
HL	Headwater level
HP	Hydraulic project
HS	Hydraulic structure
ICOLD	International commission on large dams
IDS	Information-diagnostic system
IENAUM	Institute of Engineering of the National Autonomous University of Mexico
IIGH	IInternational Institute of Geomechanics and Hydrostructures
IJHD	International Journal on Hydropower and Dams
ISRM	International society on rock mechanics
JSC	Joint stock company
LDS	Level of downstream
LRMHIM	Laboratory of Rock Mechanics of Hydroproject Institute JSC in Moscow
LUS	Level of upstream
MCE	Maximum creditable earthquake
MSL	Minimum starting level

NHL	Normal head level
NPP	Nuclear power plant
NSS	Natural stress state
PJSC	Public joint stock company
PMF	Probable maximum flood
SPP	Storage pumped plan
SR	Set of rules
SRL	Surcharged reservoir level
SS	State standard
SSS	Stress-strain state
TPS	Thermal power station

Bibliography

1. FL No. 117-FL of 07/21/1997 (as amended on 07/03/2016) "On the safety of hydraulic structures".
2. FL No. 184-FL of December 27, 2002 (as amended on July 1, 2017) "On Technical Regulation".
3. FL No. 384-FL dated December 30, 2009 (as amended on June 2, 2013) "Technical Regulation on the Safety of Buildings and Structures".
4. DGRF No. 1521 dated December 26, 2014 (as amended on December 7, 2016) "On the approval of the list of national standards and codes of rules (parts of such standards and codes of rules), the application of which complies with the requirements of the Federal Law" Technical regulation on the safety of buildings and structures".
5. DGRF No. 1303 dated November 6, 1998 (as amended on November 9, 2016) "On the approval of the Regulation on the declaration of safety of hydraulic structures".
6. SS 27751-2014. "Reliability of building structures and foundations. Key Points".
7. SS 19185-73. Hydraulic engineering. Basic concepts. Terms and Definitions. 05.23.2015.
8. SR 14.13330.2018. Construction in seismic areas. Updated edition of BR II-7-81*.
9. SR 58.13330.2012. Waterworks. The main provisions. Updated edition of BR 33-01-2003.
10. SR 20.13330.2016. Loads and impacts. Updated version of BR 2.01.07-85*.
11. SR 23.13330.2011. Foundations of hydraulic structures. Updated version of BR 2.02.02-85.
12. SR 38.13330.2012. Loads and impacts on hydraulic structures (wave, ice and from ships). Updated edition of BR 2.06.04-82*.
13. SR 39.13330.2012. Dams from soil materials. Updated version of BR 2.06.05-84*.
14. SR 40.13330.2012. Dams are concrete and reinforced concrete. Updated version of BR 2.06.06-85.
15. SR 41.13330.2012. Concrete and reinforced concrete structures of hydraulic structures. Updated version of BR 2.06.08-87.
16. SR 101.13330.2012. Retaining walls, shipping locks, fish passage facilities.
17. SS 55260.1.9-2013. Hydroelectric power stations. Hydroelectric power plants. Operational safety requirements.
18. Recommendations on checking the safety criteria for hydraulic structures of energy facilities. Order of Rostekhnadzor No. 25 dated January 24, 2013.
19. Adamovich A.N. Soil consolidation and anti-filtration curtains. M., Energy, 1980.
20. Antonov S.S., et al. Assessment of the possibility of opening a crack by contact with the base for an arch dam. *Materials of Conferences and Meetings on Hydraulic Engineering*. Energoatomizdat L., 1989, pp. 58–61.
21. Aravin V.I., Numerov S.N. Filtration calculations of hydraulic structures. M.L., Gosstroyizdat, 1955, 292 pp.
22. Argal E.S. Monolithization of concrete dams by cementation of construction joints. M., Energoatomizdat, 1987, 118 pp.
23. Arshenevsky N.N., et al. Hydroelectric stations. Energoatomizdat, 1987.
24. Asarin A.E. Dam and extreme floods. Hydrotechnical Construction, No. 8, 1993, pp. 9–11.
25. Bezukhov N.I. Fundamentals of the theory of elasticity, plasticity and creep. Higher School, Moscow, 1961, 538 pp.

26. Bellendir E.N., Ivashintsev D.A., Stefanishin D.V., Finagenov O.M., Shulman S.G. Probabilistic methods for assessing the reliability of soil hydraulic structures. SPb, Publishing House VNIIG named after B.E. Vedeneev, Volume 1, 2003, 554 pp, Volume 2, 2004, 524 pp.
27. Berezinsky S.A., Pigalev A.S. On the possible direction of improving the structures of the gravitational dam on a rocky base. Hydrotechnical Construction, No. 5, 1976, pp. 11–16.
28. Biswas A. Man and water. L., 1975, pp. 8–27.
29. Birkhoff G. *Hydrodynamics. Statement of Problems, Results and Similarity*. M., Publishing House of Foreign Literature, 1954, 184 pp.
30. Bronstein V.I. Analysis of the influence of changes in the deformability of the base and body of arched dams on their stress-strain state. *Proceedings of the Hydroproject*, vol. 68, M., 1980, pp. 44–59.
31. Bronstein V.I., Groshev M.E. Numerical modeling of the stress-strain state of high dams. Hydrotechnical Construction, No. 6, 2002, pp. 2–11.
32. Bronstein V.I. et al. Structural geometry of the arch dam of the Inguri Hydroelectric Power Station. *Materials of the Conference and Meetings on Hydraulic Engineering*, "Arch 87", L., Energoatomizdat, 1989, pp. 111–114.
33. Bronstein V.I. The state of the arch dam of the Inguri hydroelectric station and its foundation. Hydrotechnical Construction, No. 2, 1994, pp. 20–26.
34. Wentzel E.S. *Probability Theory. Textbook for Universities*. 5th ed. erased. M., Higher. school, 1998, 576 pp.
35. Vainberg A.I. Reliability and safety of hydraulic structures. Selected Issues. Kharkov, Publishing House "Tyazhpromavtomatika", 2008, 304 pp.
36. Vainberg A.I. Seismic pressure of sediment on the pressure face of a concrete gravity dam. Hydropower of Ukraine, 2017, Special issue from, pp. 51–58.
37. Vasilevsky V.V. et al. Laying particularly hard concrete mixes in the dam of the Bureyskaya hydroelectric station. Hydrotechnical Construction, No. 4, 2004, pp. 2–6.
38. Varga A.A., Remenyak M.B. System modeling of the interaction of hydraulic structures and rock masses. Hydrotechnical Construction, No. 3, 2006.
39. Vasiliev P.I., Kononov Y.I. Temperature stresses in concrete massifs. LPI, L., 1979.
40. Vasiliev A.V. et al. Laboratory studies of the downstream hydraulics during operation of the operational spillway of the Bureya HP. Hydrotechnical Construction, No. 6, 2008, pp. 36–38.
41. Velichkov S. Strengthening the dam of the Kardzhale HEP station (in Bulgarian). *Hydrotechnics and Land Reclamation*, vol. 23, No. 5, Sofia, Bulgaria, 1978, pp. 19–21.
42. Vovkushevsky A.V., Shoikhet B.A. *Calculation of Massive HS Taking into Account the Disclosure of Joints*. M., Energy Publishing House, 1981, 136 pp.
43. Volynchikov A.N., Paremud S.P. Shon La HP - the second stage of the cascade on the Da River in Vietnam. Hydrotechnical Construction, No. 6, 2006.
44. Volynchikov A.N., Mgalobelov Y.B. Justification of the reliability of the main structures of the Shon La HP in Vietnam. Hydrotechnical Construction, No. 12, 2007, pp. 23–38.
45. Volynchikov A.N., Mgalobelov Y.B., Orekhov V.V. On the earthquake resistance of the main structures of the Boguchansk HPP. Hydrotechnical Construction, No. 3, 2009, pp. 22–29.
46. Volynchikov A.N., Mgalobelov A.V., Deineko A.V. Substantiation of the designs of the lower and spill faces of a concrete dam operated in severe climatic conditions. Hydrotechnical Construction, No. 1, 2011, pp. 25–29.
47. Gaziev E.G. Morphology of cracks and stability of massifs. In the book. *"Geology of the Quaternary. Engineering Geology"*, Moscow, "Science", 1976, pp. 235–244.
48. Gaziev E.G. Stability of rock massifs and methods of their consolidation, "Stroyizdat", Moscow, 1977, 160 pp.
49. Gaziev E.G., Morozov A.S., Shaganyan V.B. The study of the strength and deformability of rock samples in conditions of volumetric stress state. *Collection of Scientific Papers of the Hydroproject*, vol. 95, 1984, pp. 83–93.
50. Gaziev E.G., Levchuk V.P., Study of the behavior of brittle polycrystalline materials in a transcendental stress-strain state. *XI Russian Conference on Rock Mechanics*, S. Petersburg, 1997, pp. 103–114.

51. Gaziev E.G. The rocky foundations of concrete dams. M., DIA, 2005, 280 p.
52. Gaziev E.G. Analysis of the modern SSS of the arch-gravity dam of the Sayano-Shushensk HEP. Hydrotechnical Construction, No. 9, 2010, pp. 48–53.
53. Gaziev E.G. Strength and deformability of brittle polycrystalline materials in bulk stress state. Hydrotechnical Construction, No. 12, 2016, pp. 2–8.
54. Geology and dams, TI-V, M., Energy, 1959–1967.
55. HS. Designer reference. M., Stroyizdat, 1983.
56. HS (in two parts). Part 1: Textbook for university students. Edited by Grishin M.M., M.: Higher. School, 1979.
57. HS (in two parts). Part 2: Textbook for university students. Edited by Grishin M.M., M.: Higher. School, 1979.
58. HS. Designer reference. G.V. Zheleznyakov, Yu.A. Ibad-Zade, P.L. Ivanov, and others; Under the total. ed. V.P. Nedrigi. M.: Stroyizdat, 1983.
59. HS: Textbook. for universities: In 2 hours, Part 1 / L. N. Rasskazov, V. G. Orekhov, Yu. P. Pravdivets and others; Ed. L. N. Rasskazov, Moscow: Stroyizdat, 1996.
60. HS: Textbook. for universities: In 2 hours, Part 2 / L. N. Rasskazov, V. G. Orekhov, Yu. P. Pravdivets and others; Ed. L. N. Rasskazov –M.: Stroyizdat, 1996.
61. HS. Textbook manual for universities. Ed. N.P. Rozanov –M.: Stroyizdat, 1978.
62. HS. Textbook manual for universities. For the editorship of A.F. Dmitriyev. Rovno Sovereign Technical University, 1999.
63. Hydraulic calculations of spillway hydraulic structures. Reference manual. M., Energy Publishing House, 1988, 541 p.
64. Gordon L.A., Gotlif A.L. Static analysis of concrete and reinforced concrete HS, M., Energy, 1982.
65. Grishin M.M. et al. Concrete dams (on rocky foundations), M., Stroyizdat, 1975, 352 pp.
66. Gupta H., et al. Dams and earthquakes, M., 1976.
67. Daily J., Harleman D. Fluid Mechanics. M.: "Energy", 1971, 480 p.
68. Durcheva V.N. Field studies of the monolithicity of high concrete dams, M., Energoatomizdat, 1988.
69. Dyatlovitsky L.I., Vainberg A.I. Stress Formation in Gravity Dams. Kiev. Naukova Dumka, 1975, 264 p.
70. Evdokimov P.D., Sapegin D.D. Strength, resistance to shear, and deformability of the foundations of structures on rocks. "Energy", M.-L., 1964. 11 p.
71. Zhilenkov V.N. Hydraulic and technological aspects of the drainage arrangement of a concrete gravity dam. News of VNIIG, t. 245, 2006.
72. Ginho M., Barbier R. Geology of dams and hydraulic structures, Gosstroyizdat, M., 1961, 355 pp.
73. Zhukov V.N. Strengthening rocky slopes and slopes with prestressed anchors. *Proceedings of the Hydroproject*, vol. 68, M., 1980, pp. 101–123.
74. Zolotov L.A., Shaitanov V.Y. Overlap of the Yangtze River in the alignment. *HEC*, No. 3, 1999
75. Ivashintsov D.A. et al. Principles of designing modern concrete dams. Hydrotechnical Construction, No. 2, 2004, pp. 6–10.
76. Kaganov G.M. et al. Ways of strengthening arched dams in wide sections. Hydrotechnical Construction, No. 8, 2011, pp. 46–50.
77. Kadomsky E.D., et al. Determination of the dimensions of the profile of a gravitational dam with an external cementation curtain. *Proceedings of the LPI*, No. 354, 1976, pp. 78–86.
78. Kalustyan E.S. Reliability of rocky foundations of high dams. Energy Construction Abroad, No. 5, 1982, pp. 17–23.
79. Kalustyan E.S. Destruction and damage of concrete dams on rocky foundations. VNIIG, 1997, 187 pp.
80. Karavaev A.V. et al. Comparison of concrete dams on a rocky base in terms of material consumption and volume of rock excavations. *Materials of the Conference and Meetings on Hydraulic Engineering*. Concrete dams work together with a rock base. L., Energy, 1979, pp. 6–13.

81. Kachanov A.M. Fundamentals of the theory of plasticity. Study Guide for State Universities. M.: Gostekhizdat, 1956. 332 s.
82. Kirillov A.P. et al. *Earthquake Resistance and Design of Earthquake-Resistant Dams. Design and Construction of Large Dams.* M., Energy Publishing House, 1985.
83. Kogan E.A. Construction of dams from rolled concrete. Analysis of the state and development prospects. Hydrotechnical Construction, No. 5, 2000, pp. 30–40.
84. Collins R. The flow of liquids through porous materials. Per. from English under the editorship of G.I. Barenblatt. M., Mir, 1962. 352 s.
85. Kuznetsov G. Results of geodetic measurements on the Inguri arch dam under construction. *Hydrotechnical Construction*, No. 3, 1983, pp. 34–38.
86. Kuzmin S.A., Deryugin G.K. On the choice of the type of conjugation of upstream high dam. *Hydrotechnical Construction*, No. 6, 2013, pp. 12–16.
87. Kulmach P.P. *Hydrodynamics of Hydraulic Structures (Main Flat Tasks).* M. Publishing House of the Academy of Sciences of the USSR, 1963, 192 p.
88. Landau Y.A. Some structural and technological solutions of dams made of rolled concrete in severe climatic conditions. *EC*, No. 8, 1994, pp. 53–55.
89. Landau Y.A. New constructive solutions of concrete dams. The dissertation for the degree of Doctor of Technical Sciences, 1997, 284 p.
90. Landau Y.A. et al. Hydropower and the environment. Kiev, Libra, 2004
91. Landau Y.A., Vainberg A.I. New solutions for interfacing concrete dams with a rock base. *Hydropower of Ukraine*, No. 1, 2007.
92. Lapin G.G., Sudakov V.B., Shangin V.S. Ways to improve the manufacturability of concrete dams. *Hydrotechnical Construction*, No. 10, 2012, pp. 2–7.
93. Lyapichev Y.P. Design and construction of modern high dams, M., Peoples' Friendship University of Russia, 2009, pp. 6–149.
94. Marchuk A.N. Dams and geodynamics. Moscow, IPP RAS, 2006.
95. Marchuk A.N. *Static Work of Concrete Dams.* Energoatom-Publishing House, 1983, 206 p.
96. Marchuk A.N., Marchuk M.A. On the state of contact of concrete with the rock under the pressure faces of dams. *Hydrotechnical Construction*, No. 6, 1989, pp. 26–30.
97. Marchuk A.N., Marchuk M.A. Tectonophysical aspects of the stress-strain state of large concrete dams. *Hydrotechnical Construction*, No. 3, 2010, pp. 31–35.
98. Marchuk A.N. About some modern trends in the construction of large concrete dams. *Hydrotechnical Construction*, No. 1, 2002.
99. Marchuk A.N. About the work of coastal adjoining arched dams. *Hydrotechnical Construction*, No. 11, 2002, pp. 10–14.
100. Marchuk A.N. The influence of tectonic stresses on the formation of the stress-strain state of concrete dams. *Hydrotechnical Construction*, No. 10, 2011, pp. 51–55.
101. Mgalobelov Y.B. Strength and stability of the rocky foundations of high dams, M., "Energy", 1979, 215 pp.
102. Mgalobelov Y.B. On the interfacing of concrete dams with a rock base. *Hydrotechnical Construction*, No. 8, 1998.
103. Mgalobelov Y.B. Comparison of reliability criteria when designing concrete dams on rocky foundations according to Russian Construction Norms and Regulations and American standards. *Hydrotechnical Construction*, No. 9, 2007, pp. 45–50.
104. Mgalobelov Y.B. Progressive solutions in the field of design and construction of concrete dams. *Hydrotechnical Construction*, No. 8–9, 2000, pp. 44–52.
105. Mgalobelov Y.B. Fam Van Hung. Justification of the reliability of the Nam Chien arch dam in Vietnam. *Hydrotechnical Construction*, Moscow, No. 1, 2008, pp. 21–35.
106. Mgalobelov Y.B. Studies of the bearing capacity of an arch dam. *Hydrotechnical Construction*, Moscow, No. 9, 2008, pp. 16–30.
107. Mgalobelov Y.B., Deineko A.V. The calculated justification for the safety of modern hydraulic structures and the peculiarities of accounting for the effects of technological equipment in an earthquake. *Hydrotechnical Construction*, Moscow, No. 7, 2010, pp. 46–50.
108. Methodology for determining the safety criteria of hydraulic structures. RAO "UES of Russia", M., 2001.
109. Guidelines for assessing the effectiveness of investment projects, M., 2000, 21 pp.

110. Guidelines for the design of optimal inserts for pairing concrete dams with a rock base, P-634-75, Gidroproekt, M., 1978, 30 pp.
111. Mirgorodsky P.P., Mozhevitinov A.L. On rational structural solutions in the rocky base of gravitational dams. *Materials of Conferences and Meetings on Hydraulic Engineering.* The work of the dam in conjunction with the rock base. Energy, L., 1979, pp. 87–90.
112. Mironenko V.A., Shestakov V.M. Fundamentals of hydrogeomechanics, "Nedra", M., 1974.
113. Mogilevskaya S.E., Vasilevsky A.G. Aging of rocky foundations of high dams: engineering-geological aspects. *Hydrotechnical Construction*, No. 6, 2002, pp. 12–15.
114. Molokov L.A., Gaziev E. G. New in the methods of research for hydraulic engineering. *EC*, No. 1 (127), Moscow, 1972, pp. 37–43.
115. Muller L. Engineering geology. The mechanics of rock massifs. "WORLD", M., 1971, 256 pp.
116. Nadai A. Plasticity and destruction of solids, Moscow, 1969.
117. Napetvaridze S.G. Seismic resistance of hydraulic structures, M., Gosstroyizdat, 1959, 216 p.
118. Natarius Y.I. Improving the seismic resistance of concrete dams, Energoatomizdat, M., 1988, 119 p.
119. Oleshkevich L.V. Lightweight concrete dams on rocky foundations, M., Stroyizdat, 1968.
120. Nuts G. et al. The stress state, strength and stability of the dam of the Katunsky hydroelectric complex. *Collection of Scientific Papers of the Hydroproject*, vol. 123, M., 1987, pp. 73–83.
121. Orekhov V.G. et al. Methods of regulating the stress state of arched dams. *All-Union Conference: Methods for Determining the Stress State and Stability of High-Pressure Hydraulic Structures and their Foundations under Static and Dynamic Loads.* M., 1972, pp. 105–112.
122. Paremud S.P., Volynchikov A.N. Progressive solutions to the technical design of the Shaun-La waterworks on the river. Yes, in Vietnam. *Hydrotechnical Construction*, No. 4, 2011, pp. 54–62.
123. Patent for the invention No. 2648186. A method of constructing a concrete dam in a river canyon. Machekhin S.V. (RU), Fink A.K. (RU), Mgalobelov Yu. B. (RU). Priority March 16, 2017.
124. Permyakova L.S. et al. Condition of the pressure front of the Sayano-Shushensky dam after completion of repair work to reduce its water permeability. *Hydrotechnical Construction*, No. 1, 2008, pp. 9–13.
125. Permyakova L.S., Reshetnikova E.N., Epifanov A.P. The filtration regime of the dam foundation of the Sayano-Shushenskaya hydroelectric station in the first years of constant operation. *Hydrotechnical Construction*, No. 4, 1994.
126. A guide for the design of arched dams. P-892-92, Hydroproject, M., 1992, 201 p.
127. Design and construction of large dams. Based on the materials of the XII (Mexico City) International Congress on Large Dams. Filtration studies of dams and their foundations. Energy Publishing, vol. 6, 1981, pp. 14–30.
128. Design and construction of large dams. Based on the materials of the XIII (New Delhi) International Congress on Large Dams. Problems of mating dams, Energoatomizdat, 1986, pp. 5–28, p. 95–141.
129. Design and construction of large dams. Based on XIII (New Delhi) international congress on large dams. Accidents and damage to large dams. Energoatomizdat, 1986, pp. 10–52.
130. Design and construction of large dams. Based on the materials of the XIII international congress on large dams. Spillways of high throughput. 1985, 141 p.
131. Prochukhan D.P., et al. Rocky foundations of hydraulic structures. Gosstroyizdat, 1971.
132. Provarova, T.P. Damage and accident of energy absorbers and concrete downstream anchorage. *Izvestiya VNIIG*, vol. 240, 2002, pp. 63–77.
133. Storytellers V.A. Moisture changes in the concrete of the Sayano-Shushenskaya dam and their influence on the stress-strain state of the structure. *Hydrotechnical Construction*, No. 10, 2012.
134. Recommendations for the hydraulic calculation of open spillways of high-pressure waterworks and erosion of a rocky channel by a discarded stream. P-80-79. VNIIG, 1979.
135. Recommendations on the methodology for compiling specialized engineering and geological models for calculating and studying rock masses. Hydroproject, P-830-85, 1985.

136. Recommendations for calculating the stability of rocky slopes, Hydroproject, P-843-86, Moscow, 1986.
137. Rechitsky V.I. Assessment of natural stresses in the coastal massif at the site of the site of the Shon La HEP. *Hydrotechnical Construction*, No. 4, 2009, pp. 55–60.
138. Romm E.S. Filtrational properties of fractured rocks "Nedra", M., 1966, 283 pp.
139. Rodionov V.D. et al. Coastal spillway of the Sayano-Shushenskaya HEP. *Hydrotechnical Construction*, No. 6, 2007, pp. 33–37.
140. Rodionov V.B. et al. Choice of the scheme of the device for the downstream of the overflow dam of the Koteshvar hydroelectric complex. *Izvestia VNIIG*, vol. 230, 1997, pp. 327–334.
141. Rozanov N.S. et al. Accidents and damage to large dams. Energoatomizdat, M., 1986, 126 p.
142. Guidance on the design of gravity and arched dams in the United States, M., Energy, 1978. Per. ed. US Burean of Reclamation, 1972.
143. Guidance on the design of landslide and landslide protective structures, Ministry of Transport, VNIITS, Moscow, 1984.
144. Savich A.I. Methodological guidelines for the use of seismoacoustic methods for assessing the deformation properties of rocks. Gidroproekt, M., 1970, 68 p.
145. Savich A.I., Lomov I.E. The main results of a set of field observations in the head part of the Inguri hydroelectric reservoir. *Collection of Scientific Papers of the Hydroproject*, vol. 96, 1983, pp. 92–104.
146. Savich A.I., Bronstein V.I. Current status and ways to ensure seismic resistance and hydrodynamic safety of large power facilities. *Hydrotechnical Construction*, No. 8–9, 2000, pp. 60–70.
147. Savich A.I., Kuyundzhich B.D. Integrated engineering-geophysical studies in the construction of hydraulic structures. "Nedra", M., 1990, 431 pp.
148. Savich A.I., Ilyin M.M., Rechitsky V.I., Zamakhaev A.M. Features of the effect of reservoirs on the rocky foundations of large dams. *Hydrotechnical Construction*, No. 3, 2003.
149. Savich A.I., Gaziev E.G. Influence of reservoirs on the behavior of the rocky foundations of high dams. *Hydrotechnical Construction*, No. 11, 2005, pp. 33–37.
150. Savich A.I., et al. Modern methods of research in hydrotechnical construction on rocky foundations. *Hydrotechnical Construction*, No. 6, 2007.
151. Savich A.I. et al. Static and dynamic behavior of the Sayano-Shushenskaya arched dam. *Hydrotechnical Construction*, No. 3, 2013, pp. 2–13.
152. Sedov L.I. Continuum mechanics. SPb., 2004.T. 1.528 s., T. 2. 448 s.
153. Semenov I.V., et al. Monitoring in the system of environmental safety of hydraulic facilities. *Hydrotechnical Construction*, No. 6, 1998, pp. 33–40.
154. Sinev V.V. et al. Design and concrete technology features of the Kousar Dam. *Hydrotechnical Construction*, No. 6, 2006.
155. Sokolov I.B., Storytellers V.A. Accounting for moisture effects on the pressure zone of concrete dams. *Hydrotechnical Construction*, No. 3, 1986, pp. 54–56.
156. Sokolov I.B. et al. Strength criteria for concrete arched dams. *Materials of Conferences and Meetings on Hydraulic Engineering*. Energy Publishing House, L., 1989, pp. 11–14.
157. Berezinsky S. A. et al.. Reference book of the designer of concrete structures of hydroelectric stations. Ed. Yu. B. Mgalobelova and Yu. P. Sergeev. M., Energoatomizdat, 1985.
158. Stafievsky V.A. et al. Condition of the dam of the Sayano-Shushenskaya YEP at the final stage of repair of concrete on the pressure face and contact zone of the rock base. *Hydrotechnical Construction*, No. 11, 2003, pp. 18–24.
159. Sudakov V.B. Technology for the construction of the dam Upper Stillwater from rolled concrete. *HECA*, No. 4, 1989, pp. 17–20.
160. Sudakov V.B., Tolkachev L.A. Modern methods of concreting high dams. M., Energo-atomizdat, 1988, pp. 118–140, pp. 229–240.
161. Sudakov V.B. et al. Modern structural and technological solutions for earthquake-resistant dams. *Hydrotechnical Construction*, No. 2, 1996, pp. 45–50.
162. Sudakov V.B. Construction of dams from rolled concrete, prospects and tasks, VNIIG, 2011, 42 pp.
163. Taycher S.I. Guidance on engineering strengthening measures on the rocky foundations of concrete dams and adjacent rock masses. Hydroproject, M., 1986, 170 p.

164. Taicher S.I., Mgalobelov Yu. B. Stability calculations of rocky coastal stops of arched dams. M., Energy, 1972, 120 p.
165. Tetelmin V.V. Analysis of irreversible processes in the alignment of the dam of the Sayano-Shushenskaya hydroelectric power station. *Hydrotechnical Construction*, No. 8, 2010, pp. 47–51.
166. Tetelmin V.V. Impact of the construction of large hydroelectric facilities on the isostatic state of the earth's crust. *Hydrotechnical Construction*, No. 11, 2009, pp. 46–50.
167. Tetelmin V.V., Ulyashinsky V.A. Elastic filtration regime at the base of high dams. *HEC*, No. 5, 2009, pp. 17–22.
168. Tetelmin V.V., Danielov E.R. The reaction of the earth's crust to the creation of large reservoirs. *HEC*, No. 3, 2014.
169. Timoshenko S.P., Goodyear J. Theory of elasticity. M., Nauka, 1975, 576 p.
170. Trapeznikov L.P. Temperature crack resistance of massive concrete structures, M., Energoatomizdat, 1986, pp. 14–62.
171. Guidelines for the design of anti-filter curtains in rocky soils, L., "Energy", 1968.
172. Tsaguria T.A. Study of the deformations of the earth's crust in the area of the Inguri hydroelectric reservoir. *KAPG Symposium on the Study of Modern Movements of the Earth's Crust*. Voronezh, 1988.
173. Tsytovich N.A., Ter-Martirosyan Z. G. Fundamentals of applied geomechanics in construction. "Higher School", M., 1981, 317 p.
174. Khrapkov A.A. et al. Investigation of the stress-strain state of the Katunsky hydroelectric dam during the construction and operational periods. *Proceedings of the Hydroproject*, No. 123, M., 1987, pp. 65–73.
175. Khrapkov A.A. Current problems in ensuring the strength of concrete dams. *Hydrotechnical Construction*, No. 6, 1994, pp. 19–22.
176. Fishman Y.A. On modern requirements for coupling gravity and buttress dams to rocky foundations. *Hydrotechnical Construction*, No. 11, 1980, pp. 12–16.
177. Fried S.D., Levenich D.P. Temperature effects on hydraulic structures in the North. Stroyizdat, L., 1978, pp. 16–41.
178. Fedosov V.E. Kapanda waterworks in Angola. *Hydrotechnical Construction*, No. 8–9, 2000, pp. 100–105.
179. Frolov B.K. Construction of the arch dam of Kelnbrein. *Energy Construction Abroad*, No. 4, 1981.
180. Schweinstein A.M. Spillways of foreign hydroelectric facilities. L. Energy, 1973, 179 p.
181. Sheinin I.S. Oscillations of hydraulic structures in liquids (reference manual on the dynamics of hydraulic structures). L., Energy, Leningrad Branch, 1967.314 s.
182. Sheinin I.S. Air curtain to protect hydraulic structures from seismic and explosive influences. *HEC*, No. 10, 1994.
183. Shulman S.G. Calculations of hydraulic structures taking into account the sequence of construction. M., Energy, 1975, 168 p.
184. Zhang Chaoran et al. Technical features of the construction of the Three Gorges hydroelectric complex. *Hydrotechnical Construction*, No. 3 and 4, 2002, pp. 49–54 and pp. 38–43.
185. Chugaev R.R. Hydraulics L., "Energy", 1975.
186. Eidelman S.Y., Durcheva V.N. Stress-strain state of the contact zone of concrete gravity dams according to field studies. *Materials of Conferences and Meetings on Hydraulic Engineering. The work of dams together with the base*. L., "Energy", 1979, pp. 121–123.
187. Arenillas M. The historical development of Spanish dams. *IJHD*, No. 3 и 4, 2006, pp. 138–142.
188. Barton N., Choubey V.D. The shear strength of rock joints in theory and practice. *Rock Mechanics*, No. 1, 1977, pp. 1–54.
189. Batraz S., et al. Design of the 100m-higt Oyuk hardfill dam. *IJHD*, No. 5, 2003, pp. 138–142.
190. Benz T. Construction of the Nant de Drance project. *IJHD*, No. 3, 2011, pp. 35–42.
191. Bieniawski Z.T. Rock mechanics design in mining and tunneling. A. Balkema, Rotterdam, 1984, p. 97–133. 1
192. Bhattacharjee S.S., Leger P. Seismic cracking and energy dissipation in concrete gravity dams. *Earthquake Engineering and Structural Dynamics*, vol. 22, 1993.

193. Brunner W. Cracking of the Revelstoke concrete gravity dam mass concrete. *Proceedings of 15-th ICOLD*, Lausanne, V.2., Q.57.R.1, 1985, pp.1–20.
194. Activities Dam Construction in China, Chinese National Committee on Large Dams, 2009, 200 p. Current.
195. Engineering and design. Gravity dam design. EM 1110-2-2200, U.S. Army Corps of Engineers, Washington, 1995.
196. Carrere A., Dussart J., Lefevre C. Foundations rocheuses de barrages en beton: Examples de traitement systematique d'amelioration des proprieties mecaniques naturelles a Go-mal Zam (Paquistan), Takamaka (Reunion), Pont de Veyrieres (France). *17th Congress on Large Dams*, vol. 3, Q.66-R19, Vienna, 1991, pp. 319–337.
197. Evdokimov P.D., Sapegin D.D. A large scale field shear test on rock. *Second Congress of the ISRM*, Beograd, 1970.
198. Engineering and design, Roller Compacted Concrete, EM 1110-2-2006, Department of the army U.S. Army Corps of Engineers, Washington, 2000.
199. FERC 2002, U.S. Federal Energy Regulatory Commission Guidelines.
200. Dunstan M.R. RCC dams in 2008. *IJHD*, No. 3, 2008, pp. 69–72.
201, Dunstan M.R. *RCC Dams – Is There a Limit to the Height Dams and Reservoirs under Changing Challenges*. CRC Press, Balkema, 2011, pp. 313–320.
202. Fujii T. Fault treatment at Nagawado Dam. *10th Congress on Large Dams*, Q.37-R9, Montreal, Canada, 1970, pp. 147–169.
203. Dams in Japan. Japan Commission on Large Dams, 2012, 509 p.
204. Dams in Switzerland. Source for World Wide Swiss Dam Engineering, 2011, 376 p.
205. Husein Malkawi A.J., et al. Computational analysis of thermal and structural stresses for RCC dams. *IJHD*, No. 4, 2004, pp. 86–95.
206. Nose M. Rock test in situ, conventional tests on rock properties and design of Kurobeg-dwa 4 dam based thereon. *8th Congress on Large Dams*, vol.1, Q.28-R82, Edinburgh, IJHD, October, 2010, pp. 20–22.
207. Hang Y., Du C. Building Jinangiao Dam. *IJHD*, October, 2010, pp. 20–22.
208. Hollingworth F. Roller compacted concrete arches dams. *IJHD*, No. 11, 1989, pp. 29–34.
209. Hoek E., Brown E.T. Practical estimates of rock mass strength. *International Journal Rock Mechanics and Mininig Science*, vol. 34, No. 8, 1997, pp. 1165–1186.
210. Hoek E., Carranza-Torres C., Corkum B. Hoek-Brown failure criterion. 2002 Edition. *5th North American Rock Mechanics Symposium and 17th*.
211. Horsky O., Blana P. The application of engineering geology to dams construction, Crech Republic, 2011, 296 p.
212. Hansen K.D., Reinhard W.A. Roller - Compacted Concrete Dams, 1991, 298 p.
213. Gaziev E., Erlikhman S. Stresses and strains in anisotropic rock foundation. *Model studies, Symposium of the ISRM, Rock Fracture, II-1*, Nancy (France), 1971.
214. Gaziev E. Criterio de resistencia para rocas y materiales frágiles policristalinos, Segunda Conferencia Magistral «Raúl J. Marsal», México, Sociedad Mexicana de Mecánica de Rocas, 1996.
215. Gaziev E. Rupture energy evaluation for brittle materials. *International Journal of Solids and Structures*, vol. 38, 2001, pp. 768–769.
216. Jonh K.W. Three dimensional analysis for jointed rock slopes. *Proceedings of the International Symposium of Rock Mechanics*, Nancy, France, 1971, Oct.
217. Joshi C.S. Optimum profile of gravity dams. *IJHD*, vol. 34, No. 9, 1982, pp. 35–36.
218. Jung S.J., Han-Uklim. Shear fracture analysis for brittle materials. *8th International Congress on Rock Mechanics*, Tokyo, vol. 1, 1995, pp. 233–235. 1
219. Kosterin N.V., Vasiliyev A.V. The multipurpose Bureya dam project in Russia. *IJHD*, No. 3, 2007, pp. 54–57.
220. Koliji A., et al. Abutment stability assessment at Hongrin Dam. *IJHD*, No. 3, 2011, pp. 56–61.
221. Gaziev E., Levtchouk V. Strength characterization for rock under multiaxial stress states. *9th International Congress of the ISRM*, Paris, 1999, pp. 101–106.
222. Guang L. RCC arch dams. *IJHD*, No. 3, 2002, pp. 95–98.
223. Landau Y.A., Mgalobelov Y.B. Structural decisions and peculiarities of static work of dams from rammed concrete. *International Symposium of RCC Dams*, China, 1991, pp. 65–75.

224. Landau Y.A., Mgalobelov Y.B. New ideas in arch dams engineering. *International Symposium of Arch Dams*, China, 1992, pp. 633–636.
225. Landau Y.A. Possibilities of increasing dam-foundation reliability by front apron design. *IJHD*, No. 7, 1993, pp. 26–28.
226. Landau Y.A. New design for composite arch dams with partial use of RCC. *IJHD*, No. 3, 1998.
227. Leguizamo High times at Miel-1. *IJHD*, No. 6, 2006, pp. 12–16.
228. Londe P. Three Dimensional Method of Analysis for Rock Slope (in French). Ann. ponts et chausses, 1965, #1, pp. 37–60.
229. Londe P., Lino M. The faced symmetrical hardfill dam: a new concept for RCC. *IJHD*, February, 1992, pp. 19–24.
230. Lucajic B., Smith G., Deans J. Use of asphalt in treatment of dam foundation leakage Stewartville dam. *Proceedings of ASCE, conference «Issues in Dam Gɔrouting»*, Col (Abr.30), USA, 1985, pp. 76–91.
231. Major dams (h>60v) under construction in 2013. *IJHD*, World atlas, 2013, pp. 16–19.
232. Martin C.D., Chandler N.A. The progressive fracture of Lac du Bonnet granite. *International Journal Rock Mechanics and Mining Science & Geomechanics Abstracts*, Vol. 31, No. 6, 1994, pp. 643–659.
233. Mgalobelov Y.B., Landau Y.A. *Non-Traditional Concrete Dams Construction in Rock Foundation*. A. A. Balkema Publisher, 1997, 266 p.
234. Mgalobelov Y.B., Fradrin B.V. Bellever the stress. *IJHD*, No. 5, 1996, pp. 38–40.
235. Mgalobelov Y.B. Evolution of methods of arch dams stability analysis, *48th US Rock Mechanics/Geomechanics Symposium*, 1–4 June 2014, Minneapolis, Minnesota.
236. Mgalobelov Y.B. Computer analysis of bearing ability of arch dam. *24 Congress on Large Dams*, 6–8 June 2012, Kyoto, Japan.
237. Mason P. Hardfill and the Ultimate Dams. HRW, November, 2004, pp. 26–29.
238. Mason P., et al. The design and construction of a fased symmetrical hardfill dams. *IJHD*, No. 3, 2008, pp. 90–94.
239. Mohr O. Zeitschrift des Vereins Deutcher Ingenieure, vol. 44, 1900, p. 1524.
240. Müller L., Das Kraftespiel im undergrund von Talsperren, Geologil und Bauwesen, 1961. 1
241. Parate N.S. Critere de rupture des roches fragile, Annales de I'Institut Technique du Batiment et des Travaux Publiques, Paris, N 253, 1969, pp. 149–160.
242. Reed G., et al. RCC for the heightening of San Vicente dam. *IJHD*, No. 5, 2003, pp. 130–136.
243. RCC dams. *IJHD*, Vol. 7, No. 3, 2000, pp. 42–63.
244. RCC dams. *IJHD*, World atlas 2014, pp. 20–45.
245. RCC dams. *Bulletin ICOLD*, No. 126, 2017.
246. RCC dams in Asia. *IJHD*, Vol. 14, No. 4, 2007, pp. 84–88.
247. Sarkaria G.S., et al. Operational efficiency of large spillways: eight Brazilian case studies. *IJHD*, No. 4, 2003, pp. 47–57.
248. Santa Clara J.M.A. The complex geology of Kariba right bank. *17 Congress on Large Dams*, Vol. 3, Vienna, 1991.
249. Savich A.I., Gaziev E.G., Rechitski V.I., Kolichko A.V. Validation of determination of rock deformation moduli by different methods. *11th International Congress of the ISRM*, Lisbon, vol. 1, 2007, pp. 517–522.
250. Savich A.I., Bronshtein V.I., Gorshkov Yu. M., Ilyin M.M., Kereselidze S.B., Remenyak M.B. Geodynamic processes at the Inguri dam area. *Proceedings of the Workshop: Geodynamical Hazards Associated with Large Dams*, Luxembourg, 1997.
251. Savich A.I., Bronshtein V.I., Ilyin M.M., Laschenov S.Y., Stepanov V.V. Monitoring of regional and local geodynamic processes in areas of high dams in C.I.S. *Proceedings of the Workshop: Geodynamical Hazards Associated with Large Dams*, Conseil de I'Europe, Luxembourg, 1998.
252. Savich A.I., Gaziev E.G., Rechitski V.I., Kolichko A.V. Validation of determination of rock deformation moduli by different methods. *11th Congress of the ISRM*, vol. 1, Lisbon, 2007, pp. 517–522.

253. Serafim I.L. New development in the construction of concrete dams. General Report, Congress ICOLD, San Francisco, 1988, pp. 23–26.
254. Serafim I.L., Del Campo F. Interstitial pressure on rock foundations of dams. «Journal of Soil Mech and Foundations Division». *Proceedings of the ASCE*, Sept. 1965.
255. Singh B., Goel R.K. *Rock Mass Classification, A Practical Approach in Civil Engineering.* Elsevier, Amsterdam, Lausanne, New York, Oxford, Shannon, Singapore, Tokyo, 1999, pp. 17–19.
256. Stojic P. Bearing Capacity of abutment and improvement of stability of left slope on Grancarevo dam. *9-th Congress on Large Dams*, vol.1, Q32-R40, Istambul, 1967, pp. 951–964.
257. Tejada L.C., et al. The design and construction of Porce II RCC Dam in Colombia. *JHD*, No. 3, 2001, pp. 51–57.
258. Terzaghi K. Measurement of stress in rock. *Geotechnique*, vol. 17, No. 1, London, 1967.
259. Toniatty N.B., Mussi J.M. Salto Caxias: Responding to Electricity Demands in Brazil, HRW, may, 1999, p. 2–15.
260. Takahashi M., Koide H. Effect of the intermediate principal stress on strength and deformation behavior of sedimentary rock at the depth shallower than 2000m. *Proceedings of ISMR-SPE International Symposium Rock at Great Depth*, Pau (France), 1989, pp. 19–26.
261. Tardieu B., et al. *Dam Shape Adaptation Resulting from Strong Earthquake Context. Dams and Reservoirs under Changing Challenges.* CRC Press, Balkema, 2011, pp. 629–636.
262. Two major RCC structures progress in Thailand. *IJHD*, No. 1, 2002, pp. 55–57.
263. The world's major dams and hydro plants. *JHD*, Yearbook, 2011, pp. 110–278.
264. Roller Compacted concrete dams. *IGHD*, World atlas and industry guide 2013, pp. 20–44.
265. World hydro potential and development. *IJHD*, World atlas and industry guide 2016, pp. 19–21.
266. US COLD, Updated Guidelines for Selecting Seismic Parameters for Dam Projects, United States Committee on Large Dams, Denver, Colorado, USA, 1999.
267. Ulusay R., Hudson J.A. The complete ISRM suggested methods for rock characterization, testing and monitoring: 1974–2006, ISRM, 2007.
268. Westergaard H.W. Water pressure on dams during earthquakes. *Proc. Am. Soc. Eng*, vol. 57. No. 9. 1931, p. 1303.
269. Westergaard H.W. Water pressure on dams during earthquakes. *Proc. Am. Soc. Eng*, vol. 59. Trans. Number 98. 1933. p. 418.
270. Win Kyaw U., et al. Myanmar's Yeywa RCC dam. *IJHD*, No. 4, 2007, pp. 77–82.
271. Zhou J., et al. San Visente: the world's largest dam heightening using RCC. *IJHD*, No. 3, 2008, pp. 73–78.

Index